This is the first scientific biography of Edward Frankland, probably the most eminent chemist of nineteenth century Britain. Amongst many other achievements, he discovered the chemical bond and founded the science of organometallic chemistry (both terms are his invention). A controversial figure throughout his life, he became a leading reformer of chemistry teaching and for nearly 40 years the government's close adviser on the purity of urban water supplies, arguably preventing a pandemic of water-borne disease.

From an apprenticeship in a druggist's shop in Lancaster, he proceeded to London to become assistant lecturer in chemistry to Lyon Playfair, and then to a PhD in Marburg under Robert Bunsen. After occupying the first chemical chair at Manchester he spent the rest of his career at numerous famous institutions in London, culminating at what became Imperial College. He was knighted in 1897.

Today a certain obscurity of reputation stems from the conspiracy of silence surrounding Frankland's origins: he was the illegitimate son of a distinguished lawyer. Frankland never gave interviews and posterity has had to guess about many of his activities. Recently, however, Professor Russell has gained access to a vast collection of his private papers, and has discovered several other major deposits, making the Frankland archive one of the largest collections of scientific papers to come to light in Britain this century. These have been fully examined in this new study which discloses, among much else, webs of conspiracy in the scientific community that demands a radical revision of the social history of Victorian science. Russell's authoritative and lively account of Frankland's achievements will be of great interest not only to professional chemists and historians of science, but also to general readers concerned with the social fabric of Victorian England.

❧ EDWARD FRANKLAND ❧

Chemistry, Controversy and Conspiracy
in Victorian England

EDWARD FRANKLAND

Chemistry, Controversy
and Conspiracy
in Victorian England

Colin A. Russell

The Open University

CAMBRIDGE
UNIVERSITY PRESS

Published by the Press Syndicate of the University of Cambridge
The Pitt Building, Trumpington Street, Cambridge CB2 1RP
40 West 20th Street, New York, NY 10011-4211, USA
10 Stamford Road, Oakleigh, Melbourne 3166, Australia

First published 1996

Printed in Great Britain at the University Press, Cambridge

A catalogue record for this book is available from the British Library

Library of Congress cataloguing in publication data
Russell, Colin Archibald.
Edward Frankland : chemistry, controversy, and conspiracy in Victorian England /
Colin A. Russell.
p. cm.
ISBN 0–521–49636–5 (hc)
1. Frankland, Edward, Sir, 1825–1899. 2. Chemists – Great Britain – Biography.
I. Title.
QD22.F65R86 1996
540′.92–dc20
[B] 95–40319 CIP
ISBN 0 521 49636 5 hardback

TO JEREMY AND MELISSA

Contents

Preface

The completion of this volume is, I have to confess, something of a relief. Edward Frankland has occupied a good deal of my energies and time for far too long. It is rather more than thirty years since I first encountered him, while writing my doctoral thesis on the rise and development of the fundamental chemical doctrine of valency. When I discovered that Frankland played a key part in that development, and moreover that he had been born a few miles from where I then lived, my interest greatly increased. It became clear to me that a biography of Edward Frankland was much overdue. As I came to discover, he was probably the most important figure in British chemistry in the last century, yet by a strange paradox relatively unknown today.

A good deal of work in local history (hitherto unknown territory to me) led to the book *Lancastrian Chemist: the early years of Sir Edward Frankland*, Open University Press, Milton Keynes, 1986. It seemed a worthwhile project on the grounds that a man's childhood and youth are likely to be his most formative years. Detailed research indicated that, in Frankland's case at least, this was profoundly true. In a sense the present volume is a sequel to that essay, but it is intended that it should also be complete in itself, so in the first couple of chapters those who have read *Lancastrian Chemist* may find some material that is familiar.

The reason for his failure to receive an extended biography became rapidly clear: the raw data were simply not available. To be sure, Frankland wrote his own recollections, *Sketches from the life of Sir Edward Frankland*, privately printed 1901. It had a slightly scandalous history, almost all copies being withdrawn within months of their release, largely on the grounds that it could have contained statements that were legally damaging, in dubious taste, hurtful to relatives, or all three. An expurgated edition was issued in 1902, rather less interesting but still an enjoyable account. But there was no way in which this could be regarded as a satisfactory treatment, partly because it was never finished, but chiefly because of the inherent limitations of any autobiography. While perfunctorily contemplating the distant prospect of possibly writing my own biographical account of Frankland something happened which transformed the whole concept in a moment.

By great good fortune we (for my wife Shirley was working with me in this), came across a vast collection of Frankland papers in private hands. Knowing these existed, I should have been foolish to the point of irresponsibility to have begun any work without first examining them all. Quickly two, three and then four other deposits, all in private ownership, became known to us, taking us halfway across the world to see some of them. All major private deposits have been microfilmed, and placed on a computer data-base; in addition many other documents have come to light in learned societies, public institutions, county record offices and so on. An account of this preliminary work on the documents has already been given ('The archives of Sir Edward Frankland: resources, problems and methods', *Brit. J. Hist. Sci.*, 1990, **23**, 175-185, by Colin and Shirley Russell). Only when this mass of new material had been digested was it deemed prudent to begin any major new assessment of Frankland's life and work. Indeed it is not going too far to say that the hitherto unpublished material has transformed our understanding not only of one man but also of the scientific and cultural environment in which he lived.

Thus the present book depends heavily on the several thousand documents that have come to light in the last ten or so years. I must repeat that most (about 90%) are in private ownership and they are not accessible to the general public. However almost all are on microfilm and in that form they can be readily consulted. Initial enquiries should be made to the Open University .

At the death of Edward Frankland his papers were divided between the four surviving children of his first marriage. They were:

> *Margaret ('Maggie')*: these have since been further divided and are currently owned by two members of her family, Miss Joan Bucknall and Mrs Myra Bucknall. There are many letters and photographs and

Maggie's invaluable *Diary,* kept over a period of over 70 years.
Fred: most of these have been destroyed many years ago, but a small residue remains in the hands of Mr and Mrs Alan Frankland.
Sophie: again many have been destroyed though a selection has been preserved by friends of one of her descendants, Mr and Mrs Patrick Campbell. They consist mainly of letters to and from Sophie in the early 1870s.
Percy: the only one of Frankland's children to follow the father's profession, Percy became Professor of Chemistry at Dundee and Birmingham. His portion of the letters is by far the largest single collection, with much of scientific interest. However even this is not as large as it once was. It is owned by Mr and Mrs Raven Frankland.

Needless to say we are greatly indebted to all the owners for permitting us to see, and then to microfilm, documents in their possession. I am deeply grateful to them all for permission to quote from their material. We owe a special debt of gratitude to Raven and Juliet Frankland for allowing access to their huge collection, for introducing us to some of the other owners, for so kindly welcoming us to their home, and for their friendship over many years.

I must also record my thanks to the Open University for supporting my research over a long time and in many ways. Their generosity has been magnificent. I am grateful to my colleagues in both Department and Faculty for support and encouragement, especially Gerrylynn Roberts and Noel Coley whose own research interests have overlapped with mine for a very long period. Others who have helped more than they know include Bill Brock and Alan Rocke. Professor Rocke has given much valuable assistance, especially in making sense of the (to me) almost illegible letters to Frankland from Kolbe, while Professor Brock helped me in the very early days to see Frankland's importance for the history of science and has since been a fount of advice and goodwill. Then, in the matter of the X-Club, I have had stimulating discussions with Jim Moore and Ruth Barton, while on that topic and related matters I have greatly profited from the work of one of my former research students, Andrew Harrison. Items of Franklandiana have come my way through the kindness of Frank James, Harold Booth, Alan Comyns, Keith Mason and John Chadwick, and other useful information has come from Ray Anderson, Alec Campbell, D G Duff, Christopher Hamlin, John Rowlinson and the late Mike Hall. I am grateful to librarians and archivists in many places: at the Royal Society, the Royal Society of Chemistry, the Royal Institution,

the Royal College of Surgeons of England, the Royal Pharmaceutical
Society, the Royal Greenwich Observatory; at the British Library
and at many central public libraries, including those in Croydon,
Kendal, Lancaster, Manchester, Newcastle, Preston; at the Public
Record Office at both Kew and Chancery Lane; at the Record
Offices of Cumbria, Essex, Lancashire, Manchester, Nottingham-
shire, Surrey; at the Auckland Public Library and the Alexander
Turnbull Library in Wellington, New Zealand; at the Deutsches
Museum at Munich, and the Chemical Heritage Foundation at
Philadelphia; at the Science Museum, the National Railway
Museum and the Geological Museum; at the Wellcome Institute
for the History of Medicine and the Institute of Historical
Research; at Lambeth Palace, the Friends' Meeting House in
London, the parish churches of St Martins-in-the-Fields and
Wimbledon, and the United Reform Church, Lancaster; at the
Marquess of Salisbury's archive in Hatfield House; and at the
universities of Birmingham, Edinburgh, Exeter, Lancaster, Liverpool,
London, Manchester, Strathclyde and Imperial College. Above all
I want to thank the staff of the libraries at the Open University and
the University of Cambridge, who, in many different ways, have
made my task immeasurably easier and more pleasant.

Finally it must be obvious that my greatest debt of thanks must
go to my wife Shirley whose long-suffering endurance has been
combined with historical expertise in her own right, and whose love
and companionship have helped in a thousand ways. We were
equally involved in the establishment of the microfilm archive, and
she has made a far more detailed study than I of the rich source
material available in the *Diary* of Maggie Frankland. So I thank her
most of all.

COLIN A. RUSSELL

Acknowledgements

Permission to reproduce material in their possession is gratefully acknowledged from

The Royal Society
The Royal Society of Chemistry, Library and Information Centre
The Royal Institution
President and Council of The Royal College of Surgeons of England
The Royal Pharmaceutical Society of Great Britain

The Public Record Office
Lancaster Public Library
Nottinghamshire Archives
British Geological Survey Archives, Keyworth
University of Birmingham Library
Syndics of Cambridge University Library
University of Exeter Library and Information Services
University of London Library
The Archives, Imperial College, London

Laporte Industries Ltd., Widnes
ICI Archives, Nobel Division

Philipps-Universität, Marburg
Deutsches Museum, München
The Alexander Turnbull Library, Wellington, New Zealand

The Marquess of Salisbury
Miss Joan Bucknall
Mrs Myra Bucknall

Mr and Mrs Patrick Campbell
Mr and Mrs Alan Frankland
Mr and Mrs Raven Frankland

Chapter notes

The following abbreviations are used for frequently cited books:

Sketches	Edward Frankland, *Sketches from the life of Sir Edward Frankland*, edited and concluded by his daughters M. N. W[est] and S. J. C[olenso], privately printed 1901, 2nd edition 1902; unless otherwise stated the second edition is intended
Experimental Researches	Edward Frankland, *Experimental Researches in Pure, Applied, and Physical Chemistry*, van Voorst, London, 1877
Lancastrian Chemist	C. A. Russell, *Lancastrian Chemist: the early years of Sir Edward Frankland*, Open University Press, Milton Keynes, 1986

The common practice is followed of indicating the *Dictionary of National Biography* as *DNB*, and the *Dictionary of Scientific Biography* as *DSB*. Also PRO = Public Record Office.

All archival material in private ownership has been copied on to microfilm at the Open University. Except for two items these are identified by a reference 'OU mf', followed by a three unit number as 01.03.0456. The first entry indicates the owner, the second the microfilm number and the third the frame number of the first page. The exceptions are (1) Maggie Frankland's *Diary*, and (2) Edward Frankland's water analysis notebooks, where the date references alone are sufficient to locate an item on microfilm; these are indicated respectively as 'MBA' and 'RFA, OU mf 09'. Such material is indicated as follows:

RFA: Archives of Mr & Mrs Raven
     Frankland   (owner reference no.: 01)
JBA: Archives of Miss Joan Bucknall (owner reference no.: 02)
PCA: Archives of Mr & Mrs Patrick
     Campbell   (owner reference no.: 03)
MBA: Archives of Mrs Myra Bucknall (owner reference no.: 04)
AFA: Archives of Mr & Mrs Alan
     Frankland   (owner reference no.: 05)

Where Edward Frankland is intended the single word 'Frankland' is generally used, except in authorship of joint papers or (in rare cases) to avoid confusion. Christian names are not usually employed in references except in the case of other members of the Frankland family.

CHAPTER I

Lancastrian
inheritance

1 Putting up the shutters

One October evening in 1845 a familiar ritual was taking place in
the ancient town of Lancaster. Autumn nights in that part of the
world can have more than a touch of chill about them and late
shoppers and work-people were scurrying home with nothing so
much in mind as the need for warmth and supper. They would
therefore have hardly noticed the actions of a young man outside
the druggist's shop in Cheapside, for they had been part of every
week-night's proceedings as long as they could remember. The
youth in question was probably the junior of three apprentices
serving their time at this establishment, and he was putting up the
shutters of his master's shop. Yet his act, simple and unremarkable
in itself, was laden with significance and symbolism for a
fellow-labourer who was in all probability watching from within
with that sense of responsibility that becomes a senior apprentice.
For him the closing of the shutters meant the end of an apprenticeship
and also the beginning of a wholly new life far away from his native
Lancaster.

The senior apprentice was called Edward Frankland.[1] He was
well known in the town for his muscular exploits with heavy goods
(he once carried a record two cwt sack of barley up a 'steep and
narrow staircase'), his efficiency behind the counter, his production
of immaculately tied parcels, his dispensing of prescriptions in
which (unlike some colleagues) he never made mistakes, and his
extra-mural activities as tooth-extractor and unofficial prescriber of

medicines, in which activity his grateful patients appeared to have greater confidence in him 'than in any of the duly qualified practitioners in the town'.[2] Now he was going, and the time had come to take a formal farewell of his master Stephen Ross.

What conversation passed between them we do not know. Frankland had quite enjoyed his work, especially towards the end, and (if the truth were told) had gained immeasurably from it. With the hindsight of another half-century he was to write a thoroughly demeaning account of his apprenticeship, 'six years' continuous hard labour, from which I derived no advantage whatever, except the facility of tying parcels neatly'.[3] However in 1845 the iron had not entered his soul, and for at least two reasons. He had not then experienced the joys and challenges of chemical research and was therefore not in a position to prescribe appropriate courses of preparation for it, and to condemn those that were inappropriate; and he had not then conceived a world view in which the simple piety of his master Stephen Ross would have been deemed reprehensible. In any case nothing could extinguish his natural exultation at the prospect of freedom and adventure. By rights he should have served Ross until his 21st birthday in January 1846, but his master had generously remitted the last three months and even Frankland, for all the jaundiced perceptions of 50 years later, had to admit his kindness and his 'real interest in the welfare of his apprentices when they left him'.[4] So it is likely that, with expressions of gratitude on one side and good wishes on the other, they parted.

The person who stepped through the shop doorway for the last time that night was of striking appearance. Silhouetted against the light he could be discerned to be a tall young man, of spare frame, with a largish head. At 5 ft $10\frac{3}{4}$ in. he was in fact rather tall for his time. He had a fair and ruddy complexion and a mass of light brown hair.[5] From his broad hands one could hardly have guessed at his extraordinary manual dexterity, though the thick glasses he always wore did not conceal an intensity of feeling and quickness of observation that marked him out from most of his fellows. But nothing in his appearance might have led a casual onlooker to recognise an individual whose restless energies and burning ambitions were to lead, or rather drive, him to the very pinnacle of his chosen profession. As he left the shop it is inconceivable that his mind was not thronged with memories from those long years that, to a young man of his vigour, must have seemed an eternity.

Fig. 1.1. The chemist's shop where Frankland was apprentice.
(J. Bucknall Archives.)

Glancing up at the shuttered windows he could see the familiar totem of a pestle and mortar displayed above the lintel. Another pharmacist in the town displays one to this very day. They were symbolic not only of a druggist's trade but also of an apprentice's drudgery. Ruefully the ex-apprentice would reflect on much larger versions of the same equipment installed on the premises, one 20lb iron pestle being rotated by a vertical rod extending through the ceiling. With apparatus of this kind he had spent hours at a time grinding mixtures as various as cocoa, 'Spanish flies' (cantharides), and mercury and lard to make mercury ointment. Not all his duties were so demanding on his considerable physical stamina. They had ranged from the selling of tea, coffee and nutmegs across the counter on a Saturday evening in 'tidy' dress, to the vending of 'all dirty and disagreeable articles', of which the three constituents of shoe blacking were typical: bone black, treacle and oil of vitriol. As he gained in seniority he was allowed to serve drugs as well as groceries and to make medicines. The precision, accuracy and tidiness required and developed by the last responsibility were to serve him well in the years ahead. But he was not to know that then.

Cheapside was an ancient thoroughfare, the mediaeval Pudding Lane. It runs south from Market Street and soon dissolves into Penny Street, an even older road in which Frankland and his parents had their residence. As he stepped out into the darkness and at once turned right Frankland may well have spared a thought for his former companions at Ross's shop. They were an odd collection, though no odder than any other group of working lads assembled to learn a trade. Had he but known it that tiny shop was to produce from the apprentices during his own half dozen years there no less than three leaders of Victorian chemistry. He often recalled them. There was Robert Galloway, 'rather gloomy and ascetic in disposition' who was later to follow in Frankland's footsteps and then become Professor of Practical Chemistry at the Museum of Irish Industry in Dublin. More congenial company had been afforded by George Maule, 'full of jocularity and spirits' whose subsequent adventures with synthetic dyestuffs made him a captain of the British chemical industry. He became wealthy enough to devote early retirement to hunting and field sports. Several other fellow-apprentices achieved more modest success in their later years.

Chiefly, however, Edward's thoughts would turn to Stephen Ross. As he rapidly passed from Cheapside to Penny Street he could make out the recently erected church of St Thomas, in the building of which Ross had been 'one of the moving spirits'. Unashamedly Low Church and uninhibitedly evangelical St

Fig. 1.2. Penny Street, Lancaster, the young Frankland lived in a house in this street. (Lancaster Public Library.)

Thomas's reflected the values of Stephen Ross that were, in later years, to be such a problem to Edward Frankland. But at this turning-point in his life he cherished no such antipathy, for he himself had recently embraced a similar form of evangelical Christianity, though in the context of nonconformity, not Anglicanism. At High Street Congregational Chapel he had embraced this faith with the enthusiasm that marked all his activities. He attended meetings, exhorted his parents and taught in the Sunday School; only one month previously he had been recorded as 'one of the most regular attenders' since January 1844.[6] Ross would have approved and encouraged, providing, of course, that it did not interfere with his duties.

Such reflections (if indeed they were entertained at this moment) would have been rapidly cut short by his arrival at no. 55, the terraced house that, since he was ten years old, had been home. Small and unpretentious, it was adequate for the modest demands of the three inhabitants: Edward, his mother and step-father. Margaret Frankland had lived in Lancaster since her son was five years old, taking in boarders for a living. One of these gentlemen was William Helm whom she married within the year. He was

Fig. 1.3. Margaret Helm, *née* Frankland, mother of Edward Frankland.
(J. Bucknall Archives.)

Fig. 1.4. William Helm stepfather to Edward Frankland. (J. Bucknall Archives.)

eleven years younger than she but, despite the disparity in age, the marriage was happy and long-lasting. Edward Frankland was devoted to them both throughout their lives, and indeed owed them a great deal. William Helm is a classic example of a trend that his step-son was to demonstrate in a more spectacular manner: upward mobility. In successive Directories or census returns he is described as :

- cabinet maker
- railway guard
- victualler
- agent
- gentleman

Cabinet-making had long been a strong tradition in Lancaster, with fine woods a staple imported commodity. The most famous firm was Gillows, though William Helm does not seem to have ever been associated with it. At this time, in 1845, he had already acquired the title of 'railway guard' but legacies of his former trade remained. Some of his furniture survives, very competent examples of careful craftsmanship. He had already passed on many of his woodworking skills to young Edward.[7]

After Helm had entered the family circle Frankland recalled that he 'was rather severe with me, and, with a thin stick, gave me many a beating which I probably well deserved'. Yet good relationships ensued and William and Margaret Helm are frequently encountered in the next 38 years of Edward's life. His mother Margaret (of whom more anon) had been the dominating influence in the development of her son's character. Despite 'a very scanty education' she taught him to read, imbued him with a love of nature and was always ready with an answer to the ceaseless flood of questions. Writing 30 or so years later he described her thus:

> She is stout and symmetrical and has had almost uniformly robust health from birth to the present time. She has brown hair which is not yet grey, fair and ruddy complexion, a nervous and sanguine temperament . . . Considering her most deficient early training my mother is a woman of most remarkable intellect and of great energy and decision of character.[8]

It is interesting that in another draft of the document from which these words were extracted several of the phrases are identical to those he actually used about himself. There can be no doubt that her son inherited from his mother more than just a certain physical likeness.

That night the little family must have had much to discuss, and doubtless did so. For not only had Edward finished his apprenticeship, he now had an opportunity that would take him far away from the town in which he had grown up, and there were many preparations to be made. Maybe in the small hours they allowed themselves to reflect on the changes that had befallen them in their adopted town. And it cannot have escaped their notice that Lancaster itself was beginning to experience changes that would be as profound as those already affecting just one insignificant family.

2 The Lancaster of Edward Frankland

At this time Lancaster[9] was far from the bustling provincial centre that might have been expected to nourish indigenous scientific talent. In fact it was just the opposite. Its 1841 census recorded less than 15 000 inhabitants. Always the poor relation of its historic rival York, it never possessed an Anglican cathedral and did not become a city until 1937. Religion in Lancaster was marked by one or two thriving nonconformist chapels, a long-established tradition of Catholic recusancy and the endowment and opening of a few new Anglican churches; the Tractarian movement from Oxford made little mark in this northern outpost (1845 was the year in which Newman made his historic pilgrimage from Canterbury to Rome). Culturally the town was beginning to awaken. The Lancaster Choral Society was founded in 1836, an early example of a movement that was to sweep many provincial towns in the 1840s. Frankland began to learn singing by the sol-fa method following a visit to the town from the singing-teacher John Hullah (1812–84) in 1843. This enabled him to join the Choral Society and take part in several concerts, including one in which the soloist was the famous contralto Maria Hawes (1816–86).[10] One part of Frankland's Lancastrian inheritance was an enduring love of music, particularly that of Handel, Haydn and Mendelssohn. Even an apprentice of Stephen Ross had some time for leisure.

Towering over the town is its historic castle, home to the Assize court where, for example, trials of the militant chartists began two years previously. Yet too much should not be deduced from this;

Lancaster experienced little industrial unrest for the simple reason
that the vast Lancashire textile industry never reached quite so far
north. There had been some expansion of cotton manufacture in
the 1830s but stagnation after the depression of 1841/2. The port
of Lancaster, once famous for its imports from the West Indies of
mahogany and other fine woods, and infamous for its slave trade,
had gone through hard times and was to experience worse in the
future. The year 1833 was the first since at least 1750 when no ship
from the West Indies called at Lancaster. The silting up of the
River Lune and the opening up of rival ports, particularly
Liverpool, were to contribute further to its industrial decline;
indeed in the very year that Frankland left his native town its
imports reached a temporary and final maximum.[11] Industrial
progress did reach Lancaster in due course, even at the domestic
level, as may be illustrated by the arrival of gas-lighting at the
Parish Church in 1844.[12] Four years earlier an Exhibition of Arts
and Manufacture displayed such prodigious novelties that it
remained open for four months. Yet mortality rates were considerably
worse than the national average, a major contributor being the
primitive means of sewage disposal, whether by drains or cess-pools.[13]
It was commonly stated that life-expectancy in Lancaster at that
time was lower than that in the worst parts of London.[14]

However in one important respect the old county town was
experiencing modernity at first hand. By the early 1840s it had
become connected to Preston by rail, irreversibly damaging the
business of the canal which since 1797 had linked the two towns; no
less than 63 packet horses were sold off by the canal company in
1842.[15] The Lancaster and Preston Junction Railway[16] terminated
at the top of Penny Street, the old station now being a home for
nurses. It must have been this company which acquired the services
of William Helm for no other line had reached the town in the early
1840s. In 1845 the Railway Mania was at its peak and plans were
afoot for a new line northwards, the Lancaster & Carlisle Railway.
The river Lune was bridged, and the 'old' station by-passed, in the
following year. For young men looking further afield for improvement
the prospects of life beyond Lancaster had never been more
inviting or more realistic. The important thing was to go!

How then did a place as economically and culturally backward as
early nineteenth century Lancaster offer any prospects of scientific
advancement? For Edward Frankland attendance at no less than

eight schools (two in Lancaster) had largely disappointed him in his quest for scientific knowledge. Efforts to learn what he called 'facts' – especially those about the natural world – were an exercise in impotent frustration. In this respect Lancaster Grammar School was the worst of all, convincing him of the utter uselessness of a classical education.[17] However his lifelong fascination for chemistry was nourished by a few aspects of his formal education, chiefly at a school in Cable Street, run by a remarkably progressive entrepreneur James Willasey. Here pupils were encouraged to make and display their own experiments, and the syllabus was 'modern' in its exclusion of classical and religious studies. Frankland would have stayed there if Willasey had been able to teach Latin, deemed by the Helms to be necessary. Unfortunately this was not the case so the Grammar School was to complete Frankland's school experience. When the *ennui* of the classics was succeeded by the tedium of apprenticeship, any hope of studying science must have seemed forlorn. Yet, apart from the very limited advantages of a disciplined approach to chemical substances, there were other hidden benefits from those years in the druggist's shop.

Somehow Frankland, and several other youths of his own age, managed to find opportunity to study in such spare time as they could find, mainly in the evenings. This happened in several ways. First there was the Mechanics' Institute founded in 1824, not as some inept attempt to maintain social control over a work-force that showed little inclination to revolt, but as an apparently genuine attempt to benefit the local mechanics and apprentices. One of the earliest in Britain, it was for a long time simply a library and by the time Frankland had joined the Ross establishment it had moved to Penny Street, within a stone's throw of his house. Even his last years at school had been relieved to some extent by the opportunity to borrow books and now he used the facility to the full. He was deeply impressed with Thomas Day's *Sandford and Merton*, and with the *Scientific dialogues* of Jeremiah Joyce, both highly popular volumes at that time. More inspiring than these was the classic *History of electricity* by Joseph Priestley. Caring nothing for merely reading about experiments he tells us that he proceeded with its help to construct an electrophorus, a Leyden jar [condenser], and a rotary generator. Further experiments in electricity and chemistry followed, mostly at home until an accidental explosion of hydrogen aroused parental wrath so effectively that from then on all home experiments were banned. The makeshift laboratory of a friend was no longer available when Frankland left school, but help was on the way.

There were in the town three medical men of progressive outlook, Christopher Johnson, and his sons Christopher and James. Concerned for the science-hungry apprentices they provided informal teaching, loan of books, and even a cottage laboratory. The latter was established by evicting a tenant from his dwelling on Green Ayre. Here apprentices would conduct experiments either from a textbook or with direct help from one of the Johnsons. Later a chemical class was established at the Mechanics' Institute, now moved from Penny Street to Back Sun Street, again with self-help as the major emphasis. Occasional lectures and other help were given by a small group of men that included the Johnsons and Stephen Ross.

This was not systematic instruction. Nor was it by any stretch of the imagination an adequate laboratory training. But it was all there was and the fact remains that it all helped to kindle in more than one man a burning passion for science and a longing to know more. When James Johnson wrote to Frankland many years later he reminded him of 'the day of small things' from which, however, great things could emerge, referring particularly to the cottage laboratory on Green Ayre.[18] In 1891 he declared that 'to the Johnsons – father and son – the late Mr. C. Johnson and the late Dr. James Johnson, he was indebted for whatever knowledge of chemistry he possessed on leaving Lancaster'. On the same occasion another member of the old chemical class, Sir William Turner, recalled: 'There was in those old days . . . a small lamp of science hung in Lancaster, and those who lit it were the family of Johnsons'.[19]

One of the most remarkable features of Lancaster at that time was its ability to nurture a number of young men later to make their mark on British scientific culture. Apart from Frankland there were Sir Richard Owen and Sir William Turner (anatomists), James Mansergh and Sir Robert Rawlinson (engineers), Edmund Atkinson, George Maule and Robert Galloway (chemists), and several others. Few achieved the academic distinction of William Whewell, Master of Trinity College, Cambridge; in 1845 he donated a collection of books to the library of the Mechanics' Institute in his native Lancaster.[20] How these 'hard progeny of the north' were enabled to grow from their unlikely soil and then to flourish in the world of science and learning is, of course, another story.

It was on the advice of one of his medical friends (Christopher Johnson) that Frankland determined to seek his fortune in London, and within a few days of leaving his apprenticeship embarked on the long journey to the capital. His destination was the laboratory of one of England's brightest young chemists, Dr Lyon Playfair, in Duke Street, Westminster. It appears that Johnson had written to both Thomas Greene, MP for Lancaster, who had a London house two doors from Playfair's laboratory, and to the Earl of Lincoln, Playfair's employer. He must have been extremely convincing for within days Lord Lincoln was writing to Greene with a promise to accept Frankland, and the MP in turn conveyed the good news to Penny Street, enclosing a letter of introduction to Playfair. There was now no reason for Edward to remain in Lancaster.

3 Journey to London

Luggage packed, arrangements made and ticket purchased Frankland left the Penny Street home and walked uphill the few hundred yards to the station. His train was almost certainly the 9.20 a.m. express.[21] His reflections can be imagined as he boarded the bright yellow carriage and the train pulled slowly out of the terminus.

As the familiar skyline of castle and priory faded from view the natural pain of parting would be diminished by recollection of the stable bonds he already had with his parents, promises to write by the new Penny Post, prospects of returning for holidays *etc*. One other consideration may have crossed his mind. He now had a girl friend, a Mary Medcalf, to whom he had actually declared his feelings and expressed his wish to marry her. Unable to offer any means of support he had refused to allow her to respond, and she for her part declined to let him speak to her parents. Her home was in Kendal but she had been on an extensive visit to Silverdale near Lancaster. Here they had recently met, but most unfortunately her mother was always present. Unable to find a way to meet in the absence of unwelcome company, Frankland decided to write before he left Lancaster and had done so in the last few days. Misfortune was doubled by the return of his letter, opened but unanswered, in an envelope bearing her father's hand. Despite this demonstration of paternal disapproval he cherished the hope that Mary would find a way to write in the next few weeks and, consoled with that thought, addressed himself to the journey to London. However she never wrote, and gradually Frankland abandoned hope in the belief that she was destined by her parents to marry 'some rich Kendal manufacturer'.[22] In fact she did not marry for another ten years, and Frankland often reproached himself for not having tried to see her after the return of his letter.

The train gradually accelerated on the level track across the eastern edge of the Fylde. Familiar sights from his boyhood crowded in upon him: rivers where he had bathed and fished, fields that he had tramped with dog and gun, the famous Lancaster to Preston canal where he had occasionally caught a packet boat, and the hills on the eastern horizon that were the eternal backdrop to many a youthful exploit. For he was fast approaching the place of his birth and earliest childhood. As the market town of Garstang came into view they may have slowed down to stop at the now-vanished station. To the west, though out of view, was the little township of Churchtown where, on 18 January, 1825, Frankland had been born. The small cottage that marked his birthplace survived until the 1960s. His mother, Margaret Frankland, was daughter of Molly Dunderdale, of good Garstang yeoman stock, and her husband John Frankland, an itinerant calico-printer working at the great print-works of Fieldings nearby. Since Edward took his mother's maiden name it was obvious to all that he was one of several illegitimate children born in the parish that year. He may have ruefully reflected on all the embarrassment and teasing caused by that situation as the train drew out of Garstang to continue its journey south.

Garstang is roughly halfway between the two ancient towns of Lancaster and Preston. The little stations of Brock, Roe Buck and Broughton flashed by as they traversed rich agricultural country dominated by the Bowland fells in the distance. The speed (which can never have exceeded 30 m.p.h.) would be further reduced on the approaches to 'proud Preston'. At this place Frankland left the train for a remarkable reason. The Lancaster and Preston Junction Railway had built a station near the canal basin a few hundred yards north of the thoroughfare of Fishergate, even though the rails continued through a short tunnel to the station of the North Union Railway who owned the line southwards. By a piece of skulduggery unusual even in Victorian days the North Union demanded the

Fig. 1.5. The birthplace of Edward Frankland, Churchtown, Garstang.
(J. Bucknall Archives.)

outrageously extortionate toll of 6d. for passengers travelling
through the tunnel under Fishergate connecting the two stations.
According to Frankland this was about 15 times the rate they were
allowed to levy. Unsurprisingly most passengers with 'through'
tickets declined to pay, left the train and walked the short distance
to pick up their train at the next station. The North Union,
defeated by its passengers' meanness, would as often as not
despatch the train south before they could reach it. Whether
Frankland suffered such a fate is not revealed, nor how concerned
he was about the possibility. He probably had other things on his
mind.

It was in Preston that his story – in one sense – began. Two
hundred yards from the former North Union station (now Preston
Station) lies a leafy square of mainly Georgian houses. In one of

these young Margaret Frankland had once been in domestic service. Her master was Edward Gorst, distinguished lawyer and Deputy Clerk of the Peace. His younger son (another Edward) was destined to follow him in that high office. Not for the first time in history a romantic attachment developed between a maid 'below stairs' and a son of the family. In this case when Margaret discovered she was pregnant she knew full well that young Edward Gorst was responsible (if that is the right word). News of his indiscretion had, at all costs, to be suppressed and Margaret Frankland was sent back home with a handsome annuity. Young Edward Frankland received his mother's maiden name and his natural father's Christian name. All his life he lived in the shadow of this family secret, with powerful reasons for suppressing enquiries into his background. It is inconceivable that such matters were far from his mind as he entered the station of the North Union Railway.

The next few miles might have had a vague familiarity as they passed the Lancashire towns of Wigan and Warrington. Edward Frankland had never ventured further south than that. Before him lay many further hours of travel through the territory of the North Union, the Grand Junction and the London and Birmingham Railways. Occasionally they would pass through small towns, hitherto only names to young Frankland, and then the strange assemblage of railway buildings (and little else) called Crewe, and Stafford with its crumbling castle. They were now traversing the line that, according to C S Lewis, ran 'through the dullest and most unfriendly strip in the island'.[23] Frankland would surely have agreed that it compared badly with his native haunts.

There was little to see except flattish fields. Yet even here was something that Frankland might have discerned as relevant to his own enterprise. Only a year or two previously the great German chemist Liebig had toured England in a bid to promote his views on the application of chemistry, most notably to physiology and agriculture. Britain, he argued, needed chemists who were no longer well-meaning amateurs but trained professionals who knew the rigours of systematic chemical analysis. They could then discover which soils would yield good crops and which, like many visible from the train, required chemical improvement. Landowners were beginning to get his message and that very year of 1845 was proving to be an important one for the history of British chemistry.

In October a long process of agitation and argument came to a head with the establishment in London of the nation's first institution devoted solely to teaching and research in that subject. The Royal College of Chemistry[24] had appointed (at Prince Albert's suggestion) a German director, A. W. Hofmann, with an academic programme based directly on the immensely innovative and successful model created at Giessen by Justus Liebig. Indeed, Liebig's was the guiding spirit behind the whole enterprise. He had been Hofmann's research supervisor and during the early 1840s had been stomping England in a personal crusade for a more disciplined and 'professional' approach to chemical science. Other institutions were beginning to teach practical chemistry, but none had the enduring effect of the Royal College of Chemistry, eventually to find reincarnation as Imperial College.[25] Its creation owed everything to a recognition of the practical importance of chemical science, not least to agriculture. It was also a natural outcome of the foundation of the world's first national institution for chemistry, the Chemical Society of London, four years previously.

Any reveries on rustic chemistry would be rudely interrupted as the train left Stafford and headed towards the Black Country. It passed through Wolverhampton, Walsall and eventually the great city of Birmingham.[26] Here was no rural paradise, but a concentration of industrial activity that in Frankland's brief experience can only have been approached in the textile city of Manchester. Birmingham's belching factory chimneys spoke at least of chemical reactions if not of chemistry, as did the work of metal-extractors, electroplaters and a multitude of kindred trades. Only a few miles away, at Ironbridge, lay the very cradle of the Industrial Revolution. The railway carriage gave to Frankland, as never before, a further view of how his subject could be applied, to the benefit of heavy industry. Such application was before long to be a subject of consuming interest to him.

As he left Birmingham he still had nearly $5\frac{1}{2}$ more hours in the train but with the coming of dusk his thoughts would focus with fresh clarity on the life awaiting him in the metropolis. Little did he know how intimately in the distant future his fortunes were to be intertwined with those of the new Royal College or with the Chemical Society. Nor could he have guessed that of all the Englishmen then alive he more than any other was to revolutionise chemistry in its theory, practice, teaching and organisation.

London was to experience an outburst of chemical activity never known there before. He would arrive just in time.

At last the capital was reached, at 8.45 p.m. if the train was on time, in an unprepossessing train-shed called 'Euston'. Among the passengers descending from their uncomfortable four-wheeled carriage was the young Lancastrian. Temporary lodgings awaited him at Bishopsgate, in the heart of the City.[27] Clutching his modest baggage he stepped out into the night.

Notes

1 Further details, and fuller references, may be found in the author's *Lancastrian Chemist: the early years of Sir Edward Frankland*, Open University Press, Milton Keynes, 1986, which deals in detail with the first twenty or so years of Frankland's life.

2 *Sketches*, 2nd ed., p.38.

3 *Ibid.*, p.21.

4 *Ibid.*, p.39.

5 Frankland, copy-letter to F. Galton, April 1874 [AFA].

6 *Men teachers' register, 7 January 1844 to 30 December 1866*, High Street Chapel, Lancaster [Lancaster Public Library].

7 In the 1841 census for 117 Penny Street Helm is already described as 'railway guard' [PRO, HO 107/503], but the electoral register shows him as occupying 'house *and shop*' until 1843: *Register of electors . . . for the Borough of Lancaster*, 1842/3 [Lancaster Public Library]. Possibly he was 'moonlighting', but in any case he was setting an example that his step-son was later to follow with such *éclat*: occupational 'pluralism'.

8 Frankland, copy-letter to F. Galton, 12 April 1874 [RFA, OU mf 01.01.0111]; this is possibly an earlier draft than that in note 5.

9 See especially A. White (ed.) *A history of Lancaster*, Ryburn Publishing, Keele University Press, Keele, 1993.

10 *Sketches*, 2nd ed., p.39; the Hawes concert was probably that on 4 September 1845.

11 M. M. Schofield, 'Outlines of an economic history of Lancaster: Part II – Lancaster from 1800 to 1860', *Trans. Lancaster Branch of the Historical Association* no. 2 (1951), 1–138.

12 'Cross Fleury', *Time-honoured Lancaster*, Lancaster, 1891, repr. 1974, p.578; the streets had been lit from 1827.

13 R. Owen in evidence to *Health of Towns Commission*, Appendix to Second Report, *P.P.*, 1845, XVII, pp.217–30.

14 See also E. Sharpe, *A history of the progress of sanitary reform in the town of Lancaster 1845–1875*, Lancaster, 1876.

15 *Time-honoured Lancaster* (note 12), p. 577,

16 M. D. Greville and G. O. Holt, *The Lancaster & Preston Junction Railway*, David & Charles, Dawlish, 1961.

17 Frankland is careful to qualify his condemnation of the Grammar School: 'I am bound to say that Whewell and Owen received their early education at this Grammar School, though not, I think, under the same master': Frankland to F. Galton (note 5).

18 J. Johnson to Frankland, 13 December [late 1870s?] [RFA, OU mf 01.03.0729].

19 *Lancaster Guardian*, 31 October 1891 (report on the formal opening of the Storey Institute, Lancaster).

20 *Time-honoured Lancaster* (note 12), p.579.

21 Bradshaw's *Guide*, 1845. The train conveyed only first class passengers but was the only one during the day with through coaches to London. His brief account of the journey (*Sketches*, 2nd ed., pp.257–8) makes no reference to an overnight trip. The first class fare was £2.7.6d.

22 *Sketches*, 1st ed., 1901, p.41.

23 C. S. Lewis, *Surprised by joy*, Fontana, London, 1959, p.25.

24 G. K. Roberts, 'The establishment of the Royal College of Chemistry: an investigation of the social context of early-Victorian chemistry', *Hist. Stud. Phys. Sci.*, 1976, **7**, 437–85.

25 An event celebrated jointly with the Royal Society of Chemistry in September 1995.

26 The present avoiding line *via* Tamworth and the Trent Valley was not then in existence and all trains between London and Lancashire passed through Birmingham.

27 The address '13, Primrose Street, Bishopsgate' is on an envelope addressed to 'Mr. Frankland' and sent from Lancaster in October 1845 [RFA, OU mf 01.02.1303].

CHAPTER 2

The road to discovery

1 Westminster, Putney and a new acquaintance

By the 1840's Britain was witnessing an unprecedented campaign
to awaken public awareness to industry's need for science. Most
prominent on the agenda was chemistry. For civil engineers
concerned with public water supplies for the rapidly growing urban
population, chemical knowledge was obviously desirable. So it was
to those involved in agriculture, whether tenants of small-holdings
or owners of great estates. Foreign competition and the threatened
repeal of the Corn Laws underlined the need for improved
crop-production, which Liebig and others had insistently connected
with use of the new chemical fertilisers. Often these same
land-owners had mining interests which chemical expertise might
also help to promote. Eventually the crusade, spearheaded by the
Prince Consort, led to the Great Exhibition of 1851 and much else
besides. But, in the mid 1840s it was spawning all manner of novel
and interesting projects, including a reconstituted section of the
Department of Woods and Forests, known as the Geological
Survey. To this small band of surveyors and geologists had been
appointed a chemist, described as 'The Pupil of Professor Liebig
and the representative of his opinions in this Country'. Indeed,
another chemist, Richard Phillips, had held office for six years, but
his health was failing, his work-load was too great, and the
newcomer was commended by no less a person than the Prime
Minister himself, Sir Robert Peel. He was appointed 'Organic
Chemist' to the Survey and given a laboratory at Duke Street,
Westminster.[1] His name was Lyon Playfair.[2] Playfair, born in 1818,
was on the threshold of a chemical career that would soon bring him

into the confidence of the Prince Consort and ultimately make him an elder statesman of British science, Liberal MP, Deputy Speaker of the House of Commons and the first Baron Playfair of St. Andrews. His diminutive stature belied his restless energies and questing ambition.

It was to Playfair's Westminster laboratory that Frankland turned his steps the morning after his arrival in London. He was immediately plunged into a totally new experience that was to redefine his goals thereafter: systematic chemical analysis. In learning analytical chemistry he was greatly helped by Playfair's young assistant Thomas Ransome, an exceptionally gifted, cheerful man who later became a manufacturing chemist. One of his sons, Cyril, became Professor of History at Leeds, and he was father of the writer Arthur Michell Ransome, who was thus grandson to Frankland's amiable colleague.[3] By dint of sheer hard work and Ransome's friendly encouragement Frankland made spectacular progress. There was a wide variety of analytical tasks to challenge him.

In the summer of 1845 Playfair had received a major assignment from the Admiralty: to undertake a survey of the British coals used for the Royal Navy.[4] Other requests followed in quick succession: could Playfair investigate means for smoke prevention? Could he look at methods for preventing gas explosions in coal mines?[5] All this was to command Playfair's attention in the year ahead, together with research on iron smelting, on the durability of stone for building purposes and other applications of chemical analysis to industry. Strange subjects for an 'organic chemist', perhaps, they reflected as much a new confidence of Lord Lincoln in science as Playfair's confidence in himself.

In 1846 the facilities and challenge of the Museum of Economic Geology (as his establishment was now called) were already too small for him, and a new opportunity arose at an institution recently established on the Surrey bank of the Thames: Putney College of Civil Engineering. This was yet another instance of the new belief that chemistry and engineering had much to do with each other and that in these exciting times knowledge should flow freely across boundaries that had hitherto kept such different branches of knowledge well apart. Without vacating his Museum post, Playfair accepted the Professorship of Chemistry at Putney College and managed to combine both responsibilities – and accept both salaries – at the same time. The Admiralty Coals Investigation was transferred to Putney from the tiny basement in Westminster. One of his assistants, John Wilson, was put in charge. But it is another of his assistants who will detain us much longer and who, in the spring of 1846, first made his way from lodgings in Lambeth[6] to the green fields of Putney. This was Edward Frankland whose progress in learning analytical chemistry had been so impressive that, at an *additional* salary of £50 per annum, he was now given the chance of passing on his new won knowledge to engineering students and of preparing the experiments for a lecture course delivered by Playfair in Putney. So, at the age of 21, Edward Frankland was appointed as Lecture Assistant to Playfair at the Putney College.

The lawns of Putney College sloped gently down to the river.[7] Once they had been gardens of the two large mansions enviably situated on the south bank of the Thames, so far from London as to be almost in the country. In 1846 they still retained their pleasant aspect, shaded by many trees, but no longer serving to isolate the inhabitants from each other or from the world outside. All 22 acres were the grounds of what must properly be called the College for Civil Engineers and of General, Practical and Scientific Education. Each July, at the annual prize day, they were thronged by the good and the great, by upwards of 80 students and by 'a great number of elegantly-dressed ladies'.[8] For the rest of the year they were the playground of the student body, a short-cut to the river and scene of numerous practical exercises in surveying, civil engineering and even lime-burning. Behind them the two former mansions, 'The Cedars' and 'Putney House', were now united by a corridor and range of study-rooms. The laundry had been converted into a sick-bay, and a coach-house to a classroom. From one set of stables, accommodation had been made for carpentry and other practical activities; from another had been fabricated a chemical laboratory and lecture room.

Putney College (as it was widely known) was one of many ephemeral institutions for technical education which rose and fell in Victoria's reign. Founded in 1839 from private philanthropy and investment, it never quite became the public institution it claimed to be, and it fell victim to defective management, both financial and academic. Student numbers gradually dwindled and the College closed in 1852. Yet despite its early demise the College did make a modest contribution to the education of the civil engineer.[9]

Fig. 2.1. Putney house, site of Putney College. (Author.)

Amongst the ancillary subjects on the curriculum was the science of chemistry (hence the lime-kiln in the garden), traditionally an appropriate pursuit for gentlemen and now seen as newly relevant to all kinds of engineering. The noble promoters of the College would probably have been extremely surprised to learn that it was the progress of chemistry rather than engineering that would give their institution a place in the annals of fundamental scientific advance. Conceivably some of them might even have been rather pleased.

When Playfair took office at Putney, for some months Frankland was able for the first time in his life to listen to a systematic course of lectures in chemistry, delivered by a master of the art, and with full emphasis upon the practical application of the science. It must have been as nectar to one whose previous experience had been limited to sporadic instruction by well-meaning medical men in their spare time, or being thrown in at the deep end in an analytical laboratory where one learned simply by example from one's fellows. Suffice to say that at the end of Playfair's course Frankland took the examination and passed it with credit, – the only written examination for which he ever sat.[10]

Frankland's connection with the Geological Survey at Westminster and the College at Putney continued for another year.[11] It is not clear how his time was divided between the institutions. By 1847 he was receiving mail addressed to him at Craig's Court (the original analytical laboratory occupied by Phillips) and at Duke Street where Playfair was in charge (when he could spare the time from other pressing duties).[12] Mineral analysis was a conspicuous feature of the Survey's work and must have occupied much of Frankland's time. It was a tedious business, but relief was at hand. Yet another of Playfair's assistants, Hermann Kolbe, showed a comradely interest in Frankland which was to have profoundly important consequences both for the young Lancastrian chemist and for science itself.

Kolbe was six and a half years older than Frankland and had already made something of a name for himself in chemical circles.[13] A native of Göttingen he had studied in the University of that city with Friedrich Wöhler and then moved to the University of Marburg where from Bunsen himself he learned the techniques of gas analysis. These were precisely the techniques needed by Playfair for his examination of fire–damp explosions in coal mines and so, on Bunsen's strong recommendation, Kolbe moved to London at about the same time as Frankland joined the Duke Street Laboratory. In fact, there was inadequate room for Kolbe and he was required to conduct all his gas analyses in his lodgings in Belvedere Road, Lambeth (somewhere very near the site of the present Royal Festival Hall),[14] just as Wilson had to decamp to Putney with the Admiralty Coals Investigation.

To assume that gas analysis was the prime interest of Hermann Kolbe, would, however, be a great mistake. True, he had already sent a paper to the Chemical Society 'On the formation of nitric acid in eudiometric combustions of gases mixed with nitrogen',[15] but other topics now became more important, particularly the action of an electric current on organic compounds. His doctoral thesis had been about the chlorination of carbon disulphide, which (amongst other things) led to the formation of 'carbon perchloride' (tetrachloride), pyrolysis of which yielded 'liquid carbon chloride' (tetrachloroethylene). From this he obtained by chlorination in the presence of water the substance trichloracetic acid, hitherto made only by the chlorination of acetic acid itself and reducible to the latter. It was thus a 'synthesis' of a genuine organic compound from non-organic materials, and deserved to rank with the classic synthesis of urea by his supervisor Wöhler in 1828. The matter may be conveniently summarised in modern symbols, thus:

$$CS_2 \xrightarrow{Cl_2} CCl_4 \xrightarrow{\text{heat}} Cl_2C=CCl_2 \xrightarrow[H_2O]{Cl_2} CCl_3.COOH$$

$$\text{organic matter} \quad CH_3.COOH$$

Such an achievement gave Kolbe a taste for organic synthesis, – of which indeed he may be said to have been the chief founder.[16] It also inserted him directly into the current controversy surrounding the electrochemical theory of the great Swedish chemist Berzelius.[17]

2 The hunt for radicals

Fundamentally this theory regarded molecules as dipolar, with the two parts united by electrical attraction. The polarities in a single binary molecule were not necessarily of equal magnitude, so a residual positive or negative polarity could enable it to link to another binary molecule, and so on. Acidic oxides (like SO_3) could combine with bases like (CaO):

CaO   SO₃
⊕ −   + ⊖   *(the circle denotes the*
                               dominant pole)
          +   --
       CaO . SO₃

The distinction between acids and their anhydrides was not yet clear. Perhaps the greatest achievement for the electrochemical theory was the attempt by Berzelius to unite in one overall scheme organic and inorganic chemistry. Organic compounds consisted of *radicals* held together in some way. Ether, for instance was an oxide of the radical ethyl. Acetic acid, once considered an oxide, had been reconstituted by Berzelius in the 1850s as a compound of methyl and oxalic acid, itself derived from water and a hypothetical oxide of carbon. Given the (false) atomic weights C$=$6 and O$=$8 he wrote:

$$\text{oxalic acid} : C_2O_3 + HO$$

Today we would halve the numbers of carbon and oxygen atoms to obtain correct C/H and O/H ratios. Acetic acid was then formulated as:

$$\text{acetic acid}: C_2H_3 + C_2O_3 + HO,$$

'C_2H_3' (or methyl) being called a *copula*, whose presence or absence did not seem to affect the affinities of the rest of the molecule. After all, oxalic acid can exist on its own as well as in combination with a whole range of hydrocarbon radicals as methyl. This theory of 'conjugation' owed much to Liebig, but was now invaluable to Berzelius because it meant that both acetic and trichloracetic acids could be formulated in similar ways (and they are very similar in properties). This had previously been difficult for Berzelius because he was unable to accept replacement of hydrogen by chlorine in the main part of a molecule. However, it did not seem to matter if this occurred in a copula. Thus we have:

$$C_2H_3 + C_2O_3 + HO, \text{ and}$$
$$C_2Cl_3 + C_2O_3 + HO$$

and these compounds, central to the Berzelian controversy, were precisely those synthesised by Kolbe.

Kolbe was not the last PhD to complete his thesis during his first job. In fact, he had moved to Marburg before completion and to a laboratory even more infused with a passion for the radical theories of organic chemistry, for Bunsen was concerned to demonstrate the radical position as strongly as possible. As far back as 1832, Liebig and Wöhler had made a whole series of compounds containing the hypothetical radical 'benzoyl', but Bunsen was to go one better. From an evil smelling liquid known to him as cacodyl and to us as tetramethyldiarsine he obtained also a series of derivatives, but in this case the radical itself was separable.[18] The fact that we recognise it to be a dimer, not a monomer, should not blind us to the sense of excitement that spurred Bunsen and his assistants to isolation of yet further radicals. The reactions were (in modern notation) those indicated on the opposite page.

It is, of course, characteristic of many inorganic compounds that they can be decomposed by electrolysis. Normally this is not true of organic substances, but one exception was by now familiar to Kolbe: the potassium salts of the aliphatic acids. So, in whatever moments Playfair allowed him from mineral and coal gas analyses, Kolbe devised a simple piece of apparatus to subject the salts to electrolysis, hoping thereby to gain a deeper insight into their constitution. His first experiments were promising, and he reported to the Chemical Society in 1847 the effects of a 'voltaic current' on a solution of valeric [pentanoic] acid. Several hydrocarbon products did indeed appear at the anode, and later on he was happy to identify one of these as 'valyl' (i.e., butyl, but it was in fact its dimer, octane).[19] Given this preliminary encouragement he turned to another strategy for confirming the presence of hydrocarbon radicals in fatty acids, and in this he enlisted the help of Edward Frankland.

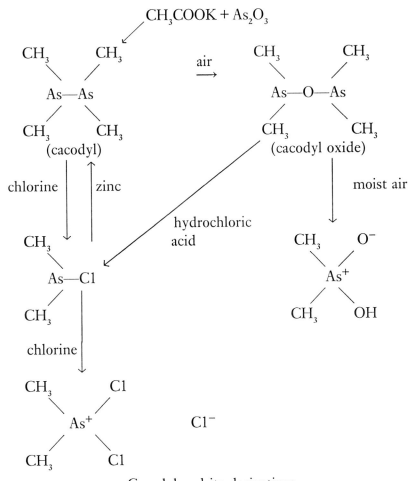

Cacodyl and its derivatives

The nitriles of valeric acid[20] and of benzoic acid[21] had recently been prepared and had been converted to the corresponding acids on hydrolysis. Their constitution, however, was highly problematic: only their empirical formulae were known with certainty. It occurred to Kolbe and Frankland, however, that they might, in reality, be cyanides. Since cyanogen could be hydrolysed to oxalic acid[22] the fatty acids in general could well be formulated like this:

radical + oxalic acid;

their nitriles would then be:

radical + cyanogen.

This formulation, however, had been made by Berzelius for the simplest acids as acetic and propionic. It so happened that the simple alkyl cyanides had recently been prepared[23] but no one knew if they could be hydrolysed to acids. If they could, the hypothesis:

nitrile = cyanide

would be confirmed, as also the general dualistic formulation of the aliphatic and aromatic acids.

Frankland observed 'to the solution of this problem Kolbe and I enthusiastically applied ourselves'. Using the method of Pelouze,[23b] Frankland treated potassium ethyl sulphate ('sulphovinate') with potassium cyanide in order to make ethyl cyanide, and then subjected the latter to boiling, concentrated aqueous potash. This reaction, of course, takes many hours to complete. Frankland at first seems to have used a distillation apparatus, repeatedly returning the distillate to the retort until no odour of the cyanide was perceptible.[24] The contents of the flask were eventually evaporated to dryness, acidified with dilute sulphuric acid and redistilled. The distillate showed all the characteristics of aqueous propionic acid. The theory was triumphantly confirmed and the result communicated to the Chemical Society on 19 April 1847.[25]

3 Cirencester

During the conduct of this research, Frankland was taking certain steps to improve his position as a chemist. The events throw interesting light on a number of aspects of career-making in mid-nineteenth century chemistry. In circumstances that are not entirely clear Frankland was offered a post as Professor of Chemistry at the Royal Agricultural College, Cirencester, founded in 1845 to teach 'the Science of Agriculture, and its various sciences connected therewith'.[26] A stipend of £100 per annum[27] would be additional to one of the same amount which he was now drawing as Playfair's Chief Assistant.[28] Playfair himself was already a shining example of how to occupy simultaneously several academic posts in plurality. Concerning the exact circumstances of this appointment, some doubt exists. According to himself, Frankland was 'offered' the post by the Principal.[29] A letter of commendation from R. Phillips, however, indicated that he was a 'candidate for the vacant post'.[30]

Playfair obviously played an important role in what followed, though possibly behind the scenes. The new Principal of the Royal Agricultural College was John Wilson, who until 1846 had been Playfair's own assistant on the Admiralty Coals Investigation; it was Wilson who offered the job to Frankland. The previous incumbent of the Chemical Chair was J. T. Way, a man who had already made his mark in agricultural analysis. Frankland rather implies that Wilson was anxious to get rid of Way, though in fact the latter was moving from Cirencester to become Consulting Chemist to the Royal Agricultural Society, in which he was succeeding Playfair (who was having to cut down on some of his many appointments). The whole affair underscores the importance of the lobby for agricultural chemistry at this time.

As part of his strategy, Frankland had decided to seek recommendations from colleagues in the Department. Writing on 28 March, Richard Phillips was able to say:

> I have great pleasure in stating my entire conviction, that should you attain the object of your wishes, you will discharge the duties of the office, not only with credit to yourself, but with great advantage to those who may receive your instructions.[31]

Andrew Ramsay (assistant geologist at the Survey) also expressed pleasure in supporting his application:

> This I can do not only from what I have heard of you from chemists well qualified to judge of your acquirements, but also from my own personal experience, having sometimes when at work in the laboratory had the benefit of your instructions.[32]

Nor were these the only commendations. From Playfair came a glowing testimony to Frankland's skill 'in all the different branches of analysis' which had so far included analyses of soils, plant ashes and organic substances.[33] The Principal of Putney College, the Rev. Morgan Cowie, acknowledged the 'great progress' made by students in practical chemistry under Frankland's guidance[34] while his old friend from Lancaster, James Johnson, wrote warmly of his ability to teach and demonstrate in chemistry and to conduct analyses.[35] Frankland's analytical prowess was also praised by Robert Hunt (Keeper of the Mining Records at the Museum of Economic Geology),[36] and his close colleague Hermann Kolbe.[37] Frankland had even attracted favourable attention from a visiting German Professor, who thought him 'distinguished among his countrymen by a true and scientific mind'.[38] Not surprisingly, Frankland was appointed to the job.

In the event this flurry of literary activity came to nothing. For some reason Frankland changed his mind and declined the post.

> Before I went down to Cirencester to commence my duties, we quarrelled about my sitting-room, which I had expected to have entirely to myself, but which I was now told I must share with several other professors.[39]

However, the Council minute of his appointment specified not merely 'the customary board . . . and one room', but also noted that 'Mr. Frankland would make the little room at the laboratory do as his sitting room'.[40] More plausibly, Frankland identified the source of his *volte face* as the advice of Kolbe. Increasingly dissatisfied with the routine inorganic analyses required of him, Kolbe longed for the freedom to pursue organic research in his former laboratory. Who better to accompany him than his recent collaborator, Edward Frankland? The pressure was on, and Frankland wrote:

> It was, in fact, owing chiefly to his graphic description of the advantages which would result from my working for a time with him in Bunsen's laboratory that I determined, to the astonishment of my friends, to throw up the Cirencester appointment. . . I decided to go over to Marburg with Kolbe in May 1847.[41]

4 Discovery at Marburg

In the second week of May, the two friends set off for the Continent, travelling by rail to Dover and steamer to Ostend. The details of the journey, his first trip abroad, were indelibly impressed on Frankland's memory: the crowds thronging Haymarket awaiting the singer Jenny Lind and impeding his journey to London Bridge; the smoothness of the sea-crossing; the 'atrocious' wines of Belgium; the river journey up the Rhine; the leafless vineyards on the banks; the ruined castles by the Rhine gorge; and the exhausting twelve hour journey from Frankfurt to Marburg by 'diligence'.[42]

The University of Marburg, founded in 1527 as the first Protestant University in Germany, once enjoyed the distinction of having also the first Chair of Chemistry anywhere in Europe. Its incumbent was Johann Hartmann in the early seventeenth century. There followed two centuries of chemical obscurity, however with the subject merely a handmaid to medical and life sciences. At length, the new chemistry of Lavoisier began to stir in Germany and in the year of Frankland's birth (1825) the Professor of Chemistry, Wurzer, took over a mediaeval monastery for his laboratory to accommodate the growing number of students. Wurzer's assistant (1839) and successor (1841) was the charismatic disciple of Liebig, Robert Bunsen. The ancient university on the banks of the River Lahn began to acquire a European reputation that owed nothing to its scenic magnificence or traditional student conviviality. Here was a new research school, comparable with the already famous one at Giessen, and promising the same kind of advantages to a young chemist prepared to work extremely hard: guidance, encouragement, example, team work and (above all) a well defined overall research programme.[43]

Warmly welcomed by Bunsen – who gave them two places in his laboratory – Kolbe and Frankland lost no time in getting down to work. First they extended their study of the hydrolysis of nitriles, and found that both methyl and amyl cyanide could be converted to acids, thus greatly increasing the credibility of their generalisation

$$nitrile = cyanide.$$

Even before their results were published[44] the French chemist, Dumas, effected the reverse change of acid to cyanide.[45] This was (in modern terms)

$$CH_3COOH \xrightarrow{NH_3} CH_3COONH_4 \xrightarrow[(-2H_2O)]{P_2O_5} CH_3CN$$

or, as Frankland put it, 'the transformation of oxatyl to cyanogen.'[46]

Thus by now, Kolbe had (he thought) prepared ethyl from propionic acid by electrolysis, and obtained propionic acid from ethyl cyanide. Nothing was more natural than to attempt to isolate ethyl from the latter compound, and this became the next research task for Frankland and himself. It was in full harmony with the objectives of Bunsen who sought, following his work on cacodyl, to establish still more firmly the radical theory of organic chemistry. Knowing the affinity between potassium and cyanide Frankland proceeded to treat ethyl cyanide with potassium metal 'with some slight expectation of isolating the radical ethyl'[47], *i.e.* in his terms

$$C_4H_5CN + K = KCN + C_4H_5,$$

or in ours

$$C_2H_5 CN + K = KCN + C_2H_5.$$

There was another reason for pursuing this course. Since the 'radicals' were expected to be gases, it was important to be able to analyse them with precision. Frankland had already learned something of gas analysis from Kolbe while in London, but now, as he said, 'I had the advantage of direct instruction from the inventor of the process, who also taught me how to make eudiometers and most of the other apparatus required'.[48]

In such instruments gases were exploded with excess air in a closed calibrated apparatus such that the volumes of reactants, and the volume of carbon dioxide produced, were known. The amount of oxygen used may be obtained by exploding with excess hydrogen *after* the main reaction and noting the change in volume. Sometimes the quantity of water produced was also obtained by noting differences before and after absorption by sulphuric acid . From the results the nature of the product could be determined, though with 'wrong' atomic weights and no recognition of Avogadro's hypothesis, the arguments appear tortuous to modern eyes, as well as often leading to incorrect conclusions. Sometimes, however, a comparative method was used in which combustion characteristics of unknown gases were compared with those of known substances, and correct conclusions were obtained.

The reaction was tried, but unfortunately the only gaseous product was 'methyl' (ethane) together with a solid they called kyanethine. Despite rigorous drying of the ethyl cyanide, traces of alcohol or water must have been present, as Frankland recognised, thus initiating both a reductive cleavage and a base-catalysed trimerization. The results, though unexpected, were of considerable interest:[49]

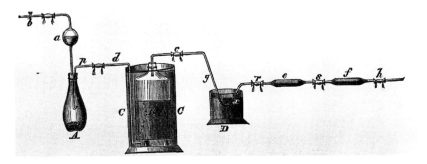

Fig. 2.2. Frankland's apparatus for producing 'methyl' [ethane] from potassium and ethyl cyanide. (*Experimental Researches.*)

Life in Marburg was not all work. The two chemists from England quickly ensconced themselves in a hotel, Der Europaïsche Hof, but dined at the nearby Ritter, relaxing afterwards over coffee and dominoes before returning to the laboratory, where they stayed until 6 or 7 o'clock. The evenings were spent in pleasant conversation, billiards and other pastimes usual in German universities. Within the first few days of their arrival they met professors from other faculties (including the Fick brothers and the philosopher F. T. Waitz) and men like Heinrich Debus whose paths frequently crossed with those of Frankland for many years to come. Although Marburg had been famous for its duels, duelling does not appear to have excited Frankland's attention.[50] On Sundays he distinguished himself from other Marburg students by attending Church, at that period being strongly influenced by the evangelical Christianity of his youth.[51] At other times there appears to have been a merry round of balls, concerts and parties. Knowing little German, Frankland must have been at something of a disadvantage, but he was quickly introduced by Kolbe[52] to a young lady who – quite exceptionally in Marburg – could speak English. Her name was Sophie Fick. Frankland's own account of his new

acquaintance (needlessly expurgated from his published autobiography) is revealing:

> Sophie made, even on first meeting, a deep impression upon me.
> Between the dances we talked a good deal to each other, and I found
> her very intelligent and well educated and capable of conversing with
> interest upon a variety of topics quite outside the usual conventional
> gossip, which is so frequently the only kind of talk which passes
> current between the sexes on such occasions. She had a graceful and
> slight figure and danced well, was lively in conversation, and her face,
> though it would not be called pretty, was very sweet and full of
> expression. She was of medium height and had very small hands and
> feet, light brown hair, good complexion and light and supple figure . . .
> We obviously liked each other from the first. It was a very pleasant
> party consisting almost entirely of professors with their wives and
> daughters. Besides dancing, one or two scenes from Shakespeare's
> play 'A Midsummer's Night's Dream', were performed by some of the
> students. Sophie was accompanied home from this party by Kolbe and
> me. We met frequently afterwards at balls, concerts, and parties: the
> last time at a ball in Pfeiffer's Garten, the evening before I left
> Marburg, when I danced the Cotillion with her, accompanied her
> home and paid her a cordial farewell.[53]

Sophie Fick was a member of a long established family from Cassel,
45 miles from Marburg.[54] Her father, F. W. Fick, was the Chief
Engineer in the State of Hesse-Cassel, and two of her brothers
taught in Marburg. Frankland had quickly made their acquaintance.
Heinrich had just become professor in the Faculty of Law, while
Ludwig occupied the Chair of Anatomy. He was rather older than
Sophie, married and was often visited by his sister.[55] It was on one
such occasion that Sophie first encountered the young English
chemist, at the dance 'in the forest on Spiegelslust about one mile
from Marburg'.[56] She was not to know that, even before his short
visit was over, Frankland had secretly determined to marry her.
With inadequate means and uncertain prospects, he wisely kept his
own counsel at that stage. They parted in 1847 'just friends'.

At the end of the three months he returned to England to a new
appointment, parting from his friend Kolbe whose sojourn in
Marburg was to last another few years. By now, Frankland had
made substantial progress as a chemist, had been a member of a
genuine research team, had experienced student life for the first
time, had tasted the delights of foreign travel and had fallen in love
with a highly intelligent girl whose impeccable ancestry, so unlike
his own, was widely recognised and respected. It is hard to

over-state the contrast between his painful years of growing up and apprenticeship in Northern Lancashire and the vistas that were now opening before him, chemically and socially. Small wonder, then, that he afterwards considered the visit to Marburg 'perhaps the most momentous journey I ever made.'[57]

Notes

1 Letter from Earl of Lincoln (H. P. F. P. Clinton, MP) to Lords of the Treasury, 15 January 1845 [PRO, Letter-books, misc., of the Geological Survey, 1844–1846, *Works* 3/6].

2 On Playfair see A. Scott, *J. Chem. Soc.*, 1905, 87, 600–5; T. Wemyss Reid, *Memoirs and correspondence of Sir Lyon Playfair*, London, 1899; *DNB*; *DSB*; also R. F. Bud and G. K. Roberts, *Science versus practice: Chemistry in Victorian Britain*, Manchester University Press, 1984, pp.86–93.

3 Arthur Ransome, *Autobiography*, ed. R. Hart–Davis, Century Publishing, London, 1985, pp.17–18. Thomas Ransome was a son of the Manchester surgeon John Ransome (1779–1837) who attended William Huskisson MP, fatally injured in the world's first railway accident in 1830. Thomas lacked financial acumen and his chemical business eventually failed. The family was related to the Ransomes of Ipswich, famous for agricultural engineering. See also *Lancastrian Chemist*, pp.144–5.

4 The matter was initiated by an enquiry from the Admiralty to Lord Lincoln (Geological Survey letter-books (note 1), 28 June 1845), duly passed to the Head of the Geological Survey, H. T. de la Beche (*ibid.*, 30 June 1845) and eventually agreed.

5 Letters from Lincoln to de la Beche, *ibid.*, 27 and 29 August 1845.

6 Frankland had soon moved from Bishopsgate to the home of his uncle John Frankland, pianoforte maker, and his wife Mary. They lived at 9 Doris Street, Lambeth, half a mile due north of the Oval.

7 For more on Putney College see *Lancastrian Chemist*, pp.145–51.

8 *Illustrated London News*, 12 August 1843, p.99.

9 From an association of old Putney students emerged the present Society of Engineers (H. P. Stephenson, letter to A. T. Walmisley, 24 January 1888, in *Trans. Soc. Eng.*, 1930, p.107).

10 Frankland's notes of Playfair's lecture-course survive [RFA, OU mf 01.04.1781].

11 His own account presents minor problems of chronology. He says he joined the Museum in October 1845, was appointed to Putney 'at the end of six months and held that post for about six months', at which time he was offered a lectureship at Cirencester Royal Agricultural College (*Sketches*, pp.64–5). That suggests late summer, 1846; this was when he met in Lancashire George Edmondson, the future proprietor of Queenwood College, who claimed he could have matched the Cirencester salary (*ibid.*, p. 66). However, manuscript evidence [see below] demonstrates that the Cirencester offer was

not made until April 1847, in which case Frankland must have met Edmondson on more than one occasion.

12 Letters to Frankland from the geologist J. B. Jukes (to Craig's Court), 1 April 1847, and from J. Bravenden of Cirencester Royal Agricultural College (to Duke Street), 14 April 1847 [RFA, OU mf 01.03.0199 and 0282].

13 On Kolbe see E. von Meyer, *J. prakt. Chem.*, 1884, **30** (n.s.), 417–66; *J. Chem. Soc.*, 1885, **47**, 323–7; *DSB;* and (in most detail) A. J. Rocke, *The Quiet Revolution: Hermann Kolbe and the science of organic chemistry*, University of California Press, Berkeley, 1993.

14 *Sketches*, 2nd. ed.p.69.

15 H. Kolbe, *Mem. Proc. Chem. Soc.*, 1847, **3**, 245–8.

16 Kolbe also showed that the product obtained from carbon disulphide and moist chlorine (trichloromethylsulphonyl chloride) could be reduced to a series of chlorinated methylsulphonic acids. For an account of all this work, with references, see J. R. Partington, *A History of Chemistry*, Macmillan, London, vol. iv, 1964, pp.504–5.

17 On Berzelius' electrochemical theory see C. A. Russell, *Ann. Sci.*, 1965, **19**, 117–45, and *idem*, Introduction, Commentary and Notes to Johnson Reprint (1972) of Berzelius' *Essai sur la théorie des proportions chimiques et sur l'influence chimique de l'électricité* (1819).

18 On Bunsen's cacodyl research see Partington (note 16), pp. 283–6, and C. A. Russell, *The History of Valency*, Leicester University Press, Leicester, 1971, pp.25–7.

19 H. Kolbe, *Mem. Chem. Soc.*, 1848, **3**, 378–80; *Lehrbuch der organischen Chemie*, Brunswick, vol. i, 1854, p. 279.

20 A. Schlieper, *Annalen der Chem.*, 1845, **56**, 1–29.

21 H. Fehling, *ibid.*, 1844, **49**, 91–7.

22 F. Wöhler, *Ann. Phys.*, 1825, **3**, 177–82.

23 (a) J. B. Dumas, *Compt. Rend.*, 1847, **25**, 383 (methyl cyanide); (b) T. J. Pelouze, *J. de Pharm.*, 1834, **20**, 399–400 (ethyl cyanide).

24 *Sketches*, 2nd. ed. p.169.

25 *Mem. Chem. Soc.*, 1847, **3**, 386–91; *Experimental Researches*, p.29.

26 *Register of the staff & students of the Royal Agricultural College from 1847 to 1897 with a short historical preface*, Cirencester, 1897; K. J. Beecham, *History of Cirencester and the Roman City Corinium*, G. H. Harmer, Cirencester, 1886, pp.207–12. The College, owing much to the advocacy of C. G. B. Daubeny, offers a remarkable parallel to that at Putney, but emphasised agriculture rather than engineering. Its ideology of scientific improvement was identical, as were its Anglican foundation and collegiate system and, indeed, many of the noble and titled subscribers. Cirencester, however, had the advantage of Royal patronage (the Prince Consort) and, after a severe shake-up in 1848, continues to flourish up to the present time.

27 *Sketches*, 2nd. ed. p.65.

28 *Ibid.*, p.66.

29 *Ibid.*, p.65

30 R. Phillips to Frankland, 25 March 1847 [RFA, OU mf 01.03.0289].

31 *Ibid.*

32 A. C. Ramsay to Frankland, 18 March 1847 [RFA, OU mf 01.03.0293].
33 Copy-letter, L. Playfair to Frankland, 19 March 1847 [RFA, OU mf 01.08.0002].
34 Copy-letter, Morgan Cowie to Council of the Royal Agricultural College, 18 March 1847 [RFA, OU mf 01.08.0004].
35 Copy-testimonial from James Johnson, 22 March 1847 [RFA, OU mf 01.08.0005].
36 Copy-letter from R. Hunt to Frankland, 23 March 1847 [RFA, OU mf 01.08.0007].
37 Copy-testimonial from H. Kolbe, n.d. [RFA, OU mf 01.08.0008].
38 Copy-testimonial (English trans.) from [J.] C. Bromeis, n.d. [RFA, OU mf 01.08.0009]. Bromeis was then teacher of chemistry and physics at the Hanau Realschule of Hesse–Cassel.
39 *Sketches*, 2nd. ed. p.65.
40 Copy of Council Minutes, 7 April 1847 [RFA, OU mf 01.03.0284].
41 *Sketches*, 2nd. ed. p.65.
42 *Ibid.*, pp. 66–8.
43 On Marburg see C. Meinel, *Die Chemie an der Universität Marburg seit beginn des 19. Jahrhunderts*, (Academia Marburgensis herausgegeben von der Phillips–Universität Marburg Band 3), N. G. Elwert Verlag, Marburg, 1978.
44 H. Kolbe and E. Frankland, *Annalen der Chemie*, 1848, **65**, 288–304; *Experimental Researches*, p.34. Frankland (incorrectly) claims this was the first use of a Liebig condenser in a reflux position (*Experimental Researches*, p.29).
45 J. B. A. Dumas, *Compt. Rend.*, 1847, **25**, 383–4.
46 *Sketches*, 2nd. ed. p.70.
47 *Ibid.*
48 *Ibid.*, p.69.
49 *Ibid.*, pp.70–1.
50 *Ibid.*, p.259.
51 *Ibid.*, p.50.
52 So his daughter Margaret later believed: Maggie Frankland, *Diary*, 17 August 1874 [MBA].
53 *Sketches*, 1st ed., 1901, pp.262–3.
54 Friedrich Fick (1783–1861), as municipal engineer he reorganised street construction in Cassel. His father Johann Christian Fick (1763–1821) had been Professor of Geography and English at Erlangen, and came from a family of Protestant emigrés from Salzburg. His nine children included Heinrich and Ludwig and the youngest and most famous son, Adolf Eugen (1829–1901), proponent of the new alliance between physiology and physics and (eventually), Rector of the University of Würzburg (*DSB*; *Neue Deutsche Biographie*, Berlin, 1960, vol. v, pp.127–9).
55 For more on the Fick family see chapter 4.
56 *Sketches*, 2nd. ed. p.259.
57 *Ibid.*, p.50.

CHAPTER 3

Queenwood

1 Towards the Millennium

Frankland's days in Marburg were brought to an abrupt and premature conclusion by a summons back to England. The call was to take up a teaching post in one of the most unusual and progressive educational institutions in the country, Queenwood College in Hampshire. He was now to get his first experience of rural life in the south of England. Queenwood College was near to the villages of Broughton and Stockbridge, just south of the Roman road linking Salisbury and Winchester. It had a distinctive, if bizarre, prehistory as might have been perceived from the inscription picked out in flint work on one of the red brick chimneys:[1]

$$\boxed{\text{C. of M. 1841}}$$

This artless symbol proclaims simply 'Commencement of the Millennium', but regrettably the institution then founded was to last not 1000 years but three. It was in fact a creation of the visionary socialist Robert Owen, intended as a grand expansion of his educational ideals in the context of communal living. Optimistically styled 'Harmony Hall', the new establishment soon raised fears of subversion, and a local magistrate was warned by the Duke of Wellington to keep an eye open for any signs of lawlessness.[2] However it was not conservative suspicions but internal mismanagement which rapidly spelt doom to this final Owenite experiment.[3]

At first the co-operative, founded in 1839, was chiefly concerned with agriculture, but lack of expertise and suitable accommodation speedily brought financial failure and much discontent. In 1841

Owen personally took command of the project and decided to add a school whose members could also help on the land. Thus was born the fore-runner of Queenwood College. The magnificent building, in whose design Owen and his two sons actively participated, was sumptuously furnished with the latest and best equipment. A visitor who called in 1842 reported a cost of over £30 000, expressing the view that 'a village of cottages, each with a garden, would surely have been more appropriate for a working community and much cheaper'.[4] A later observer reported that 'everything has been provided in the most expensive way', the Hall being 'a monument of *ill-timed* magnificence'.[5] For in fact the institution was not to last. Its remoteness from industry, the poverty of its flinty soil and the emphasis on teaching rather than production all conspired to hasten the end. By 1844 debts had been run up of £1000. As a result by August 1845 the colony had broken up and the buildings were deserted. It meant disaster for 'the more level-headed men from the north' who vainly tried to salvage the operation[6] and 'ruin of many of the poor mechanics who had embarked their savings in it'.[7] After many months of wrangling and uncertainty the trustees at last saw deliverance from their burden: the whole estate would be let for 21 years to a schoolmaster from Preston, with the eventual option to purchase.

The pedagogue in question was George Edmondson (1798–1863),[8] whose elder brother Thomas[9] was to achieve fame as inventor of the printed railway ticket. Both were sons of a Quaker trunk-maker from Lancaster and were fittingly sent to the Friends' School at Ackworth. Here they encountered an educational system that did not despise the manual arts and crafts but actually encouraged the study of natural phenomena and useful employment as gardening and spinning. Thus equipped, George left at the age of 14 to be a teacher himself and became apprentice to the reading master at Ackworth, William Singleton, who had recently opened his own school at Broom Hall, Sheffield. Here George acquired new skills in book-binding and printing as well as the outdoor arts of agriculture and surveying. No doubt he learned more than a thing or two about education from a master who maintained Quaker values to the extent of even opposing corporal punishment.

His apprenticeship over, this unusual man took the very unusual step of heading for Russia. He set sail, in 1818, as tutor to the sons of a Quaker missionary, Daniel Wheeler, whose dual achievement was to be the establishment of a Quaker meeting in Russia and the reclamation of water-logged land near St. Petersburg for the Emperor Alexander I. For seven years Edmondson laboured in those remote parts in teaching and agriculture.[10] Only once – in 1821 – did he return to England, and then for the purpose of claiming as his bride Anne Singleton[11], eldest daughter of his former master and childhood sweetheart from Ackworth days. His work 'necessitated him spending long summer days almost entirely on horseback, whilst the dark hours and deep snows of winter brought back the familiar occupation of teaching'.[12] Years later, when he was established at Queenwood, his agricultural improvements around St. Petersburg were warmly commended by a Russian lady and gentleman he chanced to meet on a train from Waterloo. By 1825, however, his time in Russia was drawing to a close as his wife's health was not equal to the climate.

In the year of Frankland's birth the Edmondsons returned to England, and George set up a school in Blackburn, Lancashire, offering its citizens the unusual opportunity of lessons in Russian. After a couple of moves within the town he settled at Lower Bank in 1827 where for 14 years he developed just the kind of practical and enlightened education that was being denied to young Edward Frankland at that very time. Amongst the facilities were a carpentry workshop and printing press. He wrote, printed, bound and published three books[13] and also issued a monthly school magazine.[14] From a single pupil in 1825 Edmondson now had a school that was bursting at the seams, and so in 1841 he removed to still more commodious premises, at Tulketh Hall, Preston.

At his new school Edmondson was able to expand his activities still further, and his fame spread so widely that pupils came from all over the country, and not all of these were from Quaker families. By 1844 he had begun to teach chemistry. A couple of years later, learning of Frankland's return to his native Lancaster for a summer holiday, he enticed him to Tulketh for a few days to receive practical instruction from the youthful chemist. The story of their first encounter has already been told.[15] Its consequences were as momentous for Frankland as for Edmondson himself. Meanwhile Tulketh Hall prospered exceedingly, applications for admission were flooding in, and still further expansion was desirable.

At just this time a call was heard from an institution in deep trouble in the remote south of England. From Hampshire came an

THE QUEENWOOD REPORTER, AUGUST 15, 1848.

Fig. 3.1. Queenwood College. (J. Bucknall Archives.)

invitation for Edmondson to decamp with his pupils and set up
school in the palatial, if desolate, accommodation once provided by
Robert Owen. The latter and his fellow-governors thought the
Lancastrian Quaker would be in sympathy with their ideals and so
could continue to propagate them in a new establishment. At first
Edmondson was reluctant to move, for in fact he looked askance at
much of Owen's ideology. After a delay of six months he was
persuaded to visit the site and returned convinced of its potential.
Now his pupils could learn not only printing, carpentry and
chemistry, but – joy of joys – could also be taught practical
agriculture. The pioneer from St. Petersburg was won over, and
early in August 1847 the great experiment began at 'Queenwood
College' (newly named after a farm on the estate). No less than 40
pupils were in residence. The first to arrive was Henry Fawcett,
later to achieve fame as the blind Post Master General. He was soon
followed by James Mansergh, future President of the Institute of
Civil Engineers, and (like Frankland) nurtured in Lancaster. After
a mere fortnight these two were elected editors of the *Queenwood
Reporter*, successor to the worthy *Tulketh Hall Mercury* and the
Lower Bank Monthly Magazine.[16] Within a few months one of the

emigrés from the north wistfully admitted in its pages that 'the country people of Hampshire are far more civil than those of Lancashire'.[17]

Some months before the school opened Edmondson had seen the need for new staff to teach the sciences and had invited his young acquaintance Edward Frankland to become his teacher of chemistry. Frankland, then serving both the Geological Survey and the College of Civil Engineering in Putney, was already seeking promotion and closed with the offer. Unfortunately, as we have seen, he had also been enticed to Marburg for the summer and, in mid-August, was savouring its heady delights to the full. With understandable pride he learned from a letter that Mr. Edmondson 'felt that they were losing by my absence'.[18] So, taking leave of Sophie, he set off the next day for Rotterdam *via* Frankfurt and the Rhine, bringing with him enough chemical apparatus to equip a laboratory at his new institution (for which he had to pay heavy customs duty in Holland). After a rough crossing to London he arrived at Queenwood on Saturday, 4 September 1847.[19]

2 Pioneers in science education

As Frankland's conveyance swept up the long drive to the stately pile known as Queenwood College he may or may not have noticed the millennial symbol high up on the chimney. He would certainly not have known that, in a most curious and unexpected way, one aspect of Owen's optimistic forecast was about to be fulfilled within his own experience. Queenwood was going to witness a truly revolutionary adventure in scientific education for the school children of England. Within his own compendious baggage lay tokens of that revolution, more substantial if less euphoric than the Owenite inscription: chemical apparatus of great variety painfully transported halfway across Europe.

Long before unpacking could be completed Frankland must have been welcomed by the presiding authorities: Edmondson, his wife Anne, their daughter Jane and his brother-in-law Josiah Singleton. He soon met his fellow tutors, including John Yeats, who appears to have been in charge of history and geography.[20] He must surely have recognised among the pupils young James Mansergh from Lancaster. But one man made the deepest impression upon him that day. This was an Irishman who had recently abandoned a rather hazardous if profitable career in surveying for railways, and, now that the 'Railway Mania' was losing momentum, sought new employment. Five years Frankland's senior he was to become a life-long friend. He was John Tyndall, soon to become eminent in both physics and the scientific politics of Victoria's reign. His immediate impression of the newcomer was distinctly favourable: 'a deeply thoughtful and instructive young man'.[21]

They lost no time in getting acquainted. Within a day they were discussing science and religion, the Orangeman noting that his sparring partner was 'an Independent'.[22] It became clear that they had much to teach each other. Frankland's eyes were opened for the first time to the fascinations of geometry as taught by a practised surveyor.

It was not long before the two friends embarked upon an extensive programme of mutual help. Tyndall instructed Frankland on mathematics and was rewarded for his pains by attendance at the latter's chemical lectures, with additional tuition in the newly equipped laboratory. From the days of his apprenticeship in Lancaster Frankland was no stranger to early rising. Both men's diaries have survived and reveal something of the passion for self-improvement that got them out of bed long before dawn, with several hours' study before the day began for lesser mortals. In their zeal for knowledge they were often aided and abetted by Josiah Singleton, who rose first, lit the fire and produced coffee for the slumberers.[23] Although addicted to rhyming doggerel for his letters, and to other minor eccentricities, Singleton was a shrewd man with a genuine interest in the well being of the young teachers.

Frankland thus records the beginnings of a typical day (19 January 1848):

> Wednesday: rose at 4 o'clock and managed to work through nearly all the simple equations contained in Bridge's *Algebra*. Gave an hour's instruction to the second chemical class . . . 9 to 10 a.m.[24]

Such exercises provided valuable stimulation for Frankland and may help to explain the huge zest with which he embarked on his programme of teaching. But they are not a sufficient explanation. For that we must surely turn to the distinctive character of Quaker education, with its emphasis on practical training and (with other Dissenting bodies) its high evaluation of science. This was specially true of Quaker 'private' schools, as opposed to their 'public' schools like Ackworth. The former were more free from the influence of school committees and the regional Quaker meetings, and therefore more able to introduce changes. Queenwood was such a 'private' school.[25] These innovative tendencies were further enhanced by the legacy of Robert Owen, both in the material facilities of the building and in the survival of his own distinctive ideas on education. These owed much to the teachings of the Swiss educational reformers Pestalozzi (1746–1827) and de Fellenberg (1771–1844), who emphasised the importance of discovery by pupils themselves. It so happened that within a few weeks of Frankland's arrival his younger colleague Yeats gave an evening lecture on Pestalozzi;[26] another colleague John Haas had taught at Fellenberg's famous school at Hofwyl in Switzerland (to which Owen's two sons had been sent).[27] To this physical and spiritual legacy of Owen must be added, finally, a growing realisation of the relevance of chemistry to agriculture, of which Frankland would be specially aware from the teaching of Liebig's British disciple Lyon Playfair.

Thus, with trepidation but some excitement, on 17 September 1847 he embarked on his first systematic course of lectures in chemistry.[28] Only once before had he ever lectured in public, and that was to an admiring audience of junior contemporaries at the Mechanics' Institute in his native town of Lancaster, just over a year previously. Here at Queenwood the results were at first 'about as bad as possible' but he seems to have made rapid progress since Singleton, 'who had not attended them [Frankland's lectures] during an interval of two or three weeks, expressed his astonishment at the progress I had made, saying he could scarcely believe it was the same lecturer'.[29] Like many another after him, being thrown in at the deep end to teach a class of boys or youths gave Frankland a facility for communication that stood him in excellent stead for the rest of his life. Whether or not his 'facility of expression' was as good as he implied there were to be plenty of future opportunities to put it to the test.

It is remarkably fortunate that a bundle of papers has survived bearing the legend 'Notes of chemical lectures at Queenwood'.[30] That this inscription is correct can be checked by the brief summary of many of these lectures given by Tyndall in his diary. They are in fact correctly labelled. In addition to a course of nearly ninety lectures (on 'chemistry') there are a few others interspersed

Fig. 3.2 John Tyndall (1820–93). (Author.)

on 'elementary chemistry'. From the contents of the main course is revealed for the first time the chemical philosophy of Edward Frankland.

Although they differ in length and detail, and most of them do not bear specific headings, their contents can be summarised as follows:

1. Origins of chemistry
2. Utilitarian value of chemistry
3. Specific gravities
4. Cohesion and repulsion
5–9. Heat
10. Chemical change
11. Acids, bases and salts
12. Nomenclature
13. Combining weights
14. Atomic theory
15. Crystals and isomorphism
16–17. Oxygen
18–19. Hydrogen
20–25. Water
26. Hydrogen peroxide; nitrogen
27. Nitrogen
28–29. The atmosphere
30–33. Compounds of oxygen and nitrogen
34–37. Carbon
38–39. Compounds of carbon and oxygen
40. Compounds of carbon and hydrogen
41–42. Coal-gas
43. Sulphur
44–47. Compounds of sulphur and oxygen
48. Hydrogen sulphide
49. Carbon bisulphide; selenium
50–53. Phosphorus and its compounds
54–55. Arsenic (including detection)
56–57. Antimony
58. Boron
59. Compounds of boron; silicon
60–61. Silicates and glass
62–67. Chlorine and its compounds
68–69. Iodine and its compounds
70–71. Bromine and its compounds
72. Fluorine compounds
73–74. Crystals and salts
75–78. Potassium; sodium; lithium
79. Potassium chlorate; gunpowder
80. Barium; strontium; calcium; magnesium

81. Cements
82. Magnesium, barium and strontium
83. Calcium sulphate
84–86. Aluminium and chromium
87. Silicates in porcelain and pottery

Several considerations immediately follow from even a cursory inspection of this list. It is obvious, for instance, that applications of chemistry are well to the fore. There is almost prophetic significance in the six lectures on water, where much attention is paid to the dissolved impurities in natural water, their estimation and removal. Silicates are chiefly of interest for their role in glass and ceramics. The mere mention of potassium chlorate triggers off a whole lecture on explosives, and so on. To a modern chemist it is clear that the absence of a periodic classification of the elements denied the lecturer a valuable expository framework. Similarly his exclusive reliance on combining weights led him to HO as the formula for water, with other erroneous formulae (like NaO) as a result. Indeed Frankland preferred equivalent to atomic weights, which is not surprising in this period before the revival of Avogadro's hypothesis through which, at last, generally agreed atomic weights were to become possible after 1860.

   That said it must be conceded that Frankland's course had surprising modernity, and, in many ways, great similarity to elementary courses in chemistry a century later. His attempts to lay down a general theoretical foundation at the beginning look forwards rather than backwards, though some earlier writers, as Berzelius, had attempted something similar in their treatises. But those were written for mature chemists, not novice schoolboys! Moreover, the young lecturer was not afraid to put before his audience developments of a most recent kind. Naturally he specially emphasises those with which he has had some kind of personal contact. Thus the wondering schoolboys would learn of Kolbe's experiments on combustion,[31] of Playfair's work with Joule on the density of ice[32] and with Mercer on amidogen,[33] to say nothing of his photometer,[34] Clegg's rotary retorts[35] and James Thompson's speculations on atomism.[36] As if to emphasise beyond all contradiction the superiority of recent chemistry Frankland prefaced his course with a lecture on the history of chemistry that stressed its 'disgraceful origins'[37] and was followed by a eulogy of its latest achievements. It is an excellent example of Whig historiography (in which past science is viewed solely through the spectacles of modern scientific understanding).

In general, Frankland follows closely in the footsteps of Berzelius (though, surprisingly, he says little of electrolysis). He writes Berzelian formulae for salts (as $CaOSO_3$) and is quite often concerned with 'elective affinity'. Yet he is not afraid to speculate:

> 'Atoms are probably only physically indivisible'.[38]
> 'May not some of the simple bodies be merely isomeric modifications of each other?'[39]
> 'Organic radicals seem to point to the compound nature of elements'.[40]

Despite the myth of Wöhler's destruction of vitalism in 1828, Frankland, like many contemporaries, freely speculated on the vital force in organic reactions 'which is not under our control'.[41] And he confessed 'catalysis is merely a cloak for our ignorance'.[42] What is astonishing is the extent to which he was prepared to take his juvenile audience into his confidence over matters of such chemical sophistication.

In conjunction with these lectures was a course of practical chemistry in a laboratory especially equipped for the purpose. On the details of the experimental course we are rather less well informed. But we do know that by 5 October Henry Fawcett had completed his first chemical preparation, 'bichromate of lead or chrome yellow',[43] that qualitative analysis was taught,[44] that some pupils were invited to assist his own private incursion into forensic chemistry by analysing the stomach contents from a suspected murder victim,[45] and that practical agricultural chemistry eventually became so prominent as to be specially mentioned in the prospectus.[46] Many years later Frankland made the claim that 'here, for the first time in an English school, experimental science was practically taught in the laboratory and in the field'.[47] Though recently repeated[48] this claim has been even more recently refuted with reference to Stonyhurst College in Lancashire, the City of London School and the earlier Dissenting Academies. A more accurate claim would be that 'Frankland and Tyndall were the first teachers of sciences and engineering in the modern sense of practical laboratory teaching',[49] though somewhat begging the question as to exactly what is 'the modern sense'.

Even in his first term Frankland was also required to teach botany. For this he found useful the collection of dried plants made in his apprentice days in Lancaster (and after his departure from

Queenwood bequeathed to Edmondson).[50] By 1848 he and Tyndall organised a joint course which placed science teaching in an even more utilitarian context. It is worth quoting their syllabus[51] in full:

The Steam Engine
Watt's single and double acting engine, cylinder, condenser, air-pump. Expansive principle; laws of Dalton and Mariotte. Motion of rotation; the crank, the eccentric, sun and planet motion, its mechanical principles. The governor; laws of vibration. Pistons, packed and metallic. Boilers; steam space, and water space, steam gauges; the indicator. Contrivances to prevent explosion; safety valves, fusible plugs. Horse-power; rules employed to determine it. Locomotion; high pressure engines; early forms of the locomotive, Trevethick's engine, walking engine; fictitious difficulty with regard to adhesion, rack rails. George Stephenson. Sub-division of flue into tubes, recent improvements, lengthening of boilers; Stephenson's patent. General Paisley.

Railways
Choosing the ground; trial levels, survey, parliamentary levels, plans and sections. Instruments; the spirit level, the theodolite, the transit instrument. Standing orders; serving notices, depositing, passage of a railway bill through parliament. Working section, cross sections, cutting and embanking, curves and slopes. Tables for earthwork; MacNeill. Bidder. Ballast, rails, chairs, keys, sleepers. Equilibrium of arches; bridges, tunnelling. Gradients; Lickey incline, Manchester and Leeds, West Riding Union, etc., diversity of opinion among engineers on this point, Robert Stephenson, Locke. Rope traction. The gauges, broad, narrow and medium. Atmospheric railways, London and Croydon, Kingstown and Dalkey. Irish Lines.

Inorganic Chemistry
The nature and properties of all the elementary substances and their compounds; the atmosphere, its constitution, the principles involved in ventilation; water; its physical and chemical properties; its supply to towns; causes of hardness in water, and methods of their removal. The chemistry of irrigation and drainage. the properties of metals and their alloys; methods employed in extracting them from their ores; their various uses in the arts. Chemical properties and preparation of acids, bases, and salts. Theory of dyeing and calico-printing. Application of chemistry to the manufacture of porcelain, earthenware, etc.

Organic Chemistry
The constitution of organic compounds. Composition and properties of saccharine and amylaceous bodies; gum, pectine, lignine, etc. The

processes of fermentation and decay; manufacture of fermented liquors, alcohol, vinegar, etc.

Agricultural Chemistry
Relation between the animal, vegetable, and mineral kingdoms. The nutrition of plants and animals. The origin of soils, and their adaptation to the growth of different species of cultivated plants. Theory of manuring, applications of chemistry to the determination of the relative value of manures, substances used in the feeding and fattening of cattle, etc. Ashes of plants. Effects and uses of fallow. Rotation of crops, etc.

Botany and Vegetable Physiology
Object and design of the science. Adaptation of plants to the circumstances in which they are placed. Elementary tissues. Organs of plants, their structure and functions. The root. Stem, leaves, and flowers. Division of plants into three great classes. Eddogens, Exogens and Acrogens. Classification of plants, Linnaean or artificial system. Natural system. Cultivation of the grasses. Cereals, tubers, etc. Application of the science to agriculture.

How far these pioneering adventures in education were successful is hard to gauge. Frankland was gratified by some of the responses,[52] but regretted the failure of the farm students to answer any questions in an examination,[53] and considered himself 'entirely unfitted' to conduct classes in elementary chemistry.[54] This was to be a shadow of things to come. Nevertheless his teaching made a considerable impression on several adults who heard him, one of whom was to become another life-long friend.

Richard Dawes[55] was rector of a neighbouring parish, Kings Somborne; he became Dean of Hereford in 1850. An enthusiast for popular science education, he gave a few lectures on agricultural chemistry at the National School in his village, using 'a small apparatus' of his own.[56] He then heard of 'a gentleman who had made the subject his professional study, and who was well qualified to give an interest to it, not only from his knowledge, but from being a good manipulator in the experiments necessary to illustrate it'.[57] The person in mind was Edward Frankland who, in the autumn of 1847, delivered six lectures in the village, attended by 'many of the gentlemen and also of the farmers in the neighbourhood'. Again Frankland's notes survive,[58] from which it is apparent that here was a more compressed version of his Queenwood course, modified to a largely agricultural theme and entitled 'Popular Chemistry':

1   Origins of Chemistry; its utility; states of matter; affinity
2. Elements; oxygen; hydrogen, water
3. Water (continued); carbon
4. Nitrogen
5. Oxides of iron; alumina; dolomite; soda
6. Carbon in plants; fertilisers; crop rotation.

A note of Frankland's requirements for demonstration apparatus leaves little doubt as to his comprehensive coverage. It does make one wonder at what hour the good farmers of Kings Somborne might have been expected to retire for the night, however. The following equipment was requisitioned for Lecture One alone:

Ground iron plates
Glass bulbs containing water
Thermometers: mercury and air
Two capsules – water and mercury
Freezing mixtures
Retort stand
Air pump
Barometer tube and mercury
Diagram of elements
Diagram of combinations of nitrogen
Glass flask and bent tube for showing latent heat of steam
Ether for freezing in vacuum
Naphtha and spirit lamps
Warm water and air pump
Chlorate of potash and sulphur
Mortar
Chalk and hydrochloric acid
Nitric acid
Sulphuric acid
Nitrate of lead and chromate of potash
Two bladders of oxygen and hydrogen
Potassium and evaporating basin
Pyrophorous
Nitrate of lime and carbonate of ammonium
Iron dust and sulphur
Voltaic battery and machine
A solution of camphor in water

Thus by lecturing, demonstrating and organising laboratory instruction Frankland gave new meaning to the phrase 'chemical
education'.

Nor was research neglected. Apart from a venture into forensic
chemistry Frankland experimented with the physiological effects
of various gases, trying them out on himself and others. It was a
topical subject. James Young Simpson had first used ether as an
anaesthetic in obstetrics only in January 1847, and chloroform the
following November. In that very month Frankland tried the effects
of breathing 'sulphuric ether' [diethyl ether] and was found by
Tyndall 'in a blissful state of inebriation . . . stammering out "well
here's a go"'.[59] Two months later he was found 'teetotally drunk'
on chloroform,[60] and in March was using it to anaesthetise a cow in
difficulties after giving birth to a calf.[61]

In less 'applied' chemistry Frankland continued on the trail of
the elusive radicals, a search begun at Marburg with Kolbe during
the previous summer. However, the intermittent nature of his
researches at Queenwood suggests a combination of heavy teaching
timetable, the delights of local excursions and some considerable
difficulty in procuring materials of the necessary purity. His first
attack (10 April) was from a new direction; instead of using ethyl
cyanide as a source of the radical he treated [diethyl] ether with,
successively, potassamide and potassium metal. But the only gaseous
product was hydrogen, and this arose from adventitious traces of
water or alcohol in the ether.[62] He then made the momentous decision
to examine the reaction between potassium and ethyl iodide, collecting
the gaseous products in a eudiometer. He argues that he 'was led to
use this compound, owing to the facility with which hydriodic acid
separates into its elements', adding that this was the first time this
important reagent was used in research.[63] It had been known since
1814, and Frankland used the original method of its discoverer,
Gay-Lussac, to prepare it,[64] by heating together a mixture of ethanol,
[red] phosphorus and iodine. After washing and separation it was
dried over calcium chloride for three days and then redistilled *in
vacuo*. On addition of potassium little seemed to happen until near
the boiling point (72°) when a violent evolution of gas took place.
The experiment was repeated in a sealed tube immersed in a heated
oil-bath for 15 minutes, and in various other ways.[65] The gas collected
was examined eudiometrically and seemed to consist of a mixture of
'methyl' and hydrogen in a ratio of 9.5:4.1. Clearly this raised more

problems than it solved, for the radical expected was *ethyl*, and whence came the hydrogen? Although Frankland does not explain himself it is fairly clear that the hydrogen must have come from water or alcohol as impurities, and many years later he realised that the main reaction yielded ethylene and ethane in equal amounts, expressed in his later notation as[66]:

$$2C_2H_5I + K_2 = C_2H_4 + C_2H_6 + 2KI$$

Then, on May 18th, came the *coup de grâce* to such experiments at Queenwood. Let Frankland himself describe the event:

> Commenced another analysis of methyl, and used a slightly larger quantity than before; on making the explosion the eudiometer was blown to atoms, one piece struck and cracked a window, another struck John [the laboratory boy] near the eye, but beyond this no damage was done. Lectured in the evening . . .[67]

Writing many years later Frankland gives a graphic account of an exploding eudiometer which was probably the same event:

> A combustible gas, consisting chiefly of ethylic hydride [C_2H_6], was mixed with oxygen, and exploded in a Bunsen's eudiometer. Expecting that the explosion would be a violent one, I pressed the open end of the eudiometer strongly down upon a plate of india-rubber with one hand, whilst with the other the knob of the Leyden jar was applied to one of the platinum wires. A deafening report followed, and the portion of the eudiometer containing the gaseous mixture (about two thirds of it) was shattered to fragments, and scattered about the room in all directions, breaking one of the windows. The tube was cut off just at the line where the mercury and gas met inside. Fortunately my hand grasped the tube below this line, and although my face was within a few inches of the part of the tube which was shattered, I escaped without even a scratch.[68]

That was not only the end of the eudiometer but also almost the end of the road for research at Queenwood, for the summer holidays beckoned and with them (as we shall see) an adventurous visit to the Continent. On his return from that journey one final attempt was to be made to isolate ethyl in Queenwood. It happened thus.

Classes at Queenwood recommenced on Thursday, 27 July. Next day the boys were all occupied writing for the *Reporter*, and so needed little supervision. Frankland began a long series of attempts to decompose chloroform with potassium but the results were ultimately inconclusive.[69] On the same day, however, he conducted an experiment of more enduring importance than any he had ever done before, and yet one that would remain unexamined and undiscussed for six months and more. Thinking that the problem with potassium was its great reactivity he resolved to substitute the much less 'positive' metal, zinc. This is pure Berzelian thinking and was put into effect only a few days before the death of the great Swedish chemist. But it was more than that. It also involved an imaginative variation on a technique developed for the potassium experiments a few months earlier. Instead of employing a sealed tube after some of the reaction had taken place he used this technique throughout. He introduced ethyl iodide into a couple of exhausted glass tubes containing zinc, sealed them and placed them in an oil bath at high temperature. The zinc began to react, a gas was evolved from the liquid which, as the pressure increased, ceased to boil. After some hours the tubes were ready to be opened but, alas, there was no eudiometer and they were stored away until such time as one might be available. That was not to happen until Queenwood days were over and a new life had begun.

3 Disharmony Hall

It is remarkable that Frankland found any time for leisure but, as in Lancaster so in Queenwood, he packed more into each day than any but the most energetic could hope to do. At weekends he would often make excursions to Southampton, the New Forest or other local places of interest. In Southampton he was glad to meet a Mr Sharpe, Manager of the Gas Works, who claimed to be the first to use gas for cooking purposes.[70] With Tyndall he made a 35 mile journey to Romsey and the famous Rufus Stone.[71] When Josiah Singleton was indisposed Frankland subjected him by turns to galvanism[72] and hydropathy;[73] Edmondson's brother-in-law survived both treatments.

During this period of his life Frankland maintained his Lancaster habit of church-going, and usually attended the Baptist chapel in nearby Broughton. Here he made friends with the elderly but infirm minister Hugh Russell[74] and his assistant A. W. Heritage.[75] They frequently appear in his *Diary* for 1848, and Frankland often writes appreciatively of their ministry or that of visiting preachers. He appears to have supported the chapel financially without resentment.[76] Visits to Anglican churches were less favourably recorded, as when he poured withering scorn on a 'Puseyite' sermon enforcing a 40 day Lenten fast on Hampshire labourers.[77] When a sceptical watchmaker questioned the Mosaic account of creation Frankland 'vigorously opposed him'.[78] A tea party with the hospitable Heritage turned out to be 'a very pleasant evening', partly because Tyndall 'managed to steer tolerably clear of his metaphysical lucubrations'.[79] During the previous Christmas holidays Frankland had attended church in Manchester[80] and while in London heard two of the most famous non-conformist preachers of the day, James Sherman and Thomas Binney.[81] On his twenty-third birthday he reflected on the year now past, regretting time wasted and asking 'has my gratitude to my Creator increased in proportion as my discoveries of his goodness have expanded?'[82] Yet he is chiefly moved by 'commotion and change' all around him, and seems to feel the ultimate good must be the acquisition of knowledge. Five days later a letter arrived from Manchester (presumably from his parents) 'relative to certain changes that have lately taken place in my religious views or rather that are likely to take place'.[83] It appears that, for all his chapel-going, problems of belief were beginning to arise, and that the theistic content of his creed was being replaced by a frantic quest for knowledge in general. Indeed it seems at times as though his search for scientific truth has itself almost religious overtones.

Such matters could be confided to his diary, shared with his mother and step-father and possibly be argued in conversation with Tyndall. They were not yet visible enough to cause much outward dissension. Unfortunately the same cannot be said of other elements of his experience which, by 1848, were beginning to make 'Harmony Hall' seem the most inappropriate name for the establishment taken over by Edmondson.

First there was a certain discomfort and cheerlessness about the place. Partly this stemmed from Edmondson's ascetic nature. He was reluctant to grant a holiday on Good Friday.[84] One April day the college was so cold and fireless that Frankland and Tyndall walked to Romsey to find 'some snug hotel'.[85] Three months before, in even colder weather, the fire in his own room so nearly filled it with smoke that Frankland was caused to protest to his *Diary* that 'some alternative must be made as I have never been accustomed to this kind of life. I always make it a first consideration wherever I take up my abode, to have a comfortable sitting room'.[86]

He later recorded the temperature at the tea table on different days as eight[87], thirty eight[88] and forty five[89] degrees Fahrenheit!

Physical discomfort was only one part of the problem. It soon became clear that discipline (or its absence) was another. In February 1848 Frankland remarked:

> I have noticed that the conduct of my pupils has been undergoing a very serious deterioration for some time past, I have now the greatest difficulty in preserving order and even those who before the Xmas vacation were noted for exemplary conduct are now become impertinent and mischievous, there certainly appears to be something wrong in the present system of treatment but what that something is is not for me to decide.[90]

With the coming of summer matters took a turn for the worse. Two boys, Henry Fawcett and Elijah Cobham, were caught bathing in a tank which supplied fresh water to the College.[91] Tyndall thought there were 'signs of mutiny among the lads'[92]. The following day food at breakfast was contaminated with soap.[93] Shortly afterwards one youth was expelled, 'he having been found beneath the bed of some of the female servants'.[94] Much of this was doubtless youthful high spirits but it is likely that Tyndall's widow, Louisa, had correctly identified another cause:

> The senior pupils – the "farmers" – were youths who came for training in the arts of farming, surveying, and engineering. Most of them were verging on manhood, some were failures from the universities, others had disappointed their parents in other vocations, and they combined to form an element of idling and unruly tendency, which would call for strong and judicious handling.[95]

Strong and judicious handling does not appear to have been greatly in evidence. For all his merits Edmondson appears weak and vacillating, and in no small degree unable to handle his own family relationships. It is here that the corporate disharmony most probably had its origin. However two other persons were most visibly involved in the disputes and wrangling that led eventually to Frankland's departure.

One was Josiah Singleton,[96] the younger brother of Mrs. Edmondson. He was employed as an assistant at Queenwood and was clearly a man of eccentric habits. He had a strong fellow-feeling for Frankland and Tyndall and, as we have seen, assisted in their early rising regime. Years later Frankland was to write to Tyndall's widow, Louisa, that 'Josiah Singleton, who was Mrs. Edmondson's brother and consequently our senior, we generally called "father"

... we both had a great affection for him'. [97] This nomenclature may have reflected the recent death of Tyndall's real father, and the perennial sense of father-deprivation that rose from Frankland's illegitimacy. There is no doubt that they considered Singleton badly treated by the Edmondsons, although the reason is not immediately obvious from Frankland's *Diary*. On one occasion Singleton gave vent to some critical remarks concerning the *Queenwood Reporter*. As this was a mouthpiece for the Edmondson family it is not surprising that his remarks 'seemed distasteful to the ladies', [98] one of whom Frankland had characterised as 'the Censoress'. [99] Singleton was not alone in his criticisms. Richard Dawes also deplored the poor composition and 'vein of puffery'[100] which ran through that otherwise estimable journal.

However it was not merely that the Edmondson ladies took unkindly to criticism. Singleton had further offended by daring 'to form an attachment with a young lady who did not quite come up to Mrs. Edmondson's ideas of 'dignity''.[101] Her husband 'tried to insinuate all sorts of things' against Singleton, making 'very ill-natured remarks about father's intended'.[102] At times Frankland becomes incandescent with indignation, fulminating at 'the most ungenerous and unchristian-like treatment'[103] meted out to his friend. But Singleton had to go, and on 1 June he departed, after a touching tribute from the senior boys and most of the staff.[104] He returned to the north, eventually opening his own school, Spring Bank, in Darwen, Lancashire.[105]

The other focus of Queenwood disharmony was the tutor John Yeats. He had formerly been at Hofwyl (where he had not been popular with the boys).[106] A rumour circulated at Queenwood that he was a spy for the Edmondsons,[107] and he gave credence to that opinion by informing the proprietor of an indiscretion by Morton, another tutor, who had taken a pupil to a public house.[108] There followed a row of classic proportions, with Singleton receiving a 'violent attack' by Yeats, and Edmondson displaying his usual symptoms of vacillation and partisanship by turns. So incensed was Frankland that he curtly returned Yeats' contribution to Singleton's leaving testimonial. To this Yeats replied that it was impossible for him to 'entertain a high regard for the private character and professional talents of Mr. Singleton, since of the one I know too little and of the other too much'. Even the pupils' address to Singleton was vetoed by Edmondson, who had previously shown it to his wife. It was to Frankland all too clearly a sign of 'deep hatred',[109] though his own assessment is unlikely to have been entirely without prejudice. However, Dawes had also expressed the view that Yeats 'would be the ruin of the establishment'.[110] What is clear beyond dispute is the unhappiness experienced by the small community at Queenwood and the inability of Edmondson to manage affairs fairly and without excessive deference to his wife. To Frankland and Tyndall it became clear that 'expatriation' was the best move,[111] though before that was to happen Anne Edmondson entered the lists on behalf of Yeats.[112] She wrote a long letter to Tyndall and Frankland, half-complaining and half-cajoling. She deplores their attitude to Yeats and regrets that it will lead to their departure. Yet she realises that this is not the only unsettling influence and refers to 'the uncomfortable state of things' which induced them 'to resort to a practice to which you were formerly unaccustomed', adding mysteriously 'you know to what I refer'. This was presumably their occasional flights to local hostelries for warmth and comfort. She wrote without her husband's knowledge (though agreed he might see the letter). It indicates more clearly than anything else her rôle in the troubled domestic politics of Queenwood.

To this lengthy epistle Tyndall and Frankland replied with one that was over twice as long.[113] They rebutted her charges of arbitrary injustice and impetuosity, rehearsed the details of their meetings with Edmondson and stoutly defended their decision to leave. Perhaps they suspected her defence of Yeats was not entirely disinterested, adding that though they were free from personal hostility to him 'we are not sure that you are equally free as regards personal attachment'. There followed more entreaties, tears, apologies and an apparently genuine attempt to redress at least some of the grievances.

At this point the turmoils of Queenwood College were to be unexpectedly, though temporarily, displaced by others yet more dramatic. They were heralded by the arrival of the summer holidays, which Tyndall and Frankland proposed to spend in Paris, partly with the intention of attending chemical lectures at some of its renowned institutions. So, on 14 June they declared their final resolution to leave Queenwood, even though four months would elapse before it could happen. Meanwhile six glorious weeks of freedom lay ahead and, early the following morning, the two friends set off for Southampton and thence to le Havre.[114] A few days in Rouen were enlivened by the presence of Edward's Lancaster friend John Scott.[115] Both here and throughout his holiday Frankland recorded in his diary minute descriptions and impressions of the ancient churches, museums and other buildings he had inspected.[116] They proceeded to Paris where, on 22 June, their ways parted, Tyndall leaving for a few days in Amiens.

No sooner had Frankland seen off his friend at the railway station than he found himself in considerable difficulty. Since the collapse of the French monarchy in February Paris had been seething with rumours of revolution and uprising. The brief months of the interim were played out in an atmosphere of suspicion of an army that was bent on avenging its humiliation in February and with unemployment and economic distress all too visible. Nor was Paris alone in its despair, for all over Europe that year, 1848, was an Age of Revolution, from Italy to Prussia and from Spain to Poland. Now it was Frankland's luck to find himself alone in Paris on precisely the day that insurrection swept the city in a wave of hate and bloodshed. Thus he describes his first awareness of real trouble ahead:

> Wending my way along the Rue St. Honoré, [I] was surprised to find almost all the shops closed. I quickened my pace, but finding the crowd to thicken as I proceeded, and seeing the soldiers running about in all directions, I struck off into a side street; but the locality being new to me, I lost my way for some time, but coming at last to the side of the river I succeeded in passing over a bridge before it was occupied by a body of troops that was just at that moment approaching. I passed through several streets, some crowded, some clear, and others occupied by soldiers; I inquired my way several times. At length I came into the neighbourhood of the Rue de la Harpe and inquired the way to the Rue Serpente, and was informed that the streets were barricaded in such a manner that it was impossible to get to it. I was turning back, when a young man said he thought I could get to my destination; I determined to make the effort, and pressing forward through a party of soldiers I soon found myself in the Rue de la Harpe, but in a part totally unknown to me. I made inquiries for the Rue Serpente from the National Guards. They pointed out the wrong direction. I pursued it for some time, and then returned, but found myself surrounded by soldiers who would not suffer me to pass. At last I over-persuaded them, and one kindly volunteered to accompany me; he escorted me until we arrived in sight of a huge barrier of paving stones raised by the insurgents. Here my guide pointed out to me the street, but could not proceed further, as we should in all probability have been fired upon by the people. As I approached the barrier I saw a

number of men armed with muskets and fixed bayonets standing behind it; they did not attempt to molest me. I crossed the barrier where another lot of such men were located at the end of the Rue Serpente, and found myself in a few more minutes safe in my own room. A quarter of an hour afterwards a high barricade was erected on the other side of the Rue Serpente which would effectually have prevented my return.[117]

In the ensuing four days the civil strife accounted for over ten thousand lives in Paris. Frankland's lively account of the conflict was derived partly from the press, partly from rumour and partly from observation from his embattled apartment. Under martial law, he was, like all others, confined to his rooms, forbidden to open doors or windows and required to illuminate his windows at night for the benefit of the military outside. When, on the evening of June 26, the worst was over and the curfew lifted he was able to join the crowds inspecting the damage, the dismantled barricades, the piles of confiscated weapons and the laid out bodies of men and horses. Frankland's own sympathies appear to have been chiefly with the republican army. He refers to the 'atrocious acts of the insurgents' as 'almost too diabolical to narrate'.[118] On the other hand he felt a characteristically warm regard for an uninhibited and amorous couple in a small boat on the river, 'how much more enviable was their lot than that of royalty'.[119]

When calm had returned to the city Frankland was able to indulge his taste for scientific discourse. Thus he heard Dumas at the École de Medicine expound on the composition of blood and milk,[120] on gelatin,[121] and on the fatty acids.[122] He heard Fremy deputise for Gay-Lussac at the Jardin des Plantes in a lecture on carbohydrates,[123] and attended other lectures on optics,[124] botany,[125] and geology.[126]. On these and other occasions he would sometimes be accompanied by G. A. Lenoir whom he had met in Bunsen's laboratory in Marburg.[127] Best of all, perhaps, he was able to purchase nine volumes of Berzelius' *Rapport annuel sur la chimie*.[128] The silent revolution chronicled therein was more profoundly to affect the personal view of Frankland (and it might be argued of posterity) than all the carnage and distress he had experienced in Paris.

On 16 July – one full month after their arrival in France – Tyndall and Frankland turned their backs on Paris and began a leisurely journey home. They called at Amiens, the site of the Battle of Waterloo, Brussels and Antwerp. From Ostend they headed for Dover and arrived in London on the evening of 23

July.[129] Following a day of visits and conversation (with Maule, Galloway, Playfair and others)[130] they returned by train to Queenwood where they 'were received with smiles'.[131] They found things a little improved but it was all too late and in October 1848 the two friends left for a new life in Germany.

Of the later history of Queenwood little need be said. Tyndall himself returned in June 1851; according to Frankland this was 'simply to wait there until he could obtain some suitable employment as Professor of Physics'.[132] He stayed until 1853, during which time he was in effect a deputy to Edmondson, whose weak government he helped to strengthen so effectively that pupil numbers more than doubled in a single year.[133] Tyndall was then succeeded by another Marburg PhD, the mathematician Thomas Archer Hirst, later to be associated with Frankland in various London adventures. Meanwhile Frankland's place was taken by his former fellow-apprentice in Lancaster, Robert Galloway[134], who came to Queenwood in September 1848.[135] On his arrival he found students studying the 'Giessen Outlines' but disliking the associated laboratory work. This was hardly surprising as only three or four hours each week were available for chemistry.[136] While at Queenwood Galloway published a text book of *Qualitative Analysis*, using his own method of group separation.[137] In 1851 he followed in Frankland's footsteps for a third time and went to Putney College, being succeeded at Queenwood by Heinrich Debus, who had previously assisted Bunsen at Marburg.[138] Edmondson died in May 1863 and his wife six months later.[139] By now the College was well into decline[140] but it continued under a new Principal (Charles Willmore), though financial difficulties eventually forced its closure.[141] On June 10, 1901 a disastrous fire destroyed the buildings and all that may now be seen above ground is the lodge and part of a wall. Yet in its day it had been the proving ground for no less than six men later to achieve professorial distinction in British institutions (Frankland, Tyndall, Debus, Galloway, T. A. Hirst, W. F. Barrett). Frankland was only there for 14 months, but his association with Queenwood left its mark upon him all his days, as was also the case for Tyndall. It was, as Frankland said, 'one of the most eventful years of our lives'.[142]

Notes

1 Though sometimes cited as 'CM 1842', the inscription with '1841' is described only three years later in C. J. H[olyoake], *A Visit to Harmony Hall!*, London, 1844, p. 7. Contemporary accounts of Queenwood College (as it became) appear also in *Illustrated London News*, 30 September 1848 and 13 January 1849.
2 A. C. Fryer, 'Queenwood College', *Owens College Magazine*, 1886, 18 (no. 2), 49–54 (50–1). The only other major account appears in D. Thompson, 'Queenwood College, Hampshire', *Ann. Sci.*, 1955, 11, 246–54, though one will also appear in W. H. Brock, *Science for all: studies in the history of Victorian science and education*, Variorum, Aldershot (in the press).
3 See F. Podmore, *Robert Owen, A Biography*, Allen & Unwin, London, 1906, p. 573.
4 Anonymous letter in *Morning Chronicle*, 13 Dec. 1842. The letter was circulated as a pamphlet four days later by the Governor of Harmony Hall, who claimed the buildings 'did not cost more than half the sum stated'. Both this letter and Holyoake's pamphlet (note 1) have been reprinted in *Co-operative Communities: Plans and Descriptions*: eleven pamphlets 1825–1847, Arno Press, New York, 1972.
5 Holyoake (note 1), p. 8.
6 Podmore (note 3), pp. 572–3. By mid August pupils were being withdrawn as quickly as possible (*New Moral World*, 1845, 13 (16 August) 488).
7 PO *Directory for Hampshire, Dorset and Wiltshire*, 1848, p. 2507.
8 *DNB*; *The Friend*, 1863, pp.152–3; D. Thomson, *Friends' Quarterly*, 1956, 10, 24–9.
9 *DNB*.
10 Edmondson's years in Russia have been described in two books by his daughter: 'J.B.', *From the Lune to the Neva*, London, 1879 (a rather fictional account in which the characters are identified by pseudonyms, as 'Doubleday' for 'Singleton', etc.); and Jane Benson, *Quaker pioneers in Russia*, Headley Bros., London, 1902.
11 Anne Edmondson was born in Lowdham, near Nottingham, to William and Anne Singleton [Lowdham Parish Records, 9 November 1800]. Her daughter Jane was born 25 years later at St. Petersburg [1851 Census, PRO, HO 107/1672].
12 *The Friend*, p. 152.
13 *A concise introduction to geography* (1837); *A course of practical geometry* (1837); and *Problems of practical perspective* (1841).
14 *The Lower Bank Monthly Magazine*.
15 *Lancastrian Chemist*, pp.152–4.
16 L. Stephen, *Life of Henry Fawcett*, London, 1885, p.9.
17 *Queenwood Reporter*, 15 February 1848, p.6.
18 *Sketches*, pp.69 and 260.
19 Tyndall, *Journal*, entry for 4 September 1847 [Royal Institution Archives, typescript].

20 The Census for 1851 identifies him as unmarried, 28 and born at Gorleston [PRO, HO 107/1672].

21 Tyndall, *Journal* (note 19).

22 *Ibid.*, entry for 5 September 1847.

23 *Ibid.*, 3 December 1847.

24 Frankland, *Diary* for 1848 and part of 1849, 19 January 1848 (typescript copy) [JBA, OU mf 02.02.1304–1484].

25 John Reader, *Of schools and schoolmasters*, Quaker Home Service, London, 1979, pp.43–4.

26 Tyndall, *Journal*, 23 October 1847.

27 *Prospectus* for Queenwood College, 1852 or later [University of London Library, MS 578]. Haas taught modern languages and was born at Marat, Switzerland, being now about 20 [1851 Census, PRO, HO 107/1672]. He came to Queenwood in May 1848 (Frankland, *Diary* (note 24), entry for 2 May 1848).

28 Tyndall, *Journal* (note 19), entry for 17 September 1847.

29 *Sketches*, p. 72.

30 'Notes of chemical lectures at Queenwood' [RFA, OU mf 01.02.1566–1858].

31 *Ibid.*, Lecture 10.

32 *Ibid.*, Lecture 21.

33 *Ibid.*, Lecture 31. During his Christmas holidays in Lancashire he spent an evening at Mercer's home at Oakenshaw, learning of his chemical and theological views (*Diary* (note 24), entry for 1 January 1848).

34 'Chemical lectures at Queenwood' (note 30), Lecture 41.

35 *Ibid.*, Lecture 41.

36 *Ibid.*, Lecture 27. Prior to his holiday visit to Mercer (note 33) he had called at Thompson's Primrose printworks at Clitheroe in company with his friend Cooper (*Diary* (note 24), entry for 1 January 1848).

37 *Ibid.*, Lecture 1.

38 *Ibid.*, Lecture 14.

39 *Ibid.*, Lecture 15.

40 *Ibid.*

41 *Ibid.*, Lecture 12.

42 *Ibid.*, Lecture 15.

43 L. Stephen (note 16), p. 10. The compound must actually have been lead *chromate*, $PbCrO_4$.

44 Frankland, *Diary* (note 24), entry for 14 January 1848.

45 *Ibid.*, 12 March 1848.

46 *Prospectus* (note 27).

47 Frankland, *Proc. Roy Soc.*, 1894, **55**, xviii; see also *Westminster Budget*, 15 December 1893.

48 D. Thompson, 'Queenwood College' (note 2), p. 246.

49 N.D. McMillan and J. Meehan, *John Tyndall*, '*X'emplar of scientific and technological education*, National Council for Education Awards, Dublin, 1980, p.26.

50 *Sketches*, p. 62.

51 *Queenwood Reporter*, 15 February 1848, pp.6–7.
52 Frankland, *Diary* (note 24), entries for 17 January and 28 February 1848.
53 *Ibid.*, 28 February 1848.
54 *Ibid.*, 9 February 1848.
55 Dawes (1793–1867), grandson of a rector of Kendal, was educated there by the blind philosopher John Gough who also taught John Dalton and William Whewell. After becoming 4th wrangler (1817) at Trinity College, Cambridge, he went on to teach mathematics at Downing College, moving to Kings Somborne in 1836; see *DNB* and D. Layton, *Science for the people*, Allen & Unwin, London, 1973.
56 R. Dawes, *Hints on an improved and self-paying system of National Education*, London, 1847, p. 30.
57 R. Dawes, *Suggestive hints towards secular instruction, making it bear upon the practical life*, 3rd ed., London, 1849, pp.vii–ix.
58 'Popular chemistry', [RFA, OU mf 01.02.1962–1990].
59 Tyndall, *Journal* (note 19), 28 November 1847. The compound was called 'sulphuric ether' because it was made using alcohol and sulphuric acid, by analogy with the products from nitric and hydrochloric acids.
60 *Ibid.*, 23 January 1848.
61 Frankland, *Diary* (note 24), entry for 11/12 March 1848.
62 *Ibid.*, 10–15 April.
63 *Sketches*, p.75.
64 J. L. Gay–Lussac, *Ann. Chim.*, 1814, **91**, 5–160.
65 Frankland, *Diary* (note 24), entry for 18 April – 16 May 1848.
66 Frankland, *Experimental Researches*, p.63; that was in 1877.
67 Frankland, *Diary* (note 24), entry for 18 May 1848.
68 Frankland, *Experimental Researches*, p.63.
69 Frankland, *Diary* (note 24), entries for 28 July – 4 October 1848.
70 *Ibid.*, 4 March 1848.
71 *Ibid.*, 30 April 1848.
72 *Ibid.*, 2 April 1848.
73 *Ibid.*, 5/6 April 1848.
74 Hugh Russell (*c.* 1785–1849) was a Scotsman who spent 40 years of his ministry at this historic chapel (*Baptist Manual*, London 1850, p.43).
75 A. W. Heritage (1815–1871): *Baptist Handbook*, 1872, pp.226–7.
76 Frankland, *Diary* (note 24), entries for 13 February and 26 March 1848.
77 *Ibid.*
78 *Ibid.*, 17 February 1848.
79 *Ibid.*, 7 April 1848.
80 *Ibid.*, 2 January 1848.
81 *Ibid.*, 9 January 1848. Sherman (1796–1862) was Minister of Surrey Chapel, and Binney (1798–1874) of the Weigh House Chapel; see *DNB*.
82 Frankland, *Diary* (note 24), entry for 18 January 1848.
83 *Ibid.*, 23 January 1848.
84 *Ibid.*, 21 April 1848.
85 *Ibid.*, 9 April 1848.

86 *Ibid.*, 25 January 1848.
87 *Ibid.*, 26 April 1848.
88 *Ibid.*, 27 April 1848.
89 *Ibid.*, 1 May 1848.
90 *Ibid.*, 23 February 1848.
91 *Ibid.*, 6 May 1848.
92 Tyndall, *Journal* (note 19), 7 May 1848.
93 Frankland, *Diary* (note 24), entry for 8 May 1848.
94 *Ibid.*, 16 May 1848.
95 Thompson, 'Queenwood College' (note 2), p.248.
96 Born 3 January 1811, typescript *Dictionary of Quaker biography* [Friends Meeting House Archives]. He died 10 November 1889 in Southport, Probate records [Somerset House]. His only publication appears to have been *A book of texts containing an epitome of Christian faith and duty*, Darwen and London, 1866, reflecting his practice as a schoolmaster of encouraging pupils to learn one Biblical text each day; and a paper read before St. John's Church Guild, Darwen, on 1 June 1882, 'Rest' [British Library, MSS 4371.e.34(9)].
97 Frankland, letter to Louisa Tyndall, 20 April 1894 [Royal Institution Archives].
98 Frankland, *Diary* (note 24), entry for 15 February 1848.
99 *Ibid.*, 25 January 1848; this was presumably Anne Edmondson.
100 Frankland, *Diary* (note 24), entry for 12 August 1848.
101 *Ibid.*, 1 June 1848.
102 *Ibid.*, 13 June 1848.
103 *Ibid.*, 21 June 1848.
104 *Ibid.*, 31 May 1848.
105 *Ibid.*, He married the following year a Miss Martha Walker of Manchester, (Tyndall, *Journal*, 18 January 1849). He wrote in optimistic terms to Frankland and Tyndall from Spring Bank: J. Singleton to Frankland and J. Tyndall, 21 August 1848 [RFA, OU mf 01.02.1307]. According to local *Directories* he remained there until at least 1881.
106 Frankland, *Diary* (note 24), entry for 20 August 1848.
107 *Ibid.*, 31 May 1848.
108 *Ibid.*, 30 May 1848.
109 *Ibid.*, 31 May 1848.
110 *Ibid.*, 17 September 1848.
111 *Ibid.*, 21 May 1848.
112 *Ibid.*, 6 June 1848: transcript of letter dated '6–5–1848'.
113 *Ibid.*, 7 June 1848: transcript of letter dated 'June 8th 1848'.
114 *Ibid.*, 15 June 1848.
115 *Ibid.*, 17 June 1848.
116 *Ibid.*, 23 June 1848.
117 *Ibid.*, 23 June 1848. A detailed account of Frankland's experience of the June insurrection appears in his *Sketches*, pp. 456–69. Most of that material has since been published (1976) in a *Document Collection* for Open

University Course A321, 'The Revolutions of 1848', pp. 37–40. For this reason the present chapter gives a brief account of his impression, though the original *Diary* (note 24) contains further material as yet unpublished.

118 Frankland, *Diary* (note 24), entry for 26 June 1848.

119 *Ibid.*, 2 July 1848.

120 *Ibid.*, 4 July 1848.

121 *Ibid.*, 8 July 1848.

122 *Ibid.*, 11 July 1848.

123 *Ibid.*, 8 July 1848.

124 *Ibid.*, 8 July 1848.

125 *Ibid.*, 10, 14 July 1848.

126 *Ibid.*, 12 July 1848.

127 This was presumably G. A. Lenoir, a fellow student in Bunsen's laboratory (*Sketches*, p.81, where he is misrepresented as 'Herr Senior') who had recently confirmed the existence of pentathionic acid (*Annalen der Chem.*, 1847, **52**, 253).

128 Frankland, *Diary* (note 24), entries for 10, 12 July 1848.

129 *Ibid.*, 23 July 1848.

130 *Ibid.*, 24–5 July 1848.

131 *Ibid.*, 25 July 1848.

132 *Westminster Budget*, 15 December 1893.

133 A. S. Eve and C. H. Creasy, *The life and work of John Tyndall*, Macmillan, London, 1945, p.34.

134 Robert Galloway (1822/3–1896) was born at Cartmel (then in Lancashire), was a fellow-apprentice of Frankland in the Lancaster pharmacy shop, and his immediate superior. He followed in Frankland's steps to Queenwood and Putney and eventually became Professor of Applied Chemistry at the Museum of Irish Industry in Dublin. See 'Cross Fleury', *Time-honoured Lancaster*, Lancaster, 1891, pp.303–4; and Anon., *J. Chem. Soc.*, 1896, **69**, 733–4.

135 Frankland, *Diary* (note 24), entry for 9 September 1848; Galloway appears to have needed little persuasion to come since he did not get on well with Playfair to whom he was an assistant (*ibid.*, 24 July 1848).

136 R. Galloway, *Education, scientific and technical; or, how the inductive sciences are taught, and how they ought to be taught*, London, 1881, pp.258–9, 346. The 'Giessen outlines' were presumably *Instruction in chemical analysis (Qualitative)*, by C. R. Fresenius 'with a preface by Professor Liebig', ed. J. L. Bullock, 2nd ed., London, 1846 (Fresenius being responsible for elementary laboratory instruction at Giessen).

137 R. Galloway, *A manual of qualitative analysis*, London, 1850. This is essentially a simplified version of Fresenius (note 136), but introduces the term 'Groups' (starting with Group V for elements with insoluble chlorides).

138 On Debus (1824–1916) see J. C. Poggendorf, *Biographisch-literarisches Handwörterbuch zur Geschichte der exacten Naturwissenschaften*, Leipzig. He later became Professor of Chemistry at the Royal Naval College, Greenwich.

139 He left less than £5,000, and she less than £7,000, Probate records [Somerset House].
140 A letter from a newly appointed science master, W. H. Wilcox, reported that in 1893 'the Chemical Laboratory is in a very bad state, so that it will be some time before I get it in good order': W. H. Wilcox to J. T. Wood, 22 April 1893 [Nottinghamshire Archives, M24/326].
141 An octogenarian inhabitant recounted to the author in 1981 how the Queenwood boys came down to buy plum-cake in the village once a week, and how the College's debts to local shopkeepers were partly paid in kind (knives, forks, etc.).
142 Frankland, *Diary* (note 24), entry for 2 October 1848.

CHAPTER 4

New worlds in Germany

1 Return to Marburg

When Frankland and Tyndall sailed for Germany on 18th October 1848 it was anything but a desperate bid to escape the quarrels and privations of the dour regime at Queenwood. Still less was it an unpremeditated dash by impetuous youth. Six months ago Frankland had sent to his friend Debus certain enquiries[1] regarding the university at Marburg where he had already spent three whirlwind months the previous summer. Within days a reply was on its way to Queenwood, urging him to come as soon as possible and assuring him of a welcome from Bunsen and all his friends[2]. The message was augmented by another letter from the same university, when Hermann Kolbe urged him 'be quick and come soon'[3]. On the other hand it was far from the kind of shrewd opportunism one expects from a research student today, selecting carefully from a wide range of possible options. So familiar in the later nineteenth century was the spectacle of English students going to German universities that it is easy to forget that, apart from a few hardy pilgrims to Giessen, in the 1840s it was a considerable novelty. In the case of Marburg itself no Englishman had ever graduated from its ancient university, the only one in the tiny state of Hesse-Cassel. So Frankland and Tyndall were embarking on a course far more adventurous and uncertain than anything that can be imagined today. Why then did they do it?

The principal magnet which drew the budding researchers to Marburg was the charismatic figure of Robert Wilhelm Bunsen. Unlike his fellow countrymen who followed the Giessen trail,

Frankland had already experienced much kindness and help from Bunsen and had, of course, made friends among the Professor's junior staff. Had he not been enticed to Marburg the previous summer by Kolbe it is doubtful whether he would now be bent on a research degree in a university that (unlike Giessen) was inadequately supported by the government and largely unknown in Britain. [4]

Thus with a cab so full of baggage that they had to walk from Tyndall's lodgings to Blackwall Station, the two friends embarked on their two day voyage to Rotterdam. They journeyed by steamer down the Rhine, Frankland recording their impressions in some detail: the 'pretty' scenery of the Dutch flat lands; the Gothic splendours of the cathedrals in Dusseldorf and Cologne; the majestic channel of the Rhine beyond Coblenz. From Frankfurt they were compelled to travel by road to Marburg (which was not linked by rail until 1850). Frankland travelled overnight, leaving Tyndall, who did not care for such nocturnal conveyance, to arrive the following day (25 October). [5] They quickly settled into their lodgings from which they would frequently sally forth to enjoy convivial dinners at the Gasthaus Ritter.

The striking view that greeted their eyes had changed little since Frankland last saw it over a year ago, but it was new to Tyndall whose delight was immediate and evident:

> A very picturesque town Marburg is. It clambers pleasantly up the
> hillsides, and falls as pleasantly towards the Lahn. On a May day,
> when the orchards are in blossom, and the chestnuts clothed with their
> heavy foliage, Marburg is truly lovely. [6]

Now, with autumn nearly over, the view of Marburg from a nearby hill was, in Frankland's words, 'a most lovely prospect'. [7] It was to be only a few days before the landscape was covered with snow, [8] and another six weeks before Frankland was to be found skating on the frozen Lahn. [9] Nearby the forests were to offer plenty of scope for excursions, especially on Sunday. Sometimes Frankland would elect to cross the mountains on foot, in the company of friends, in order to attend a social function in a village not too far away. Whether it was a grand ball with the Cotillion being danced into the small hours, or a musical evening in a professor's home, or throwing 'a select party' in his own apartments, Frankland plunged into the social life of town and university. Ever appreciative of the opposite sex he approvingly records parties of 'pretty damsels' on festive occasions. [10] Several of these young ladies caught his eye in a special way, as a certain 'highly accomplished young lady' whom he met at a ball at Kirchhein:

> I was not slow in engaging Miss N. . . of Wetter as my partner in the Cotillion, during which dance we became still better acquainted with each other; the dance continued until 5 a.m. after which I had the pleasure to accompany Miss N . . . to the house of her friend; the regret at parting was mitigated by the anticipation of a future meeting.[11]

Then there was 'the sprightly Fräulein von Gehren', who was 'pretty and lively, but too short to be deemed handsome', though she danced well and was a great favourite with the officers. Special notice was taken of Fräulein Grebe, with whose brother he read English and German and whose society he could enjoy because he understood her to be already engaged.[12] However, they exchanged bouquets[13] and were often seen together. On one occasion:

> Her eyes lent their expressive brilliancy to the scene and my position really became dangerous as I sat by her at supper, although I have always had a horror of celebrated beauties.[14]

In the end Frankland did not have to choose between 'the sparkling expressive eye of a Grebe and the exquisitely moulded foot of a Gehren'[15] for Sophie Fick was to claim his heart. Yet of that young lady there were to be no sightings[16] until late the following summer, when he received 'a gracious glance' as she passed his window.[17] Only her portrait was evident at her brother's house, for a bereavement kept her in Cassel for many months, it seems, and later Frankland was disappointed not to meet her even though she was known to be in the town.

However the social life of Marburg compassed more than excursions, picnics and parties in attractive female company. Many were the riotous gatherings with fellow students for taking tea, dining or drinking late at night (though Frankland rarely appears to have imbibed more alcohol than he should). Among his closest friends were, of course, Tyndall (from Queenwood) and Debus (from his first visit to Marburg). But he also enjoyed the company of several others later to make their mark on science.

One of these companions was William Leonard Faber, an American whom Frankland christened 'Yankee' and who eventually returned to his own country as a metallurgist and mining engineer. He travelled extensively in Germany, studying local techniques and attending the universities of Marburg, Giessen and Freiburg. He wrote several times to Frankland in the early 1850s, complaining

Fig. 4.1. The building at the University of Marburg that formerly housed
Bunsen's laboratory. (Marburg.)

more than once of his long absence from home, expressed as a
disgust for Germany (except its scenery).[18] He settled in America
shortly afterwards, collaborating with James C. Booth, a professor
in Philadelphia. Faber translated the *Tables for Qualitative Chemical
Analysis*[19] by Heinrich Will which he had encountered at Giessen,
and jointly with Booth edited an American translation of Regnault's
Elements of Chemistry.[20] Among the entries in that work attributed
to 'W. L. F.' are notes on Frankland's studies on 'ethyl',[21] the
research by Frankland and Kolbe on amines,[22] and gas manufacture
in Manchester.[23] Eventually the itinerant mining engineer settled
down to the quieter life of journalism, and in 1860 wrote to
Frankland inviting him to be a European correspondent of his local
paper *The Prospect*.[24] Whether Frankland agreed is not known. In
Marburg Faber was a frequent companion and helped Frankland in
the translation of his dissertation.[25]

A visitor to Marburg whom Frankland had previously encountered
in England was Faber's friend Conrad Bromeis (1820–61), who had
graduated at Marburg and then taught chemistry and physics at the
Realschule in nearby Hanau. He was to concentrate on chemical

technology,[26] became a correspondent of Frankland,[27] and acted for him as a referee.[28] He and Faber helped Frankland to 'see in' the New Year of 1849.[29]

Nevertheless for all the pleasure afforded by good company Frankland felt at times a stranger in a foreign land. He was a good correspondent and often wrote to friends back home; one Christmas meal reminded him nostalgically of 'the good old times and dear Grandmother at Churchtown'.[30] He was once worried by farmers' dogs 'probably on account of the strangeness of my attire',[31] and would have found it hard to shake off the image of English people cherished by the Marburgians. The only other Englishman apparently known to the inhabitants was a lunatic named Mr Magnus who was led about by his keeper[32] and one day encountered at an inn by Frankland, 'quite mad'.[33] Nor would the impression be dispelled by Tyndall's unusual practice of 'hanging by his legs from the branches of the trees and performing all sorts of queer antics'.[34]

Yet he was strongly attracted to many distinctively German values and ideals. He was often to be found reading Goethe's *Faust*, usually with friends, and was a rapt listener to the disquisitions by Waitz on Kant and other German philosophers. He was sufficiently confident to commence German entries in his *Diary* as early as November 1848, and soon picked up enough of the spoken language to understand lectures and converse adequately on social occasions. One characteristically German sentiment struck deep chords within him: 'the German division of Society into classes, which is here effected according to <u>mind</u> and not <u>money</u>.' Here was a sentiment with which this impoverished Englishman could readily identify, and which concurred with what he had learned in Lancaster days from the self-improvement classes sponsored by the Johnsons and others. On his return to England it would become for some time a *leitmotif* in his struggle for social acceptance.[35] His warm admiration for the German ideal of enlightened scholarship was fully shared by his friend John Tyndall:

> The German student, and Germans in general, so far as I have seen them, are kindhearted, loveable men. They are thoughtful men, able men to boot; their voice is gone out into all lands. In England, German thought is silently executing its high mission; slowly and steadily ripening, into results which will one day be as palpable as others, of historic fame, which have originated in this land of reformers.[36]

Such sentiments will surface again and again in the later life of Edward Frankland.

One other aspect of distinctively German culture impinged sharply on Frankland during his year at Marburg: the political instability of the German states, and particularly Hesse-Cassel whose government vacillated over reform and over possible links with Prussia. A 'tremendous row between the peasants of Kölbe and Marbach' called out the Bürgergarde, and 'everyone expected revolution, but all ended in two men with broken heads being taken into custody'.[37] More serious was the sight of people arming themselves after receiving news of the barricades at Dresden, for a Prussian bombardment of Marburg was expected any day. It did not happen, though Frankland was requested to fight for the crown jewels! He 'politely declined'.[38] Such experiences, like the Paris uprising, were all part of the educative process embarked upon by the young man from Lancashire. But in his eyes all were peripheral to the primary object of the whole adventure: to learn chemistry with Bunsen.

2 Research in Bunsen's laboratory

On the day after their arrival Tyndall was introduced to Bunsen, and Frankland commenced work in the laboratory. The coming year was to be of the first importance for him and for chemistry. The laboratory in which Frankland once again found himself was no longer enlivened by the presence of his friend Hermann Kolbe, for the latter had moved to an editorial post at Braunschweig a few days before Frankland's arrival.[39] However Bunsen was as active as ever, Tyndall was never far away as he was promptly set to work on quantitative analysis, and Debus was as good as his word in welcoming the stranger. Attendance at lectures[40] would soon extend his acquaintance with students and professors. In addition to Bunsen's lectures (9 to 10 a.m.) Frankland enrolled for a course in physics delivered by Professor Gerling[41] during the previous hour and another one by Stegman[42] on algebraic geometry in the afternoon (less successful as he could not keep awake in the over-heated room). Later he attended other lectures on botany and mineralogy, a series on Kantian philosophy by his previous acquaintance, the youthful Professor Waitz[43], and courses on organic chemistry and voltaic electricity from Bunsen himself.[44]

The shape of things to come was plain from Frankland's first day in the laboratory. Neither equipment nor chemicals taken for granted today were available then. Chloroform, procured by some unspecified means, had to be washed with water, (presumably) dried, and distilled a few days later. Not deterred by his experiences at Queenwood he bought an ounce of potassium, then still a fairly rare commodity. Money was scarce and every purchase had to be an investment. As for apparatus we first find him at the laborious task of graduating and calibrating a eudiometer, and then preparing apparatus for distillation and for organic analysis by the methods of Liebig.[45] He appears to have been left largely to his own devices. In those early days Bunsen's only recorded interactions with him appear to have been a demonstration of his own method for preparing ozone (by electrolysis of chromic acid solutions)[46] and an evening conversation in which the great man received copies of memoirs by Frankland and Kolbe, and conversed on volcanic minerals and rock oil formation.[47] Frankland's debt to Bunsen must surely have been chiefly in the inspiration needed for a hunt for 'radicals' and for the invaluable new techniques of gas analysis. On these two topics he waxed lyrical in a letter to Tyndall many years later, giving his opinion of Bunsen's contributions to science. There is more than a hint of autobiography in the following:

> The beneficial influence which the discovery of Cacodyl has exerted upon the development of sound views in organic chemistry, and the pursuit of rational paths of research, can scarcely be over-estimated, since the support which it gave to the theory of compound organic radicals, enabled that theory to exist and to bring forth abundant fruits, throughout a period in the history of science when there was but feeble evidence of its truth, and until those researches, both of supporters and antagonists, finally established the fundamental accuracy of that theory.

On the importance of Bunsen's contributions to gas analysis Frankland continued:

> Previous to his labours in this field, eudiometrical determinations were of a very crude and unsatisfactory character; but his improvements in the construction of eudiometers enabled every operator to construct these instruments for himself with facility, and possessing an accuracy of measurement which could not be exceeded in the workshop of the philosophical instrument maker. These improvements, together with his new methods of applying absorbents to gaseous mixtures, have imparted to eudiometrical determinations a degree of accuracy never exceeded and rarely equalled in other departments of analytical chemistry. It is also perhaps worthy of notice that the subsequent

improvements in gaseous analysis have been chiefly if not entirely
made by the pupils of Bunsen.[48]

Of these pupils he mentioned only Regnault. Modesty can have
been the only reason for excluding his own.

As with many a subsequent PhD student the first few months of
research seemed to lead nowhere. Instead of embarking on a guided
exploration of new fields he began by attempting to continue work
begun at Queenwood. He returned to his ambitious project of using
potassium to extract the 'formyl' radical from chloroform (hence
his early acquisition of the two reagents). Experiments in England
had usually ended in explosion, the same result attending his early
attempts in Marburg. However a sealed tube from this reaction
carried out at Queenwood survived the journey and arrived on 2
November.[49] In his *Diary* he assumed 'the K must at a certain
elevation of temperature instantly and explosively combine with
chlorine',[50] though he later attributed the explosion to the introduction
of potassium hydroxide on the potassium's surface, thus generating
carbon monoxide from the chloroform and hence the explosive
potassium carbonyl $[(KCO)_6]$.[51]

Undeterred by these difficulties Frankland proceeded to demonstrate a characteristic that became prominent in all his subsequent
work. He devised, and constructed, alternative apparatus using his
very considerable manual skills. First, he tried heating the reagents
together in a flask, condensing out the chloroform vapour in a
cooled glass bulb, and collecting the residual gas (hopefully 'formyl').
Regrettably little was obtained due to the formation of potassium
chloride on the surface of the metal, so inhibiting further reaction.
Next, he reverted to another Queenwood technique and made two
'very strong tubes' as reaction vessels, immersed them in an oil-bath
at 100°C., and left them for some time. On return he found everything
was 'shattered to atoms', though fortunately no one had been present
at the time.[52] A further variation led to another explosion, 'piercing
about 10 to 15 holes through the window placed for the protection
of the observer', who was, by extreme good fortune, so unscathed as
to go on to write out some mathematical notes.[53] As if that were not
sufficient warning Frankland proceeded on the following day to
reconstruct the apparatus and encase it in a wooden box whose
open side faced a blank wall. This time there was neither explosion
nor apparent evolution of gas.[54]

Despite further reverses Frankland continued his quest for

'formyl'. Occasionally he did obtain a gaseous product which he subjected to gas analysis, though his figures imply large amounts of gas not containing carbon (possibly this was hydrogen).[55] But he was clearly getting nowhere and, as autumn merged into winter, the change of scenery, the approach of Christmas and the new delights of intellectual conversation were distractions that he could hardly avoid and he came to the reluctant conclusion that 'so far as original research was concerned, I consider that the first three months of my time in Marburg were wasted'.[56]

The festive season over, Frankland's thoughts reverted to another class of organic compounds that had yielded impressive results in his earlier quest for radicals, the fatty acids. So, in conformity with his notions of the relation between cyanides and acids, on 8 January he tried to make formic acid by the alkaline decomposition of potassium cyanide. He was delighted with the evolution of 'eine grosse Masse' of ammonia, and possibly encouraged by the result to visit Bunsen that evening in order to discuss his doctoral examination.[57]

The following day saw a return to his by now familiar exploitation of the intense reactivity of the alkali metals; it will be recalled that these had been known for barely 40 years and were still extremely expensive, so Frankland's use of them indicates remarkable confidence in their effectiveness and in his own skill in handling them. His first experiment with carboxylic acids and alkali metals was the action of sodium on pure valerianic acid.[58] It gave off a vapour which Frankland identified by its odour as 'the radical discovered by Kolbe (C_8H_9).' He then posited similar changes in these and other members of the same series:

formic acid	H,C_2O_3,HO	$+2Na$	$=H_2$	$+2CO+2NaO$
acetic acid	C_2H_3,C_2O_3,HO	$+2Na$	$=C_2H_3$	$+2CO+H+2NaO$
butyric acid	C_6H_7,C_2O_3,HO	$+2Na$	$=C_6H_7$	$+2CO+H+2NaO$
valerianic acid	C_8H_9,C_2O_3,HO	$+2Na$	$=C_8H_9$	$+2CO+H+2NaO$
caproic acid	$C_{10}H_{11},C_2O_3,HO$	$+2Na$	$=C_{10}H_{11}$	$+2CO+H+2NaO$
oenanthilic acid	$C_{12}H_{13},C_2O_3,HO$	$+2Na$	$=C_{12}H_{13}$	$+2CO+H+2NaO$
caprylic acid	$C_{14}H_{15},C_2O_3,HO$	$+2Na$	$=C_{14}H_{15}$	$+2CO+H+2NaO$
margaric acid	$C_{32}H_{33},C_2O_3,HO$	$+2Na$	$=C_{32}H_{33}$	$+2CO+H+2NaO$

His formulae are of course based on the old atomic weights $C=6$ and $O=8$, and the 'radicals' are actually dimers, so in modern terminology the products from acetic and valerianic acids (for example) would have

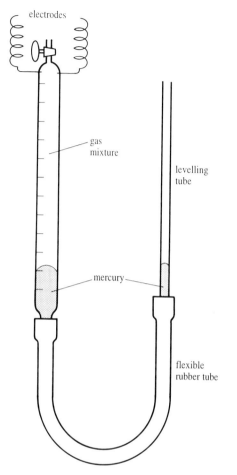

electrodes

gas
mixture

levelling
tube

mercury

flexible
rubber tube

In such instruments gases were exploded with excess air in a closed calibrated apparatus such that the volumes of reactants, and the volume of carbon dioxide produced, were known. The amount of oxygen used may be obtained by exploding with excess hydrogen *after* the main reaction and noting the change in volume. Sometimes the quantity of water produced was also obtained by noting differences before and after absorption by sulphuric acid. From the results the nature of the product could be determined, though with 'wrong' atomic weights and no recognition of Avogadro's hypothesis, the arguments appear tortuous to modern eyes, as well as often leading to incorrect conclusions. Sometimes, however, a comparative method was used in which combustion characteristics of unknown gases were compared with those of known substances, and correct conclusions were obtained.

Fig. 4.2. Simplest form of eudiometric gas analysis used by Frankland at Marburg. (Author.)

been respectively C_2H_6 and C_8H_{18}. The real problem, he confided to his diary, was that 'all the water atoms were decomposed & therefore hydrogen gas has developed'. He then made a decision of great importance to his subsequent work. Abandoning the highly reactive sodium he resolved to 'study the preparation of dry methyl iodide and its decomposition by zinc'.[59]

Why he should have done this the diary does not say, though it is highly probable that the same reasons applied as those that led him to replace potassium by zinc at Queenwood. The choice of methyl iodide is rather more problematic, especially as an unopened tube from the ethyl iodide/zinc reaction had survived its long journey from Queenwood and awaited attention. He later suggested that it must have seemed 'best to begin with the most simple term of the series'[60] and possibly that was so. Methyl iodide was a newer compound than its ethyl analogue, having been made by Dumas and Peligot[61] from the methyl alcohol [methanol] that they had discovered the previous year.[62] Doubtless using their method Frankland prepared the compound on 27 January. Then, between a four-hour reading with Tyndall and Faber of Goethe's *Faust*, and a study on his own of six chapters of Exodus, Frankland distilled the methyl iodide in his laboratory in the afternoon of Sunday, 29 January. But an unexpected setback awaited him. To purify the methyl iodide from all impurities, including water, was 'the work of three weeks'.[63]

Understandably his thoughts turned with greater urgency to that sealed tube from his luggage. On 18 February he opened one end of the tube under water, with an inverted receiver to collect any gas that might be present; it is not difficult to recapture something of the excitement:

> The rush of gas was so impetuous that some of the liquid and solid contents of the tube were ejected into the water, although the tube was held perpendicularly; the quantity of gas from so small a tube was enormous . . . On pouring a few drops of water into the tube, a violent effervescence took place and much gas was evolved.[64]

The gas was collected in tubes over water. It was soluble in alcohol, inflammable, and of strong ethereal smell. Analysis suggested that ethyl [butane] had indeed been found, together with about 20% each of methyl [ethane] and ethylene. Further preparations were started, in one of which needle-like crystals and a limpid liquid were noticed in the tube before opening. A few other metals were tried instead of zinc, but only tin appeared to undergo appreciable reaction. Meanwhile a method was devised to obtain 'ethyl' without its accompanying hydrocarbons. Frankland had noticed that, when the tubes were opened, there was an initial rush of gas, followed by a slower evolution as though a liquid in the tube was now gently boiling. He correctly assumed the latter was 'ethyl', since butane is easily liquefied under pressure. Accordingly he collected the later gaseous product separately and purified it further by diffusion through a porous septum of gypsum. The 'ethyl' was shown to be generally unreactive, but to react with chlorine in daylight and with bromine on heating. All this, and much more on similar lines, occupied Frankland's time between the date he commenced writing his dissertation (22 April) and his doctoral examination (30 June).

To modern minds a doctorate awarded on the results of less than a year's work may seem of little value. Yet that was the normal practice in Germany at the time, as was the *viva voce* examination of the candidate by every professor of the Faculty who was minded to ask questions. In fact Frankland got off fairly lightly, with only botany, mathematics and crystallography added to chemistry as subjects for inquisition. He was even spared a disputation in the Great Hall, with the added problem of having to deliver it in Latin. Doubtless mercy was being shown to a foreigner, for he was the first Englishman to graduate in Marburg. The examination took three hours, and the process of installation a further hour and a half, proceedings being wound up by an oration from Bunsen on the volcanic phenomena of Iceland.[65] Perhaps it was an appropriate subject to conclude a hot summer's day in Marburg. Frankland was awarded the degree of PhD *cum laude*. His dissertation, *Ueber die Isolirung des Aethyls*, was a slim volume occupying 45 pages of print.[66]

This was not quite the end of research in Marburg for another month remained to him. Having explored the ethyl question to his own and his examiners' satisfaction he now returned to the subject he had intended to begin with, the isolation of methyl. To some extent the urgency had departed as what purported to be 'methyl' had turned up as a by-product in the reaction between ethyl iodide and zinc. Nevertheless it would be good to obtain it in pure condition and, once again, his thoughts turned to sealed tubes that had been put away for the time being. These contained the products from the action of zinc on methyl iodide from earlier in the year. As he said 'out of these tubes some very wonderful and unexpected things came'.[67] His *Diary* entry for 12 July begins laconically enough: 'Set Ozone apparatus agoing',[68] for that substance was still a matter of great interest to Bunsen. A fortnight later Frankland concluded that it was HO_4. [69]But the *Diary* immediately continues for the same day:

> Opened 2 tubes in which C_2H_3I [methyl iodide] had been decomposed by Zn. A large quantity of gas issued from the tubes, and appeared to be acted upon by the water which exhibited a strong effervescence. On afterwards breaking off the tube-end and pouring water on the residue a violent action took place accompanied by a brilliant greenish blue flame.[70]

That flame was one or two feet high, and was followed by a relatively slow evolution of 'a gas of a most insupportable odour' which on ignition deposited a black stain on a cold surface. This was shown to be zinc, though Bunsen on entering the laboratory from his lecture room next door at first assumed it to be arsenic (a common impurity in commercially available zinc), with dire consequences for the inhabitants. The agitated professor, who rushed to the laboratory windows and threw them wide open, was calmed only by Frankland's demonstration that the black stains on porcelain were soluble in hydrochloric acid, and therefore could not be arsenic but were probably zinc. Frankland himself was prostrated for the afternoon, but later attributed his malaise not to chemical poisoning but to lumbago.[71]

The volatile zinc compound was shown on analysis to be a totally new kind of compound, called by Frankland 'zinc methyl' [dimethylzinc]. Of its significance more will be seen later. On August 4 Frankland left Marburg for a long journey home that would take him first to Giessen and then Berlin.

3 Interlude in Giessen

About 18 miles to the south of Marburg lay the little town of Giessen, in Hesse–Darmstadt. Here was the state's only university, a small institution that had, however, one enduring claim to fame. Since 1839 it had incorporated one of the world's first research schools in chemistry[72], founded in 1824 by Justus Liebig. Here was developed the first major technique for the quantitative analysis of organic compounds (by oxidation to CO_2 and nitrogen), and here began one of the earliest programmes of laboratory training as a prerequisite for experimental research. Many were the chemical pilgrims to that shrine of practical organic chemistry. Their names read like a roll-call of leading 19th century chemists, and by 1849 had included the following who all were to occupy university chairs in Britain: Thomas Anderson (Glasgow), James Blyth (Cork), Benjamin C. Brodie (Oxford), William Gregory (Edinburgh), William A. Miller (King's College, London), Lyon Playfair (Edinburgh), Edmond Ronalds (Galway), and Alexander Williamson (University College, London). Others included Robert Kane (future Dean of the Royal College of Science in Dublin), together with Thomas Richardson and Walter Crum (well known chemical manufacturers on the Tyne and Clyde respectively). Among German chemists there was A. W. Hofmann, son of the Giessen laboratory architect and pupil of Liebig from 1843 to 1845, and F.

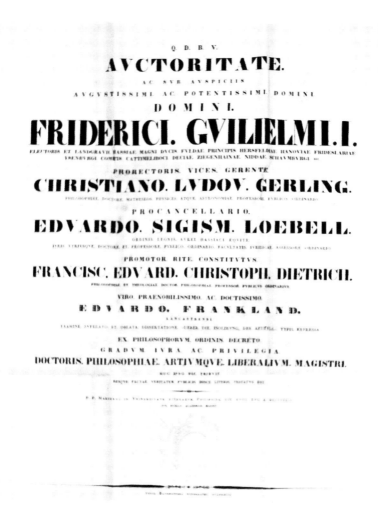

Fig. 4.3. Frankland's PhD diploma from the University of Marburg.
(R. Frankland Archives.)

A. Kekulé who was already in residence at this time. Small wonder, then, that Frankland sought (and obtained) an interview with Liebig and an offer of a place in his laboratory during the winter term.[73]

By the time Frankland came to Giessen there were few research students in the laboratory and Liebig himself had largely withdrawn from experimental work in favour of writing activities. Nevertheless he gave to the young Englishman a characteristically warm welcome and introduced him at once to his assistants Strecker[74] and Fleitmann.[75]

His work in the laboratory was a natural continuation of the research at Marburg. It was virtually confined to one topic: the isolation of the radical 'amyl' by the action of zinc on amyl iodide. The reaction is much slower than is the case of the simpler alkyl iodides, even when heated in sealed tubes; replacement of pure zinc by zinc amalgam led to a faster reaction. In due course Frankland was able to isolate and examine three products exactly analogous to those obtained from methyl or ethyl iodides. They were [in modern terms] pentane and pentene, and a fairly involatile liquid supposed to be 'amyl' but actually decane. The results were of course published in Liebig's own journal, *Annalen der Chemie und Pharmacie*, as well as in the *Journal of the Chemical Society*.[76] During this laboratory work Liebig was able to give a good deal of attention to his English visitor, to the latter's great profit. Frankland also attended Liebig's lectures, more polished presentations than those of Bunsen.

The winter of 1849–50 proved to be exceptionally hard in Giessen, the temperature at times being −30°C; a kettleful of water left on a red hot stove at night was a solid mass of ice in the morning. This led to a night of great anxiety after a visit from Heinrich Fick and a friend Dr Lucius. They and Frankland had a merry time at the Einhorn Hotel, eating and drinking to unusual excess. Frankland himself had to be carried up to bed in the hotel by the waiters and Lucius also stayed at that establishment. However Fick wandered out into a snowstorm and was feared lost in the sub-zero temperatures. Fortunately he was found next morning, – in bed in Frankland's lodgings. As Frankland observed afterwards 'This was the only time in which I even approached intoxication, and the effect experienced on the following day was not likely to make me wish to repeat the experiment'.[77]

However Christmas was soon upon him and all experiments were completed by 23 December. He took his leave with promises from Liebig to assist him in England, and requests to return in order to use his glass-blowing skills for the benefit of the Giessen laboratory. After a short holiday in Cassel, Frankland turned his face towards Berlin where he had secured an invitation from Heinrich Rose to work in his laboratory. On arrival he was introduced to the good and the great (and derived much satisfaction in reporting it to his parents).[78] But it was not to be. Before the Christmas vacation was over a letter arrived from England which turned his fortunes again. Lyon Playfair was retiring from his chair

Fig. 4.4. Edward Frankland about 1850 (apparently in graduation robes). (*Sketches.*)

at the Putney College of Civil Engineering and he offered the post to Frankland. Accordingly he departed hot-foot for England, calling at Marburg and Giessen on the way, and arrived just before term commenced. The salary was not good, and the future of the college insecure, but it did provide a footing in England. And now something else had happened that made job prospects a great deal more important: he had become engaged to Sophie Fick.

4 Sophie

A man's diary is often more important for what it omits than for what it includes. In reading Frankland's *Diary* for 1848/9 one cannot avoid being struck with the virtual absence from its pages of the person who might have been expected to be the most conspicuous. This is the young lady whom, on his last visit, he had secretly resolved to marry, Sophie Fick. It may be recalled that Sophie, though a resident of Cassel, had been a frequent visitor to Marburg where two of her brothers, Heinrich[79] and Ludwig[80], occupied university chairs. Now, on his return, Frankland quickly renewed his acquaintance with the two Fick professors, but of Sophie herself there was not a trace. In fact a bereavement kept her in Cassel for many months. However she had 'an intimate friend' in Marburg, Fräulein Fulda, with whom Frankland was frequently seen 'because it was from her alone that I could gain intelligence of Sophie'.[81] Otherwise he had to be content with a glimpse of her portrait in the house of her brother (presumably Ludwig), which 'to say the least at first glance produced an electric effect' He went on to add:

> Should it be my fate to enjoy much of her society in Marburg, my situation will probably become imminently perilous and probably she alone can save me from perpetrating an engagement; perhaps it is fortunate for me that the bereavement that has occurred in the family will prevent me for a time from seeing her.[82]

He was probably right. This apparently hard-headed view shows remarkable strength of character and sense of responsibility that goes far to dispel any suggestion that his intentions were less than serious. On a couple of excursions to Cassel itself he failed to see her,[83] but on the very day his dissertation was handed over he glanced out of his window and, to his immense delight, 'was favoured with a gracious glance from Miss F.'[84] Whether her visit to Marburg has been deliberately timed to coincide with the end of his doctoral studies it is impossible to be sure; but it must remain a possibility. Thereafter she remained invisible until a week after the *viva* examination Then one afternoon they met in Pfeiffer's Garten, the scene of their parting 22 months before. Artlessly he reported 'an hour's pleasant conversation with the lovely Miss F. – worth all the girls in Marburg put together'.[85] His long wait was over.

The rapid development of this relationship Frankland recorded in the first edition of his *Sketches,* but as that account has been withdrawn from the public domain it may be interesting to have Frankland's impressions in his own words :

> We were delighted to see each other again, and our former pleasant intercourse was renewed as if our last meeting had occurred a week,

instead of more than a year, ago; and this although, as she subsequently told me, she had heard that I had no longer any interest in her, but was devoted to, if not engaged to, Fräulein Grebe. Before we parted some arrangements were made with the Fick family for an early picnic at Hanse Haus, a restaurant on a hill just outside on the other side of the river Lahn. From the moment of our meeting in Pfeiffer's Garten I became soon strongly impressed with the idea that my future relation with Sophie must be decided immediately. But how? the answer to that question perplexed me immensely. I had no immediate prospect of being able to marry or obtaining any remunerative work. In fact I had nothing to live upon except £70 per annum allowed me by my parents. Would it not then be heartless and cruel to try to persuade Sophie to link her fortune with a penniless student? On the other hand, I was hopeful and ambitious. I had already acquired somewhat of a reputation as a discoverer of ethyl, a discovery which, at this time, was exciting much interest in the chemical world, and was communicated to many learned societies and announced in many journals and periodicals. When in an optimistic mood, I could not believe that this belauded discovery, together with the training I had received in chemistry and allied sciences culminating in my university degree, would not enable me to obtain some professional appointment with a 'living wage'. On the other hand, when in a pessimistic mood, I remembered numerous instances in which discoveries of far greater value and interest had, so far as their authors were concerned, been passed over without any substantial reward. My optimistic moods, however, were more frequent and more prolonged than my pessimistic ones, and so it came about that, after a sleepless night spent in anxious thoughts, I resolved to ask Sophie to be my wife on the next occasion of our meeting. This was at the Hanse Haus picnic on July 29, 1849, where, separating her from the rest of the party, I proposed to her in the simplest words, and she accepted me. I explained to her that we could not marry immediately and that our engagement might be a long one. She wished it to be kept secret for the present, because she was convinced that it would be met with much opposition from her mother, to whom, therefore, she wished to tell it personally on her return to Cassel, which would be in a month or two. We finally arranged that we each should tell it to our most intimate friend. She told it to Auguste Schaumberg and I to Dr. Kolbe.

This secrecy was both painful and inconvenient for us, but we contrived to meet at parties and dances as often as possible, and we arranged for some English reading at the house of Fräulein Fulda, where we three read 'Jack Hinton'. On one occasion we contrived to get Fräulein Fulda out of the room, I sending her for a jug, and seizing that opportunity I got my first kiss.[86]

Doubtless the next visit to Cassel was anticipated with some foreboding. There could well be parental opposition of unpredictable

Table 4.1. *The Fick dynasty. Showing names in text and main academic connections*

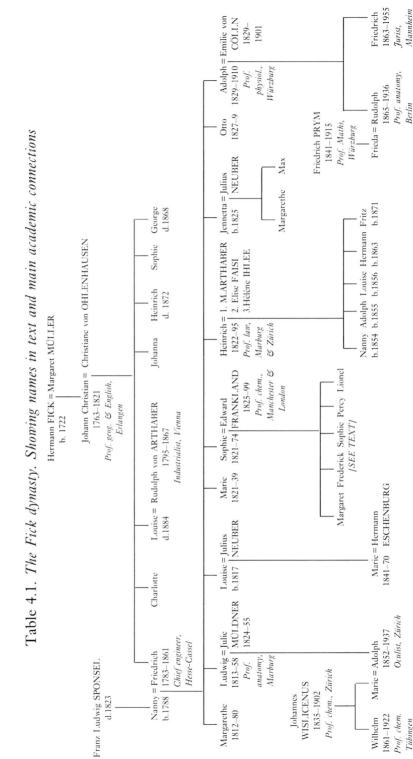

[Sources: *Neue Deutsche Biographie*, H. Fick (note 79), Family papers]

magnitude. For Edward Frankland it would involve, even if all went well, a new set of family relationships quite different from any he had experienced before. Sophie's family origins could be traced far back to Salzburg, from which city the Ficks were expelled in about 1660 on account of their Protestant faith. After a century or so of wandering in the area around Bayreuth they settled in Erlangen where Sophie's grandfather, Johann Christian Fick, became professor of geography and English language. His son Friedrich moved to Cassel, becoming chief municipal engineer for the tiny state of which it was the capital, Hesse-Cassel. Sophie was the fourth of his nine children (one of whom died in infancy). In one respect the Fick family attitudes would commend themselves to the young English chemist: a respect for academic values, so characteristic of the rising German middle class. Three of Sophie's brothers were (or became) university professors and several other members of the family acquired academic connections through marriage. The abbreviated family tree[87] (Table 4.1) illustrates this point and includes many of the people who later played a part in the life of Edward Frankland.

As Sophie's visit to Marburg drew to a close the prospect of meeting her formidable family had to be faced:

> On her departure alone for Cassel I was very anxious to accompany her in the diligence, which she on no account would allow, as she feared the discovery of our secret, so I had to go there alone the following day. This was on a Saturday, and we had arranged that we should meet on the following day in the Roman Catholic Church, and at the conclusion of the service hear what sort of reception I should meet with from her parents. She and her sister Margarethe sat down below and I in the gallery. They saw me, and from their glances upwards I saw that her sister was already in the secret. On coming out of the church her mother (who was also in the church) and her sister were allowed to go ahead, whilst Sophie and I followed them at a convenient distance. Sophie then informed me that her parents were willing, and that if I called after dinner (3 o'clock) her father would be alone and would see me. I called at the appointed time, I presented myself on the *zweite Etage* and was ushered into a room where sat old Oberbaurath Fick alone.
>
> I had never seen him before, and introduced myself as Dr. Frankland. He received me very cordially, and after the usual courtesy, I said I had come to ask his permission to marry his daughter Sophie. He replied that from the account of me which Sophie had given him, and from what he had heard of me for some time past from

Fig. 4.5. The Fick family into which Frankland married; Sophie is third from the left. (J. Bucknall Archives.)

his son in Marburg, he had no hesitation in giving me permission. He said his daughter Sophie had been a very good daughter to them, and was especially regarded as 'das häuslichste Mädchen' amongst his daughters, and he was confident that she would make me a good and useful wife. He then introduced me to Sophie's mother, who cordially seconded her husband's decision, and welcomed me as her future son-in-law. Sophie afterwards told me that her mother had said after this interview that, although she had a prejudice against Englishmen, she would be quite content to let her go with me anywhere in the world. I was then introduced to Margarethe and Janetta (Frau Neuber), and spent a pleasant afternoon with them and stayed for supper. After supper I took leave, and Sophie lighted me downstairs with a candle. I kissed her and bade her good-night at the front door.[88]

Doubtless his acceptance by the family must have come as a great relief. Sophie's oldest brother, Ludwig, had opposed their engagement on the grounds that Sophie's delicate health would give her only a year or so to live, but he was evidently over-ruled. As Frankland later recalled with heavy sarcasm 'I have ever since placed little reliance on doctors' opinions'.[89] There followed a mineralogical tour of Riechelsdorf in the company of Bromeis, a brief return to Cassel and Sophie, and then the memorable term at Giessen. His next reunion was to be at Christmas, after an overnight journey of such intense cold that two hours before the hotel fire were required before he was thoroughly warm.

> Now, however, I was off like lightning to my Sophie, who had been waiting for me two hours with intense impatience; she had contrived that we should meet *alone*. The meeting scene I will not attempt to describe, for you can much better imagine it for yourselves; after it was over we joined the rest of the family (father, mother, sister, and two brothers, the one from Marburg and the other from Berlin). After the mutual greetings were over Sophie and myself retired into another room to enjoy ourselves by ourselves, for we had of course many things to talk over after so long a separation. In the evening we all (except mother, who had a stiff neck) went to Sister Jeanette's to see what the Christkindchen (Christ's angel) had got for us at the Christmas tree. I have before told you something about this beautiful custom, which prevails throughout all Germany. A room is devoted to this purpose, and in it is planted some days previously a young fir-tree, reaching to the ceiling; the branches are covered with all sorts of fruit, confectionery, and other ornaments, and stuck full of wax tapers; upon the topmost twig is placed the miniature figure of an angel flying. A few evenings before Christmas-eve the angel, dressed in white, enters the house with great solemnity, and inquires very particularly how the children have behaved since last Christmas, mentioning each by name; if they have been tolerably good, the angel praises them, and promises to remember them at the Christmas tree. As Christmas Eve approaches, the children become almost frantic in anticipation of their presents. The presents, which are made to old as well as young, are all directed and laid at the foot of the tree, or somewhere in the room, and no one has the slightest idea what the presents are until 6 o'clock on Christmas Eve, when the tapers on the tree are lighted by one of the old members of the family, and after a bell has been rung three times, the door is thrown open, and the children, wrought up to the highest pitch of curiosity, rush in with unbounded delight, and soon pounce upon the presents intended for them, and this curiosity and pleasure is also shared in by the other members of the family, for each of whom there is sure to be something. I had taken care to bring with me from Giessen

a few presents, viz. a whist-box for mother, a pocket work-case for my own Sophie, two volumes of Moore's poems for sister Margarethe, and some toys for each of Jeanette's four children. These I sent to Jeanette, with directions about them, as soon as I got to Cassel, in order that she might have time to arrange them before evening. Accordingly, at 6 o'clock, after the due observance of all the ceremonies, we all rushed into the room. For myself were three presents: first, an album containing entries from all the members of the family, presented by Jeanette; second, a water-colour drawing of the landscape from the drawing-room window in Father's house, executed and presented by my Sophie; and third, a worked worsted *Polster*, to enable one to sleep when travelling, also executed and presented by Sophie, the comforts of which I have already found on my journey to Berlin.

After all the presents had been distributed we had a first-rate supper, drank several bottles of wine, and were very merry until a late hour. After taking Sophie home I went to my hotel, and was soon fast asleep, as I had had no rest the night previous; the next day, Christmas Day, and by a curious coincidence the birthday of both my fathers[90], was of course a great holiday, and it was determined to celebrate this double birthday by a grand supper in the evening, to which were invited Dr. and Mrs Neuber (my brother-and sister-in-law, with whom we were the previous evening), and Prof. Hildebrant (M.P. for Marburg) and lady, so that altogether the party consisted of eleven persons. Both healths were drunk, with due honours and loud applause, in bumpers of Rhine wine. Sophie's birthday is on January 9, my mother-in-law's on December 3. Neither of these birthdays is, however, celebrated: the first because of Sophie's twin-sister, who died of typhus fever in her twentieth year, and the second because it is the birthday and deathday of Dr. Fick's wife, who died of consumption a few weeks after my arrival in Marburg; I gave you at that time an account of her death. But I find I am becoming too *prolix*, as Uncle Richard would say, and will therefore cut a long story short by merely saying that after spending eight days of the most unmingled happiness with my Sophie, being with her, *mostly alone*, from 10 a.m. to 10 p.m. (for I had all my meals there except breakfast), I started on Wednesday morning, the 2nd inst., for Berlin by railway.[91]

This description of Christmas in Germany was in a letter to his parents, sadly excised from the second edition of his *Sketches*. In the same letter he wrote of one last contact with Sophie that remained before he left Germany for England:

Yesterday afternoon my brother Adolf arrived from Cassel, and brought me a long letter from my beloved Sophie, enclosing one for mother, about which she says the following:

'Yesterday I wrote the enclosed note to thy dear parents; I would gladly have said more, but found so much difficulty in saying what I have said, on account of not being able to express myself properly in English, that I am afraid thy beloved parents will not be able to read it for laughing at my many mistakes; this note will prove to them how little part I took in the translation of the former one. I hope, love, thou wilt give them a short commentary upon it, that they may be able to undertake it, as I hope that thou at least, dearest, canst read it.'

Of course I had no difficulty in understanding it, because the mistakes arise from employing the German instead of the English idiom, but fearing lest one or two expressions might not be quite clear to you, I have enclosed them along with the true English in parenthesis. My late visit to Cassel has convinced me more than ever of the invaluable treasure I possess in that girl; she is so simple-hearted, kind, devoted and affectionate, and her love for me is scarcely credible; her whole existence is so completely swamped in mine, and she is so fearful lest she should lose my affection, that I am really afraid lest she should die during my absence in England, for during five weeks previous to my visit to Cassel she was quite ill, owing to nothing but anxiety on my account; the day before my arrival, however, she was suddenly quite well again, and continues in excellent health and spirits during my stay.[92]

But now another year of separation lay ahead, broken only by a month's visit to Germany during the summer holidays of 1850. After that they were not to meet until February 1851 when Sophie arrived in England for her wedding to Edward Frankland.

Notes

1 Frankland, *Diary* for 1848 and part of 1849, entry for 21 May 1848 (typescript copy) [JBA, OU mf 02.02.1304–1484].
2 H. Debus to Frankland, 4 June 1848 [RFA, OU mf 01.02.1391].
3 H. Kolbe to Frankland, 1 August 1848 [RFA, OU mf 01.02.1297].
4 C. Meinel, *Die Chemie an der Universität Marburg seit beginn des 19. Jahrhunderts* (Academia Marburgensis herausgegeben von der Phillips-Universität Marburg Band 3), N. G. Elwert Verlag, Marburg, 1978.
5 *Sketches* 2nd. ed., pp.94–5.
6 J. Tyndall, *New fragments*, New York, 1892, p.233.
7 Frankland, *Diary* (note 1), entry for 5 November 1848.
8 *Ibid.*, 12 November 1848.
9 *Ibid.*, 30 December 1848.
10 *Ibid.*, 12 January 1849; the damsels arrived by 'sledgefuls' on that occasion.
11 *Ibid.*, 26 December 1848.
12 *Sketches,* first edition, p.270.
13 Frankland, *Diary* (note 1), entries for 21 and 22 May 1848.
14 *Ibid.*, 9 March 1849.
15 *Ibid.*, 5 March 1849.
16 *Ibid.*, 28 May, 6 June 1848, *etc.*

17 *Ibid.*, 17 June 1848.

18 W. L. Faber to Frankland, 20 March, 1 June, 26 June, 11 December 1850, 1 February 1851 [RFA, OU mf 01.02.1376; 01.02.1345; 01.02.1317; 01.02.1354; 01.02.1327.

19 H. Will, *Tables for Qualitative Chemical Analysis*, trans. W. L. Faber, Philadelphia, 1852.

20 H. V. Regnault, *Elements of Chemistry*, trans. T. F. Betton, and ed. with notes by J. C. Booth and W. L. Faber, Philadelphia, 1852; many subsequent editions.

21 *Ibid.*, vol. 2, pp.568–71.

22 *Ibid.*, vol. 2, p.625.

23 *Ibid.*, vol. 2. p.793.

24 W. L. Faber to Frankland, 30 January 1860 [RFA, OU mf 01.04.1222].

25 Frankland, *Diary* (note 1), entry for 9 May 1849.

26 Meinel (note 4), pp.227–9.

27 Letters from Bromeis to Frankland [RFA, OU mf 01.08.0009, 01.03.0279 and 01.03.0233].

28 [Undated] references, from their context 1847 and 1850 [RFA, OU mf 01.04.1840 and 01.03.0911]; each suggests a previous acquaintance with Frankland in England.

29 Frankland, *Diary* (note 1), entry for 31 December 1848.

30 *Ibid.*, 18 December 1848.

31 *Ibid.*, 25 May 1849.

32 *Sketches*, p.106.

33 Frankland, *Diary* (note 1), entry for 4 May 1849.

34 *Sketches*, p.265.

35 Frankland, *Diary* (note 1), entry for 30 December 1848; he was content to note 'a difference of opinion about kissing German ladies'.

36 'Wat Ripton' [J. Tyndall], 'The German student', *Preston Chronicle*, 24 February 1849.

37 *Ibid.*, 11 March 1849.

38 *Ibid.*, 9, 10 and 11 May 1849.

39 This is implied clearly in Kolbe's letter (note 3), but placed a year earlier by Alan Rocke (A. J. Rocke, *The Quiet Revolution: Hermann Kolbe and the science of organic chemistry*, University of California Press, Berkeley, 1993, p.69).

40 *Sketches* 2nd. Ed., p.96.

41 Christian Ludwig Gerling (1788–1864) was Director of the mathematical-physical institute and, in 1849, Dean of the Faculty of Philosophy (F. Gundlach, *Catalogus professorum Academicae Marburgensis: Die akademischen Lehrer der Phillips-Universität in Marburg von 1527 bis 1910*, G. Braun, Marburg, 1927).

42 Gundlach (note 41).

43 Franz Theodor Waitz (1821–1864), Professor of Philosophy from 1848 (*ibid.*).

44 *Sketches*, 2nd. Ed. p.96.

45 Frankland, *Diary* (note 1), entries for 26 October – 2 November 1848.

46 *Ibid.*, 28 October 1848; Frankland said this later led to his own isolation of metallic chromium for the first time (*Sketches*, 2nd. Ed. p.98).

47 Frankland, *Diary* (note 1), entry for 3 November 1848.

48 Letter Frankland to Tyndall, 22 January 1857 [Royal Institution Archives, Tyndall Collection, 9/E3.8].

49 Frankland, *Diary* (note 1), 2 November 1848.

50 *Ibid.*, 9 November 1848.

51 *Sketches*, p.84.

52 Frankland, *Diary* (note 1), entries for 6 and 7 November 1848.

53 *Ibid.*, 8 and 9 November 1848.

54 *Ibid.*, 10 and 11 November 1848.

55 *Ibid.*, 25 November 1848; thus in one experiment 2 volumes of gas consumed 1.25 volumes of oxygen and yielded 0.994 volumes of carbon dioxide, consistent with a mixture including CO and H_2 in a ratio of 2:3.

56 *Sketches*, p.99.

57 Frankland, *Diary* (note 1), entry for 8 January 1849.

58 *Ibid.*, 9 January 1849.

59 *Ibid.*, 19–24 January 1849.

60 *Sketches*, 2nd. Ed. p.99.

61 J. B. A. Dumas and E. M. Peligot, *J. de Pharm.*, 1834, **20**, 548.

62 *Idem, Annalen der Chem.*, 1835, **15**, 20.

63 *Sketches*, p.99.

64 Frankland, *Diary* (note 1), entry for 18 February 1849.

65 *Sketches*, pp.99 – 106.

66 Frankland, Dissertation, *Ueber die Isolirung des Aethyls*, Marburg, 1849 [RFA, OU mf 01.03.0926–0949].

67 *Ibid.*, p.106.

68 Frankland, *Diary* (note 1), entry for 12 July, 1849.

69 *Ibid.*, 25 July 1849.

70 *Ibid.*, 12 July 1849.

71 *Sketches*, 2nd. Ed. pp.107–8; he suffered periodic bouts of lumbago in later life.

72 J. Morrell, 'The chemist breeders: the research schools of Liebig and Thomas Thomson', *Ambix*, 1972, **19**, 1–46.

73 *Sketches*, 2nd. Ed. p.110 (4 August).

74 A. F. L. Strecker (1822–1871), later professor in Oslo, Tübingen and Würzburg.

75 F. F. T. Fleitmann (1828–1904) went to study with Liebig in 1848; he became an industrial chemist and metallurgist, specialising in nickel extraction (*Neue Deutsche Biographie*, Berlin, 1960, vol. 5).

76 'Isolirung des Amyls', *Annalen der Chem.*, 1850, **74**, 41–70; 'Isolation of the radical amyl', *J. Chem. Soc.*, 1850, **3**, 30–52.

77 *Sketches*, 1st ed., p.276.

78 Frankland to William and Margaret Helm, 8 January 1850, in *Sketches*, 1st ed., pp.278–88 (shorter version in 2nd ed., pp.267–73).

79 Alexander Heinrich Fick (1822–95) became Professor of Law at Marburg (from 1847) and at Zurich (from 1851): Gundlach (note 41), p.162; *Neue Deutsche Biographie*, Berlin, 1960, vol. 5; Hélène Fick, geb. Ihlee, *Heinrich Fick, Ein Lebensbild*, Zurich, 1897.

80 Franz Ludwig Fick (1813–58), Professor of Anatomy at Marburg; his wife

was Julie Müldner (1824–55), their son Adolph Gaston Eugen Fick
(1852–1937) becoming a famous oculist, his own son being the architect
Roderick Fick (1886–1955) (Grundlach (note 41), p.205; *Neue Deutsche
Biographie*, Berlin, 1960, vol. 5).

81 *Sketches*, 1st ed., p.269.
82 Frankland, *Diary* (note 1), entry for 17 December 1848.
83 *Ibid.*, 28 May, 6 June 1849.
84 *Ibid.*, 17 June 1849.
85 *Ibid.*, 7 July 1849.
86 *Sketches* (1st ed.), pp.271–3.
87 Gundlach (note 41); *Neue Deutsche Biographie*, Berlin, 1960, vol. 5; and *Diary*
of Maggie Frankland, 1869, end-papers [MBA].
88 *Sketches* (1st ed.), pp.273–4.
89 *Ibid.*, p.307.
90 *I.e.* his father-in-law Friedrich Fick and his step-father William Helm.
91 Frankland to William and Margaret Helm (note 78), pp.280–3.
92 *Ibid.*, pp.286–7.

CHAPTER 5

Fundamental discoveries in chemistry

1 Putney revisited

A feature of the history of science that often attracts attention is its sheer unpredictability. That is to say, not that valid generalisations can never be made and tested, but rather that new and unforeseen effects may suddenly become apparent and the course of subsequent history turn out in quite unexpected ways. This was certainly true for the next part of Frankland's life. Who, for example, could have predicted that one of chemistry's fundamental laws would have come to light, not in the famous laboratory of the great Bunsen, but in a very minor English institution, racked with social division and financial uncertainty and very soon to be forced to close its doors? Yet a rich irony of the story is that at such a place Frankland not only discovered the phenomenon of valency but also effectively established that branch of chemical science that we know today as organometallic chemistry. These two fundamental pieces of research form the major theme of this chapter,[1] but first we must look briefly at the institution in which they took place.

When Frankland left Germany in early 1850 he had gained a good knowledge of the German language and, through his engagement to Sophie Fick, an entrée into a respected German family. He had also acquired a PhD, a growing reputation, at least in Germany, for being the 'discoverer of Ethyl' and the acquaintance of many leading European scientists. In the critical month after his doctoral examination in Marburg he had inadvertently stumbled upon a new kind of chemical compound, a discovery that was to have repercussions through the whole of chemistry in the years ahead. It was infinitely more important than his alleged isolation of 'ethyl' which was, of course, nothing of the kind but merely n-butane.

Now a new door was open to him, or rather perhaps an old door was reopening. While waiting in Berlin for the commencement of term in January 1850 he received an unexpected letter from his former boss, Lyon Playfair, at the Putney College of Civil Engineering.[2] It announced that

> the Professorship of Practical Chemistry at Putney, lately held by Mr. A. Phillips[3], is vacant and I have the sanction of the Principal to offer it to you . . . Now this is an opening such as may not occur again so fair in 10 years and of course it is worth while your taking it. I do not think you will make a fortune by the position, but it gives you a lab^y free of cost and a probable income suff^t to keep you out of debt. Besides, I may tell you in confidence, but on this subject I make no promises, nor should it enter your calculations except as a <u>possibility</u>, – it is probable that I may receive another Professorship[4] before July in wc case you would doubtless succeed me in the whole thing. But this is an opening to bring you before the public such as may not occur again for years and you would be unwise not to accept it. In case you do so you must <u>instantly</u> set off and come here or you will lose much of your chances of progress. Decide by return of post . . .

Perhaps it was the confidential disclosure in this imperious communication that persuaded Frankland to forsake the possibility of several weeks' research at Berlin for the more dubious prospect of teaching in a college that he already knew to be plagued by weak leadership, student unrest and an ambiguous place in the provision of public and private technical education.[5] Term was due to start on Tuesday, 15 January, and Playfair had suggested that he should arrive at the very latest by the end of that week. There was, therefore, no time to lose. In fact Frankland was not able to leave until that very Tuesday and can only just have met his deadline, since *en route* he enjoyed two days in Cassel, went to a ball in Marburg and spent a night in Giessen celebrating the twentieth anniversary of Liebig's appointment; he was not a man to waste good opportunities![6]

Outwardly Frankland's second spell at Putney was fairly uneventful. He took lodgings nearby, where he was visited by friends such as T. A. Hirst[7], with whom he would spend an occasional literary evening (as when the two friends paid an unannounced call upon Thomas Carlyle at his Chelsea home,

Fig. 5.1 Lyon Playfair (1818–84): Frankland's first chemical supervisor.
(Author.)

found him away, and repaired to a hotel for a poetry reading of
Tennyson[8]). He allowed himself few holidays in this year at
Putney, though did use the summer vacation to visit his mother and
step-father at their new home in Disley, Cheshire, and to spend a
month or so with Sophie at Cassel.[9]

Frankland did not, as is often suggested, succeed Playfair
immediately, but appears to have begun as a kind of superior
assistant to the Professor. Only after the summer of 1850 does he
seem to have been in command, and even then he was often referred
to as 'Lecturer', which was really a better description of his

functions. From the start Frankland's life was dominated by teaching. It had to be so, for all the incumbent's income was derived from student fees. Generally laboratory fees were £10.10.0 (ten guineas) per term for a six-day week; lectures cost £3.3.0 per term, except that they were free to those who attended laboratory classes.[10] By 1850 student numbers were falling, so the alarmed college authorities tried to halt the decline with a change of name, 'The College for Practical and Scientific Education, Putney'.[11] A surviving class book contains student records from 1847 to 1851, and discloses that in July 1850 only 26 students sat the General Examination.[12] At the same time the numbers taking Practical Chemistry and Metallurgy were reported as 73 (and included a Mr Maule, presumably the George Maule of later dyestuffs fame and a former fellow apprentice of Frankland).[13] However in the following term the chemistry class was attended by only 19 persons. The Class Book also includes Frankland's accounts, estimating the total fees due for his year at Putney as £210.19.5$\frac{1}{2}$, whereas he actually received £191.3.6. One defaulter was the son of a Col. Cameron of Christchurch, to whom Frankland wrote on three occasions to demand his fee of £6.6.0; the outcome is unknown.[14] From these sums he had to pay for consumable items in the laboratory so his residual income amounted to only about £100, as compared with the sum of £300 that he estimated as necessary if marriage were to be undertaken.[15]

A set of Frankland's lecture notes[16] has survived, dated from 1 May to 5 October 1850. Their titles were as follows:

> 3 May. Compounds of C with O (carbonic acid only]
> 4 May. Carbohydrogens [hydrocarbons]
> 7 May. Manufacture of gas
> 15 May. Sulphur [element only]
> 25 May. Compounds of S with O
> 29 May. Compounds of S and H
> 5 June. Compounds of P and O
> 8 June. Compounds of P with H
> 12 June. As
> 19 June. AsO_5; compounds of As and H [and S, P and Cl]
> 22 June. Detection of arsenic
> 26 June. Antimony [including oxides]
> 29 June. Antimony and hydrogen

25 Sept. Silicon [and boron]
 5 Oct.   Chlorine

Clearly a course designed to cover the most important non-metals and their compounds, this is basically descriptive chemistry but with much emphasis on industrial applications. Most remarkable, perhaps, is the concentration upon arsenic, anticipating the rôle of forensic chemistry in the spate of arsenic poisonings in later Victorian England.

A chemistry examination paper for July, 1850, confirms a strong 'applied' emphasis, with questions on the manufacture of sulphuric acid and its uses in calico-printing; dyeing; agriculture; gas diffusion and ventilation; and three questions on coal-gas.[17] It is not significantly different from previous chemistry papers. The corresponding paper for physics (also taught by Frankland) places a similar emphasis on applications, including atmospheric electricity (two questions); construction of mercury thermometers; principles underlying compressed air engines; use of polarised light in sugar determinations; and the employment of electricity as a motive power.[18] These reflect the syllabus for chemistry and physics,[19] a comprehensive course of 150 lectures intended to be delivered over two years and strongly emphasising Frankland's growing conviction about the utility of applied science.

A prospectus for a course on metallurgy is conceived in the same pragmatic spirit:

> The object of this course is to make the Student practically acquainted with Chemical Manipulation and Analysis, the Assaying of Metallic Ores, and the various other Chemical determinations which constantly occur in the Arts and Manufactures. When he has become expert in qualitative and quantitative analysis he will have the opportunity of applying his knowledge to the investigation of Soils, Ashes of Plants, Rocks, Minerals, Waters for the supply of Towns, &c.; and will have to conduct Assays of all the more commonly occurring Metals, as also determinations of the economic value of Coal and other fuel, for which the College Laboratory and Assay-Room present every convenience.[20]

Such an ambitious teaching programme, undertaken by one man alone, would leave most modern lecturers breathless. It suggests an almost prophetic understanding of the possibilities of technological education, together with a rare combination in the lecturer of poised self-confidence and almost manic energy. Yet even at Putney teaching was not allowed to supplant research, and a brief

PUTNEY COLLEGE.

EXAMINATION.—JULY 1850.

PHYSICS.

1. What is meant by sensible, latent and specific heat? What is the latent heat of water and of steam?

2. Describe the principles upon which the indications of the thermometer depend; and state how you would construct a mercurial thermometer.

3. What is the coefficient of expansion of gases and vapours for each degree of Fahrenheit's scale? Apply this to the solution of the following queries. If 1000 cubic feet of air, at 300° F., be cooled to 32° F., what will then be its volume? If 1000 cubic feet of steam, at 212° F., be heated to 1000° F., what volume will it then occupy, the pressure remaining the same?

4. Why would there be no economy in substituting alcohol for water in our steam-boilers? and why does a given amount of heat convert the same quantity of water into steam under whatever pressure the steam may be generated?

5. When air is suddenly expanded to twice its former volume, what diminution of temperature takes place? How does this diminution of temperature affect engines worked by compressed air?

6. What is the undulatory theory of light, and what are the principal facts which support it?

7. Explain the phænomenon of double refraction in Iceland spar and other crystals.

8. What do you mean by the polarization of light; and how can polarized light be applied to determining the value of saccharine juices for the manufacture of sugar?

9. What is the velocity of light, and how can it be determined?

10. State the laws relating to the intensity of light.

11. How do you explain the phænomena of electricity in general?

12. State minutely how you would protect a large irregular building from lightning. What precautions are necessary in the employment of conductors?

13. Explain the action of atmospheric electricity as exhibited in thunder-storms. How can a charged cloud produce the most destructive effects over a wide extent of country without parting with any of its own electricity?

14. What evidences have we of the mutual connection and convertibility of heat, light, electricity and the chemical force? How do these evidences bear upon the employment of electricity as a motive power, and the recent alleged invention of Mr. Payne, Worcester, U.S, for the inexpensive production of heat and light by means of electricity?

Fig. 5.2 Putney College Physics examination paper July 1850. R. Frankland Archives

report by Frankland refers to his analysis of ores, of gases by Bunsen's method, and (most significantly for the future) of water from the nearby Thames.[21] He also continued the coal investigation started by Playfair on behalf of the Admiralty.[22] He later referred several times to these investigations at Putney on calorific powers of coal,[23] some samples of which he had obtained from the nearby Wandsworth Brewery.[24]

These investigations, however, were entirely subsidiary to the main thrust of Frankland's research, the continuation of his work on the 'zinc alkyls'. From these pioneering studies was to come the inauguration of a great new branch of chemical science, organometallic chemistry.

2 The foundation of organometallic chemistry

It is perfectly true that an organic compound containing a metal[25] had been known since 1827 when Zeise[26] isolated the salt named in his honour, potassium trichloro(ethylene)platinate(II) monohydrate, or

$$K^+[C_2H_4PtCl_3]^-. \ H_2O,$$

(though elucidation of its π-electron structure had to wait for over a century). And we have already seen that by the early 1840s Bunsen had done extensive work on the cacodyl series of compounds containing the element arsenic (then sometimes regarded as a metal). Yet it remained entirely appropriate for an early German treatise on organometallic compounds[27] to bear the following dedication:

Dem Andenken
Sir Edward Franklands
gewidmet

for, as many historians have recognised, the true founder of organometallic chemistry was Edward Frankland himself.[28] This was so because it was Frankland who (a) first recognised the existence of a new *series* of compounds,[29] (b) gave to them the name by which they are universally known today, 'organometallic',[30] and (c) made it a matter of sustained effort to produce them in as many varieties and as large quantities as possible.[31]

It is worth recalling that his first experiments on zinc and alkyl halides at Marburg (or even at Queenwood) led only to the isolation of *hydrocarbon* products, mistaken at first for the prized 'radicals'. They included the hydrocarbon ethane, of which Frankland is thus the discoverer, though its recognition is often overshadowed by his more famous discovery of dimethylzinc. At that time, of course, Frankland regarded ethane as the radical 'methyl'. His analysis of its gravimetric composition cannot be faulted, but as he did not accept the hypothesis of Avogadro he had no means of knowing whether or not the empirical formula was also the molecular formula. The same was of course true of his slightly earlier discovery of butane ('ethyl'). Thus a reaction that we might summarise as

$$2C_2H_5I + Zn \rightarrow ZnI_2 + C_4H_{10},$$

Frankland wrote in German[32] and English[33] publications as, respectively,

$$\left. \begin{array}{c} C_4H_5J \\ Zn \end{array} \right\} = \frac{C_4H_5}{ZnJ} \quad \text{and} \quad C_4H_5I + Zn = C_4H_5 + ZnI,$$

where $C = 6$ and Zn (or Zn) $= 32.5$, and where empirical and molecular formulae were not distinguished.

It was not long before Gerhardt saw in Frankland's results a further illustration of the law of *homology*, the existence of families of closely related compounds or homologues. In that case the so-called radicals were probably homologues of marsh–gas, necessitating a doubling of Frankland's formulae for his 'methyl', 'ethyl', *etc.*[34] Frankland kept to his position, however, and in his paper on amyl compared his products with hydrogen, which is fairly unreactive in the free state, but active *in statu nascenti*, giving rise to many compounds, such as 'hydrurets' like 'light carburetted hydrogen', $C_2H_3 + H$, *etc.*[35] Hofmann attempted unsuccessfully to resolve the issue by experiment and took up the dubious implication that hydrogen was unreactive.[36] At least it combined readily with chlorine, which is more than could be said for Frankland's 'radicals'. After a dispassionate survey of both sides of the question Hofmann came down, reluctantly it seems, 'in favour of the lower formulae'. Frankland at once replied, dealt with Hofmann's objections to his contention, and announced some eudiometric results of his own, a few of which are extremely hard to explain.[37]

So at least he admitted years later, saying that his experimental data were then 'nearly the only evidence in support of the non-identity of the two series of hydrocarbons', and suggesting a repetition of the experiments.[38] The crux of the problem – the issue of whether and when formulae should be doubled – was pointed out by Brodie in a paper that clearly set out the issues, and concluded on the side of Gerhardt.[39] But it was not until 1864 that Schorlemmer finally established the identity of Frankland's radicals with the paraffins.[40]

The following seems to have been the sequence of events leading to Frankland's recognition of organometallic compounds prior to his year at Putney:

- first studies of the action of zinc on ethyl iodide, begun on 28 July 1848, at Queenwood;[41] the sealed tube was not opened until next year in Marburg.

- preparation of 'ethyl' from ethyl iodide and zinc; it was discovered on March 4/5 1849 and announced in a paper[42] submitted to the Chemical Society on 17 June 1849 (though received after 18 June), and in Frankland's Dissertation.[43]

- formation of an unidentified white crystalline body as an intermediate, reported in the above paper to the Chemical Society. It was subsequently shown to be EtZnI.[44]

- isolation of 'methyl' from methyl iodide and zinc shortly afterwards, and also a new kind of compound altogether (dimethylzinc[45] in our terms). This was reported in a second paper, read to the Chemical Society on 5 November 1849.[46]

- the analogous preparation of diethylzinc[47] seems to have rapidly followed for, although Frankland's *Diary* contains no mention of it, *both* dialkylzinc compounds were reported in this paper to the Chemical Society.

- preparation and examination of 'zinc amyl' in Liebig's laboratory during the late autumn of 1849; this was published in February 1850.[48]

There were (and are) formidable difficulties in handling dialkylzinc compounds, for they are spontaneously inflammable in air, immediately decomposed by water, of unpleasant odour and (as we now know) toxic in varying degrees. Yet Frankland was undaunted by the difficulties and made a study of organometallic compounds a major research objective. One of the most pressing problems, their great chemical reactivity, he addressed by a series of appropriate techniques, including scrupulous drying of apparatus and reagents and consistent use of inert atmospheres such as hydrogen.

For the next few years Frankland was initially driven by a strategy that seems strange, even misconceived, today. It derived directly from his understanding of the nature of these new compounds. This controversy as to the nature of Frankland's 'radicals' raged strongly, but for only a few months. Its immediate effect, however, was to reinforce Kolbe's paper of the same period,[49] to keep alive the concept of a radical, and to stimulate Frankland still further to experiment with organometallic compounds in the hope of making yet more 'radicals'. Indeed, the organometallic compounds themselves were regarded as radicals. As he wrote of one in 1849: 'It is highly probable that this body which for the present I propose to call Zincmethyl, plays the part of a radical, combining directly with oxygen, chlorine, iodine, *etc.*'[50] There was thus the double attraction of discovering more radicals of the cacodyl type, and of using them to produce 'alcohol radicals'. In the same paper he predicted some 20 more compounds of the kind known today as metal alkyls. He saw the possibility of using methyl, ethyl and amyl alkyls to replace 'negative elements', as chlorine and oxygen, with 'positive elements' and 'thus ascend the scale of homologous bodies at pleasure'.[51] While in Germany he had carried out occasional experiments with alkyl iodides acting on several different metals, but without definite results. Therefore when Frankland left the Continent for England in the winter of 1849/50 he came with every stimulus to pursue further his work on organometallic compounds. How this came about is best described in his own words:

> I was particularly anxious to try the effect of light in producing or favouring the decomposition of the iodides of the alcohol radicals by various metals. The chemical laboratory at Putney was particularly well situated for carrying out such experiments. It was an isolated building, situated in the middle of an extensive lawn, sloping down to the Thames . . . and had a flat place upon the roof, on which experiments in the open air could be conveniently carried on. Having provided myself with a concave platinised reflector to concentrate the sun's rays, I proceeded, as the spring advanced, to expose iodide of ethyl in contact with various metals to intense solar light. When it was desired to exclude the simultaneous action of heat, the sealed tubes containing the iodide and metals were placed under water, coloured blue, by a solution of ammonio-sulphate of copper.[52]

The importance of light is interesting in this context for it is now believed that many of these reactions proceed by a radical–ion mechanism in which a homolytic fission of a carbon–halogen bond is photochemically induced.[53] In a paper communicated to the Royal Society on 10 May 1852, Frankland reported many interesting findings, including the action of zinc and methyl iodide at 150° C to give a gas, perfectly resembling 'the methyl procured by Kolbe from the electrolysis of acetic acid', some crystals, and a colourless liquid that was shown to be zinc methyl. Descriptions followed of zinc ethyl and amyl, and (by a remarkable achievement) of methyl mercury iodide.[54] Before this, however, Frankland had much to report on new organic compounds of the metal tin.[55]

He first described the preparation of 'iodide of stanethylium', 'C_4H_5SnI', from tin and ethyl iodide in strong sunlight, or by heating at 180°. On successive treatments with alkali, dilute hydrochloric acid and zinc this gave rise to a yellow oil, 'C_4H_5Sn', which he termed 'stanethylium'. It boiled, with decomposition, at 150°, and combined with air to form an oxide, and with hydrochloric, hydrobromic and hydriodic acids to form, respectively, chloride, bromide and iodide of stanethylium. The relationships that Frankland discovered may be summarised thus:

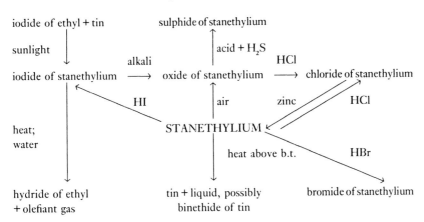

Here, then, was another substance like zinc ethyl capable of rapid combination with the non-metals; more important still,

> stanethylium perfectly resembles cacodyl in its reactions, combining directly with the electronegative elements and regenerating the compounds from which it has been derived.[56]

Although Frankland's method had to wait until 1911 for further detailed investigation,[57] his work has been much quoted in the modern literature, and it is clear that most of his observations are still impeccable. The reaction between ethyl iodide and tin has not, however, been reproduced with photochemical induction alone.[58] But his observation that a temperature of 180° is needed conforms with a known phase change in the metal, at 161°, from β-tin to the more reactive γ-tin.[59] In modern terms the initial reaction is

$$2EtI + Sn \longrightarrow Et_2SnI_2,$$

probably involving a catalytic or photochemical loosening of the carbon–iodine bond to form a transient intermediate EtSnI, followed by a carbene-type uptake of a molecule of ethyl iodide to yield the diethyltin diiodide.[60] Frankland's 'stanethylium' was certainly not simple diethyltin; as long afterwards as 1968 it was being argued that such a substance had not yet been prepared.[61] Such a molecule would be expected to be inherently unstable, comparable with the carbenes, and has so far been found only as a transient intermediate in processes as the polarographic reduction of diethyltin dichloride.[62] It would therefore appear that Frankland's 'stanethylium' was a polymeric molecule $(Et_2Sn)_x$, where x was probably about 6. Such materials have been prepared[63] as pyrophoric oils or waxes which readily decompose on heating into tin and tetraethyltin, the latter being obviously Frankland's 'binethide of tin':

$$(Et_2Sn)_x \longrightarrow \tfrac{1}{2}xSn + \tfrac{1}{2}xEt_4Sn.$$

Other new bodies reported in the same paper include 'zinc amyl' and 'mercuric methiodide'. In years to come Frankland was to report further advances in organometallic chemistry but none of them eclipsed in importance those in his seminal paper of 1852. Partly this was because of other theoretical issues raised at the same time [see Section 3, below] and partly because the 1852 paper laid down general principles for preparing and handling such difficult materials.

Two final qualifications need to be made. First, there was another publication, by Löwig, almost at the same time dealing with organic compounds of tin, and this described the production of [polymeric] diethyltin from ethyl iodide and a sodium/tin alloy, together with its conversion into diethyltin dihalides.[64] And secondly, as has been noted, Frankland was wrong in assuming formation of monomeric diethyltin, a fact of some importance in view of the later use he made of his product in theoretical argument.

This research, largely done at Putney, brought him rapid recognition. His neologism *organometallic chemistry* was soon in common use; in 1853 Thomas Anderson of Glasgow thanked him for his 'very interesting paper on the organo metallic compounds',[65] and the term featured frequently in Frankland's correspondence with the Royal Society,[66] from whom he received grants for organometallic research in 1850, 1852, 1855 and 1858. Frankland's papers on the subject submitted to the Royal Society received most favourable referees' reports throughout the 1850s.[67] In 1857 Frankland was awarded their Royal Medal for 'the isolation of the organic radicals of the alcohols' and for 'researches in the metallic derivatives of alcohol'.[68] W. A. Miller confessed that he had got the history of organometallic chemistry wrong in his *Elements of chemistry* but would set the record straight in the next edition.[69] As for individual friends, his former assistant W. J. Russell spoke with some understatement of diethylzinc as 'a most interesting compound'[70], and the veteran inventor John Mercer wrote of Frankland's 'beautiful discoveries of the organo-metallic bodies', adding that

> all who love the science of chemistry ought to be grateful to you for filling with such new and valuable matter so many of the blank pages in the great chemical volume.[71]

As the century progressed the zinc alkyls retained their usefulness as synthetic tools, though after 1912 they have been largely replaced by Grignard reagents which are much easier to handle and have certain other advantages. However they still find a use in syntheses such as the Reformatsky reaction (using α-halogeno-esters and related compounds). They also played an important part in the creation of a modern atomic weight system by Cannizzaro, announced in his momentous *Sketch of a course of chemical philosophy* of 1858, where he brings together evidence from vapour densities and other hitherto unrelated fields as specific heats. He writes:

> It being proved from the density of zinc ethyl vapour, and from its specific heat, that the complete molecule of zinc ethyl contains a single atom of zinc combined with two ethyl radicals . . . no one can deny that there will be prepared compounds containing a single atom of zinc combined with two different monatomic radicals.[72]

In other words, 'zinc ethyl' is Et_2Zn and $Zn = 65$ (not 32.5 as Frankland had taken it). What Cannizzaro did not say in this seminal paper, but did admit elsewhere, was the crucial rôle of the zinc alkyls in his intellectual pilgrimage. Writing to Ludwig Mond after Frankland's death he remarked 'that the discovery of Zinc-ethyl and Zinc-methyl of Frankland, *had been the starting-point* of my theory on the atomic weights of metals' [my italics], adding that it was his own demonstration of the atomic weight of zinc and other metals (using densities of their volatile compounds) that led to the general acceptance of the theory of Avogadro, and that 'this was also the opinion of Frankland' who 'was amongst those who most appreciated and exaggerated the part which I had in the reform of the formulas of mineral compounds'.[73]

Frankland's methylmercury iodide has since acquired unenviable notoriety as the first of many organomercury compounds known to cause great environmental damage, from the unfortunate deaths attributed to his own laboratory in the mid-1860s (p. 251) to our own contemporary problems of mercury residues in sea-water fish. Of the other compounds Frankland discovered, those containing an alkyl-tin bond have been of most practical value, compounds of the type Bu_2SnX_2 and Bu_3SnX finding significant industrial rôles as biocides, PVC stabilisers *etc.* in recent years. However the most lasting and significant development from the work in organometallic chemistry by Edward Frankland lay in the area of fundamental chemical theory known as valency.

3 Foundations of the theory of valency

One of the most fundamental doctrines of chemistry has been that of valency (valence), second only perhaps to the atomic theory upon which it is founded. Its early history has been well-charted[74] and few would now deny that Frankland was one of its major architects.[75] As the following chapter will relate, much controversy surrounds the subsequent course of events but there can be little doubt about the origins of valency theory. They lay in Frankland's own work in organometallic chemistry. The year was 1850.

As he exposed his tubes of alkyl halides and metals to the sunlight remarkable changes began to occur and a whole new tract of organic chemistry became gradually revealed. More important still was the gradual dawning of a new awareness of a fundamental regularity in nature that no one else had noticed:

> I had not proceeded far in the investigation of these compounds at Putney before the facts brought to light began to impress upon me the

existence of a fixity in the maximum combining value or capacity of
saturation in the metallic elements which had not before been suspected.[76]

The paper in which these ideas were first announced was
communicated to the Royal Society on 10 May. Its publication was,
however, delayed by a year as it 'was inadvertently laid aside in his
private drawer by the Secretary (Professor Stokes)'.[77] This delay
has created the impression that it was during the *next* period of his
life (at Manchester) that Frankland arrived at 'the law of valency'.
However he explicitly stated that

> the conception of that law was first forced upon my attention by my
> study of Organometallic bodies when holding the Chair of Chemistry
> in the Putney College of Civil Engineers three years previously.[78]

The great reactivity of his organometallic compounds offered an
obvious parallel to the behaviour of cacodyl, and this was duly
noted. Now Kolbe had regarded cacodyl as 'arsenic conjugated
with two atoms of methyl', where conjugation did not alter the
'essential chemical character' of a substance; cacodyl still retained
characteristics of arsenic itself.[79] If he were right, Frankland's new
compounds must be deemed to have similar character, with the
'alcohol radicals' conjugated on the zinc, tin, *etc.* In that case,
therefore, 'stanethylium' should closely resemble elementary tin,
as indeed cacodyl resembles arsenic.

> If therefore we assume the organo-metallic bodies above mentioned to
> be conjugated with various hydrocarbons, we might reasonably expect
> that the chemical relations of the metal to oxygen, chlorine, sulphur,
> *etc.* would remain unchanged.[80]

But in fact *this was not so*, even for cacodyl, and now the new
organometallic compounds reiterated the difficulty. With the
arsenic compounds one would expect two stages in oxidation by
analogy with the behaviour of elementary arsenic:

cacodyl ⟶ protoxide of cacodyl ⟶ further product (X)
and cacodylic acid

arsenic ⟶ protoxide of arsenic ⟶ arsenic acid
and arsenious acid

(or in modern terms, $As \longrightarrow As_2O_3 \longrightarrow As_2O_5$).

In fact the second product (X) for cacodyl was not found, even
when cacodylic acid was distilled with arsenic acid. Moreover,
protoxide of cacodyl dissolves readily in hydrochloric acid, whereas

Frankland believed that protoxide of arsenic [As_2O_3] had no well-defined basic character.

'Stanethylium' was as difficult. Like tin, it should, if conjugated, give rise to a protoxide and peroxide, but Frankland was 'quite unable to form any higher oxide than that described'.[81] Only on removal of ethyl would another equivalent of oxygen go in. Stibethyl posed the same problem.

So it was that the facts forced themselves upon Frankland with such directness that he was impelled to write:

> When the formulae of inorganic chemical compounds are considered, even a superficial observer is struck with the general symmetry of their construction; the compounds of nitrogen, phosphorus, antimony and arsenic especially exhibit the tendency of the elements to form compounds containing 3 or 5 equivalents of other elements, and it is in these proportions that their affinities are best satisfied; thus in the ternal group we have *NO_3, NH_3, NI_3, NS_3, PO_3, PH_3, PCl_3, SbO_3, $SbCl_3$, AsO_3, AsH_3, $AsCl_3$, etc.*; and in the 5-atom group *NO_5, NH_4O, NH_4I, PO_5, PH_4I, etc.* Without offering any hypothesis regarding the cause of this symmetrical grouping of atoms, it is sufficiently evident, from the examples just given, that such a tendency or law prevails, and that, no matter what the character of the uniting atoms may be, the combining power of the attracting element, if I may be allowed the term, is always satisfied by the same number of these atoms.[82]

In the light of this regularity Frankland was in fact contemplating a reconciliation between the two rival views of organic chemistry that had been the cause of much bitter conflict over the last decade or so. These were the radical theory already encountered, and a view proposed chiefly by the French chemists Dumas, Laurent and Gerhardt that became known as the type theory. According to Gerhardt's latest formulation of the theory, actually advanced just *after* the publication of Frankland's paper, organic compounds could be grouped into four fundamental families or 'types'.[83] These were based on the simple molecules hydrogen, hydrochloric acid, water and ammonia. They were intended to be viewed as unitary bodies, being represented in the following way as a matter of convenience and in order to display relationships between individual members of the 'family' and its simplest member:

$$\left.\begin{matrix} H \\ H \end{matrix}\right\} \qquad \left.\begin{matrix} H \\ Cl \end{matrix}\right\} \qquad \left.\begin{matrix} H \\ H \end{matrix}\right\}O \qquad \left.\begin{matrix} H \\ H \\ H \end{matrix}\right\}N$$

Thus, as Williamson had already shown,[84] alcohols and ethers could both be profitably regarded as members of the water type:

$$\text{alcohols} \quad \left.\begin{matrix} R \\ H \end{matrix}\right\} O \qquad\qquad \text{ethers} \quad \left.\begin{matrix} R \\ R' \end{matrix}\right\} O$$

There was no structural significance to these formulae; they showed how a molecule reacted rather than how it was constructed (then a highly problematic concept).

The daring achievement of Frankland was to take this general view, envision a whole range of new inorganic 'types', and relate his new organometallic compounds to these. The metal's oxygen, sulphur or chlorine compounds would then be seen as the true molecular types, and his new substances as their derivatives. But in so doing he was sailing dangerously near the wind. For he had already committed himself to a radical view of chemistry (in both senses of the term), with entities like ethyl and methyl pre-existing within a molecule. Such assurance was anathema to the type chemists whose types were not merely being multiplied but also given something suspiciously like an internal constitution. Yet in reality Frankland was taking each view as an *analogy* and saying, in effect, that *both* analogies might at times be appropriate. Whereas the radical theory (or one version of it) regarded the 'conjugate organic radicals' as analogous to the metal, when it came to considering their uptake of negative elements like oxygen or the halogens it might be better to compare them with these inorganic 'types'. Accordingly he gave the following table[85] (overleaf). With one slight and doubtful exception, all known cases were covered in this way. After one or two brief comments, and a suggestion for changes in nomenclature, he brought his classic paper to a close.

It could be said that the Theory of Radicals, winding its tortuous way through the dark forest of organic chemistry, with innumerable checks and many fundamental errors, emerged vindicated by the work of Frankland. But its triumph was relatively short-lived. For not merely were Frankland's 'radicals' soon to be shown to be not true radicals but dimers, but the supposed similarity in reactivity between cacodyl and the zinc alkyls was quite spurious, cacodyl being dimeric and dialkylzinc monomeric. Stanethylium was much later shown to be an oligomer, and the activity of the different organometallic compounds came from as wide a variety of causes as

Inorganic types   *Organo-metallic derivatives*

$As \begin{cases} S \\ S \end{cases}$ $As \begin{cases} C_2H_3 \\ C_2H_3 \end{cases}$ cacodyl

$As \begin{cases} O \\ O \\ O \end{cases}$ $As \begin{cases} C_2H_3 & \text{oxide of} \\ C_2H_3 & \text{cacodyl} \\ O \end{cases}$

$As \begin{cases} O \\ O \\ O \\ O \\ O \end{cases}$ $As \begin{cases} C_2H_3 & \text{cacodylic} \\ C_2H_3 & \text{acid} \\ O \\ O \\ O \end{cases}$

$Zn \quad O$ $Zn \quad C_2H_3$

$Zn \begin{cases} O \\ O_x \end{cases}$ $Zn \begin{cases} C_2H_3 \\ O_x \end{cases}$

$Sb \begin{cases} O \\ O \\ O \end{cases}$ $Sb \begin{cases} C_4H_5 \\ C_4H_5 \\ C_4H_5 \end{cases}$

$Sb \begin{cases} O \\ O \\ O \\ O \\ O \end{cases}$ $Sb \begin{cases} C_4H_5 \\ C_4H_5 \\ C_4H_5 \\ O \\ O \end{cases}$ and $Sb \begin{cases} C_4H_5 \\ C_4H_5 \\ C_4H_5 \\ C_4H_5 \\ O \end{cases}$

                                                binoxide of   oxide of
                                                stibethine   stibethylium

$Sn \quad O$ $Sn \quad C_4H_5$

$Sn \begin{cases} O \\ O \end{cases}$ $Sn \begin{cases} C_4H_5 \\ O \end{cases}$

$Hg \begin{cases} I \\ I \end{cases}$ $Hg \begin{cases} C_2H_3 \\ I \end{cases}$

their differing kinds of structures. Yet in this paper was an admission that the electrochemical (radical) viewpoint was only one of several, and an evocation of the type theory as the climax of the argument.

An important point has been made by Ladenburg:

> In the way in which the development actually took place, the influence of Kolbe, and more particularly that of Frankland, upon the supporters of the Gerhardt-Williamson school (Wurtz, Kekulé and Odling) can hardly fail to be recognised. Both schools were required, in order to raise the significance of the formulae to what it subsequently became.[86]

From this reconciliation emerged one thing far greater than either: the notion that elements have a definite combining power. For all the obscurity of the reactions of the new compounds and of what we now call their structures that simple idea emerged with crystal clarity.

Notes

1. Some of the material in the last two sections is derived from the author's *The History of Valency*, Leicester University Press, 1971.
2. L. Playfair to Frankland, 8 January 1850 [RFA, OU mf 01.02.1340].
3. J. Arthur Phillips (1822–1887), formerly of the École des Mines, Paris [*Prospectus* for 1848, British Library, MSS 8364.de.24]; he is sometimes confused with Richard Phillips (1778–1851), Playfair's assistant at the Museum of Economic Geology and President of the Chemical Society, 1849–50; in his time the post was little more than a demonstratorship.
4. Playfair was possibly referring to his appointment as a Commissioner for the Exhibition to be held in 1851; he already held a Chair at the School of Mines and was, like several of his disciples (including Frankland) not averse to holding several posts at the same time.
5. On Putney College see also *Lancastrian Chemist*, pp.145–51.
6. *Sketches*, 2nd ed., pp.273–4.
7. T. A. Hirst, *Diary*, 6 October 1850, reprinted in W. H. Brock and R. M. MacLeod, *Natural knowledge in social context: the journals of Thomas Archer Hirst F.R.S.*, Mansell, London, 1980.
8. *Ibid.*, 2 October 1850.
9. *Sketches*, 2nd ed., p.276.
10. Printed Syllabuses for 1850 [RFA, OU mf 01.03.0922 and 0923].
11. *Ibid.*
12. Putney Class Book [RFA, OU mf 01.06.0029 – 0106].
13. Putney College *Report*, 1850 [British Library, MSS 8364.de.41].
14. Copy-letters, Frankland to W. E. Cameron on 13 September 1851, 25 October 1852 and 1 November 1852 [RFA, OU mf 01.06.0109, 0275 and 0280].

15 *Sketches,* 2nd ed., p.276.

16 Lecture notes from 1 May to 5 October 1850 [RFA, OU mf 01.02.1918–1961].

17 Putney College *Report,* 1850, pp.32–3: Chemistry examination paper, July 1850 (note 13).

18 Physics examination paper, July 1850, two copies, the first of which has Frankland's marking scheme; interestingly the question that gained the most marks ($12\frac{1}{2}$%) was that on electricity as motive power [RFA, OU mf 01.02.1892–3].

19 Syllabus for chemistry and physics (Michaelmas Term, 1850) [RFA, OU mf 01.03.0923].

20 Prospectus for practical chemistry and metallurgy (Michaelmas Term, 1850) [RFA, OU mf 01.03.0922].

21 Putney College *Report,* 1850 , pp.89–90 (note 13).

22 *Sketches* (2nd ed.), pp.64, 118–19.

23 Copy-letter, Frankland to J. Russell & Sons, 4 July 1853 [RFA, OU mf 01.06.0320].

24 Copy-letter, Frankland to unknown recipient, 30 July 1853 [RFA, OU mf 01.06.0351].

25 J. S. Thayer, 'Historical origins of organometallic chemistry, Part I, Zeise's salt', *J. Chem. Educ.,* 1969, **46,** 442–3.

26 W. C. Zeise, *Ann. Phys.,* 1827, **9,** 632 (a single page brief report).

27 A. E. Krause and A. von Grosse, *Die Chemie der metall-organischen Verbindungen,* Verlag von Gebrüder Borntraeger, Berlin, 1937.

28 C. A. Russell, 'Edward Frankland, founder of organometallic chemistry', *Chem. Brit.,* 1982, **18,** 737–8; S. Radosavljevic and A. Leko, 'The path of development towards the first organometallic compounds', *Hemijski Pregled,* 1971, **12,** pt. 2, 36–45 [in Serbo-Croat].

29 As in the title of his paper 'On a new series of organic bodies containing metals and phosphorus', *J. Chem. Soc.,* 1849, **2,** 297–9; *Annalen der Chem.,* 1849, **71,** 213–16.

30 As in the text of his paper 'On a new series of organic bodies containing metals', *Phil. Trans.,* 1852, **142,** 417–44; the term was sometimes hyphenated (*organo-metallic*) in the early years.

31 Thus in applying for a grant from the Royal Society he specifies the high expense of buying the alcohols, bromine and iodine needed for making large quantities of zinc alkyls: Copy-letter, Frankland to C. R. Weld, September 1851 [RFA, OU mf 01.06.0190].

32 Frankland, dissertation, *Ueber die Isolirung des Aethyls* [RFA, OU mf 01.03.0926–0949].

33 *J. Chem. Soc.,* 1850, **2,** 263–90; another version appeared as *Annalen der Chem.,* 1849, **71,** 171–213. On the basis of this paper Frankland's preparation of the first organozinc compound is wrongly ascribed to 1848 (J. Boersma in G. Wilkinson (ed.), *Comprehensive organometallic chemistry,* Pergamon Press, Oxford, vol. 2, 1982, p.826); but the paper makes no specific reference to such a compound and in any case each version dates from 1849, not 1848.

34 C. Gerhardt, *Compt. rend. des trav. de chimie,* 1850, **11,** 233–36. He concludes

with passion: 'Du dualisme, que reste-t-il encore debout? Je le demande aux ésprits de bonne foi'.

35 Frankland, *J. Chem. Soc.*, 1851, **3**, 30–52 (50); in the issue for February 1850.

36 A. W. Hofmann, *ibid.*, 1851, **3**, 121–34.

37 Frankland, *ibid.*, 1851, **3**, 322–47.

38 Frankland, *Experimental Researches*, 1877, p.65.

39 B. C. Brodie, *J. Chem. Soc.*, 1851, **3**, 405–11.

40 C. Schorlemmer, *ibid.*, 1864, **17**, 262–5.

41 Frankland, *Diary*, 28 July 1848 (typescript copy) [JBA, OU mf 02.02.1304–1484].

42 Frankland, 1850 (note 33).

43 Frankland, dissertation (note 32).

44 Ethyl zinc iodide is polymeric in the solid state (P. T. Moseley and H. M. M. Shearer, *Chem. Commun.*, 1966, 876); on heating it yields diethylzinc and zinc iodide, with which it is in equilibrium in ethereal solution, though the monomer may be stabilised in addition compounds with 2,2'-dipyridyl *etc.* (J. J. Habeeb, A. Osman and D. G. Tuck, *J. Organomet. Chem.*, 1980, **185**, 117–27).

45 Dimethylzinc is now more usually obtained by alternative methods, as by the action of trimethylaluminium on zinc(II) acetate: A. L. Galyer and G. Wilkinson, *Inorganic Syntheses*, 1979, **19**, 253–6.

46 Frankland, 1849 (note 29). This paper of 1849, in its two versions, must constitute the foundation document of organometallic chemistry as it has the first explicit reference to the two dialkylzinc compounds.

47 Diethylzinc is still prepared from these reagents, Frankland himself later using a copper pressure vessel (*Phil. Trans.*, 1855, **145**, 259–75); a more recent method allows a zinc/copper couple to react with a mixture of ethyl iodide and bromide: C. R. Noller, *Organic Syntheses*, Coll. Vol. 2, 1943, pp.184–7.

48 Frankland, 1851 (note 35).

49 H. Kolbe, *Annalen der Chem.*, 1850, **75**, 211–39.

50 Frankland, 1849 (note 29a), p.298.

51 Copy-letter, Frankland to C. R. Weld, September 1851 (note 31). This intention was repeatedly re-affirmed in the years ahead, as in 1852 (note 30), 1855 (*Rep. Brit. Assoc. Adv. Sci., (Trans. of Sections)*, 1855, **25**, 62; *Phil. Trans.*, 1855, **145**, 259–275), and 1856 (*Phil. Trans.*, 1856, **147**, 59–78). In the event, most advances of the kind proposed were to be carried out by others. Conceivably one reason for Frankland's insistence was an awareness of the grant-generating potential of such a programme.

52 *Sketches*, 2nd ed., p.115.

53 J. W. Nicholson, J. A. Douek and J. D. Collins, *J. Organomet. Chem.*, 1982, **233**, 173–83.

54 Reactions of alkyl halides and pure mercury are notoriously sluggish, though the iodides (as here) are the least reluctant. Production of Frankland's methyl mercury iodide was a remarkable tribute to his experimental skill and persistence.

55 See J. W. Nicholson, 'The early history of organotin chemistry', *J. Chem. Educ.*, 1989, **66**, 621–3.

56 Frankland, 1852 (note 30), p.423.

57 B. Emmert and W. Eller, *Ber.*, 1911, **44**, 2328–31.

58 Nicholson (note 55).

59 Nicholson *et al.* (note 53).

60 A. G. Davies and P. J. Smith, in G. Wilkinson (ed.), *Comprehensive organometallic chemistry*, Pergamon Press, Oxford, vol. 2, 1982, p.546.

61 G. E. Coates and K. Wade, *Organometallic compounds*, Methuen, London, 3rd ed., 1968; the only true alkyl derivatives of Sn(II) are those with very large substituents or those with π-electron systems (as cyclopentadienes).

62 R. C. Poller, *The chemistry of organotin compounds*, Logos Press, London, 1970, p.152.

63 H.-P. Ritter and W. P. Neumann, *J. Organomet. Chem.*, 1973, **56**, 199–208; U. Blaukat and W. P. Neumann, *ibid.*, 1973, **63**, 27–39.

64 *Annalen der Chem.*, 1852, **84**, 308–35 (report on organometallic researches of C. Löwig); Löwig (1803–1890) was a professor at Zurich.

65 T. Anderson to Frankland, 13 January 1853 [RFA, OU mf 01.03.0236].

66 Copy-letters, Frankland to the Royal Society, September 1851; 29 March 1854; 1 September 1855 [RFA, OU mf 01.06.0110; 01.08.0088; 01.04.2063].

67 Referees' Reports from Hofmann and Playfair (1852), Graham and Hofmann (1855), Miller (1857) and Brodie (1859) [Royal Society Archives, RR 2/69, 70; 3/107, 108, 109; 4/103].

68 W. Sharpey to Frankland, 9 November 1857 [RFA, OU mf 01.02.0298].

69 W. A. Miller to Frankland, 25 March 1857 [RFA, OU mf 01.02.0241]; the chief point of contention appeared to be Miller's view that organometallic tin compounds are 'not so simply accounted for on Frankland's view', a statement omitted altogether in the second edition; and a suggestion that Frankland's 'ingenious' view 'may probably admit of further extension', later modified to an admission that 'indeed it contains the germ of the theory of polybasic elements': W. A. Miller, *Elements of chemistry, theoretical and practical*, London, vol. 3, 1st ed., 1857, pp.214 and 216; 2nd ed., 1862, p.221.

70 W. J. Russell to Frankland, 21 March 1855 [RFA, OU mf 01.03.0259].

71 J. Mercer to Frankland, 23 December 1860 [RFA, OU mf 01.03.0220].

72 S. Cannizzaro, *Sketch of a course of chemical philosophy* (1858), Alembic Club Reprint, Edinburgh, 1910, p.52.

73 S. Cannizzaro to L. Mond, 11 October 1899 [ICI Archives, Nobel Division, Millbank].

74 W. G. Palmer, *A history of the concept of valency to 1930*, Cambridge University Press, 1965; C. A. Russell, *The history of valency* (note 1).

75 C. A. Russell, 'The influence of Frankland on the rise of the theory of valency', *Actes of 10th International Congress of the History of Science*, Ithaca, 1962, pp.883–6.

76 Note 30; the paper was also published by the Chemical Society, *J. Chem. Soc.*, 1854, **6**, 57–74.

77 *Sketches*, 2nd ed., p.187.

78 Typescript speech to the Royal Society in acceptance of the Copley Medal,
    30 November 1894 [RFA, OU mf 01.04.0037].

79 H. Kolbe, *Handwörterbuch der Chemie*, Braunschweig, 1848, vol. 3, pp.177,
    185, 442.

80 Frankland, 1852 (note 30), pp.439–40.

81 *Ibid.*, p.440.

82 *Ibid.*, p.440.

83 C. Gerhardt, *Ann. chim. phys.*, 1853 [3], 37, 331.

84 A. W. Williamson, *Rep. Brit. Assoc. Adv. Sci., (Trans. of Sections)*, 1850, **20**,
    65; on Williamson (1824–1904) see *DNB*, *DSB*, E. Divers, *Proc. Roy. Soc.*,
    1907, A78, xxiv-xliv, and G. C. Foster, *J. Chem. Soc.*, 1905, **87**, 605–6.

85 Frankland, 1852 (note 30), pp.441–2.

86 A. Ladenburg, *Lectures on the history and development of chemistry since the
    time of Lavoisier*, trans. L. Dobbin, Alembic Club, Edinburgh, 1905, p.233.

CHAPTER 6

Frankland and the development of valency

1 Constituents of a theory of valency

Frankland was to spend only one year at Putney. He then returned north to Manchester (1851) where he was to remain until he moved in 1857 to spend the rest of his working life in London. The circumstances of those changes will be noted later, but it is pertinent to remark that valency remained a continuing preoccupation long after the Putney days were over, and indeed conditioned much of his later thinking on other matters. It is therefore convenient at this point to consider his rôle in the further development of valency[1], venturing on occasion well beyond the 1850s. We shall note three distinctive constituents of a theory of valency, all of which can be attributed to Frankland.

1.1 A new synthesis: radicals and types

The 1852 paper was, of course, an attempt to unite the radical and type theories, most obviously by the stratagem of using a quasi-type notation to convey an insight arrived at by a programme of research conceived almost entirely within the framework of dualistic radical theory. Frankland was, by his own admission, concerned to isolate the first 'radicals', and indeed for a long time thought he had done so. Yet in fact his synthesis was far more than a matter of representation or of language. An early insight comes from a document written before the *Phil. Trans.* paper was published.

The Edinburgh chemist William Gregory (1803–1858), pupil of Liebig, and author of a widely-used textbook *Outlines of Chemistry*,[2] had written to Frankland for the latest news of his researches on the zinc alkyls for inclusion in a new edition of the book (part of which duly appeared in 1852 with a new title[3]). In particular he had picked up Frankland's reference to the initial solid product from ethyl iodide and zinc and the fact that he had failed to identify it; was it perhaps a mixture of zinc ethyl and iodide of zinc?[4] In his reply[5] Frankland confirms Gregory in that opinion (which was actually mistaken):

> You are quite right in your conjectures respecting the crystalline compound formed by the action of zinc upon C_4H_5I; the crystals are simply ZnI saturated with the C_4H_5Zn which last distils off when heat is applied and leaves pure ZnI behind.

As for the more general enquiry he wrote [my italics]:

> I am not aware that I have made any observations not yet published about the radicals &c that are worthy of notice; I have some notions that C_2H_3Zn, C_4H_5Zn, C_4H_5Sn, C_2H_3Hg &c including kacodyle are *not radicals in the strict sense of the term, but are compounds of metals with the respective organic groups in which the group actually replaces an equivalent of O, S, H &c.* This opinion is formed upon the composition of the compounds of the bodies above mentioned compared with the corresponding compounds of the metals themselves. Thus with As we have

$$As \begin{cases} O \\ O \\ O \end{cases} \qquad As \begin{cases} C_2H_3 \\ C_2H_3 \\ O \end{cases} \text{ oxide of cacodyl}$$

$$As \begin{cases} O \\ O \\ O \\ O \\ O \end{cases} \qquad As \begin{cases} C_2H_3 \\ C_2H_3 \\ O \\ O \\ O \end{cases} \text{ cacodylic acid}$$

and with Sn

$$Sn \quad O \qquad Sn \quad C_4H_5$$

$$Sn \begin{cases} O \\ O \end{cases} \qquad Sn \begin{cases} C_4H_5 \\ O \end{cases}$$

> but I have not yet sufficiently supported these views by
> experiment and have only now just come to follow the subject further.

From this letter several important facts emerge. First, the italicised passage indicates a new perception of radicals which excludes substances as reactive as the zinc alkyls. Second, his comparison of the new substances 'with the corresponding compounds of the metals themselves' is entirely in the spirit of the type theory where family resemblances were so strongly stressed. Third, his use of type formulae in a private letter suggests that their appearance in his published paper was more than a matter of mere convention or good public relations with the 'typists'. Fourth, the tentative nature of Frankland's theorising *at this stage* is plain from his final sentence.

Eight years later he wrote to Lyon Playfair (now ensconced in the chemical chair at Edinburgh) in response to a magisterial letter[6] proposing a mode of representing organic compounds in terms of parent oxides. Frankland's letter[7] drew heavily on an address recently given (7 June) to the Chemical Society[8] and is a concise exposition of his views:

> Your letters have interested me much, & your views regarding the peroxide type allow many bodies to be much more simply formulated than they can be according to the methods now in use. Independently however of its affording a simple notation I confess your arguments in favour of the truth of your theory are very strong & might probably have convinced me were I not accustomed to look at chemical compounds through the medium of organometallic bodies, which have led me to views differing somewhat, though perhaps not fundamentally, from your own. One or two of my ideas on the subject you will see in the next number of the Chemical News in a short abstract of my lecture to the Chem. Soc. on Organometallic bodies. I will therefore content myself with a very brief statement here.
>
> I believe that each element has the power of attracting around itself a certain maximum number of molecules. An element so united with its maximum number of molecules is in a state of Chemical Saturation. Of course a biatomic molecule here plays the part of two molecules &c.
>
> An element can however form compounds containing a smaller number of molecules than that necessary for saturation; each compound of this kind represents a stage of stability. Thus antimony has a stage of stability of SbR_3 whilst its state of saturation is SbR_5. Arsenic has two stages of stability, viz. AsR_2 & AsR_3.
>
> So-called radicals are nothing more than bodies one or two stages short of Chemical saturation. When a body is one molecule short, either of chemical saturation or of a stage of stability[,] it is uniatomic

as $As(C_2H_3)_2$ and $As(C_2H_3)_4$. It is obvious that a body which is below a stage of stability can have a double atomic character, thus cacodyl is uniatomic in oxide of cacodyl & tetratomic in cacodylic acid. Again Arseniomonomethyl (AsC_2H_3) is biatomic in arsenious dioxymethide $(As(C_2H_3)O_2)$ & quadratomic in monomethylarsenic acid $(As(C_2H_3)O_4)$.

After illustrating his point with a number of examples from organic chemistry Frankland remarked:

> It is not however a matter of indifference which of these modes of notation is adopted. The simplest mode of representation is always that in which the element of the highest polyatomic character is made (if I may be allowed the expression) the moulding element . . . Pray excuse this rough, imperfect & dogmatic statement of the still somewhat crude ideas which have been for many years floating in my mind.

Radicals are now reduced to 'nothing more than bodies one or more stages short of chemical saturation' (which would not have pleased Berzelius or even Kolbe), while Frankland's 'stages of stability', representing the possibility that one atom might have more than one combining power, undermined a central feature of classical type theory (and would have infuriated Kekulé). Yet by means of modifications like these Frankland was able to unite radicals and types in a bold new synthesis that went well beyond its stylised representation in the 1852 paper.

1.2 Tetravalent carbon atoms

Although Frankland's early work was in the borderland between inorganic and organic chemistry its obvious application was to that same territory inhabited by organometallic compounds, and, by extension, to their simpler inorganic counterparts. The 'great leap forward' of applying valency concepts to organic chemistry is not usually associated with Frankland but rather with A. S. Couper (in two almost totally forgotten papers [9]) and F. A. Kekulé (in a series of papers [10] from 1857–8). It is sufficient to note here that Kekulé is usually credited with having first applied valency to organic chemistry by enunciating the concept of the tetravalent carbon atom. It happened as follows.

In a couple of papers on fulminates [11] Kekulé introduced a *fifth* type, based on methane, and known as the 'marsh-gas type', writing compounds as follows (using the $C = 6$ convention):

C_2	H	H	H	H	marsh-gas
C_2	H	H	H	Cl	methyl chloride
C_2	H	Cl	Cl	Cl	chloroform
C_2	(NO_4)	H	H	(C_2N)	[hypothetical] fulminic acid

Meanwhile his more famous paper on polyatomic radicals[12] used the concept of types to develop the notion that 'the number of atoms of an element or radical that combines with one atom . . . depends upon the basicity or relationship-size [*Verwandtschaftsgrosse*] of the component parts', elements with basicities of 1, 2 or 3 being respectively exemplified by the types HH, OH_2 and NH_3. To this he added a notable footnote:

> Carbon, as may be easily shown, and as I shall explain in detail later, is tetrabasic and tetratomic; *i.e.*, one atom of carbon = C = 12 is equivalent to four atoms of hydrogen.

The major exposition of his theory came in the seminal paper of 1858 'on the constitution and metamorphoses of chemical compounds and on the chemical nature of carbon'.[13] Acknowledging his debt to Williamson, Odling (who in 1855 had argued for a CH_4 or 'marsh-gas' type[14]), Gerhardt and Wurtz (but *not* Frankland), Kekulé deemed it necessary 'to go back to the elements themselves', and argued that radicals are combinations of atoms in which affinity units are joined together. This reconciliation of type and radical theories was quite as definite as Frankland's, though a little later. Kekulé went on to note that the carbon atom 'always combines with four atoms of a monatomic, or two atoms of a diatomic, element', which regularity 'leads to the view that carbon is tetratomic (tetrabasic)'.

In subsequent discussions it is not surprising that the discovery of the tetravalent carbon atom has nearly always been ascribed to Kekulé. Yet this is quite wrong for not only does it overlook the proposal of Odling, it also ignores important work by Frankland and his associate Hermann Kolbe[15].

In the few years immediately following his move to Manchester (1851) Frankland continued to contemplate the prospect of replacing hydrogen atoms in organic compounds by alkyl, using zinc alkyls for the purpose. In 1855 he reported his own addition of diethyl zinc to nitric oxide [to give EtZn-O-NEt-NO, though he regarded it as 'replacing oxygen by ethyl'] and a similar addition by his pupil

Hobson of dimethylzinc to sulphur dioxide.[16] He was overjoyed when his former pupil James Wanklyn later discovered ethylsodium and, by addition to carbon dioxide, obtained (after hydrolysis) the simple organic compound propionic acid.[17] He wrote warmly to congratulate him.[18] Kolbe, meanwhile, regarded his friend's work with some suspicion, indicating an aberration from the truth enshrined in electrochemical dualism. In the first part of his *Lehrbuch* Kolbe strenuously rejected much of the current ideas of the 'typists', including the doubling of hydrocarbon 'radicals' of the kind that he and Frankland had isolated. But he parted company with Frankland by retaining the theory of copulae.[19] However what must have been a considerable correspondence between the two friends happily led to 'a substantial agreement'.[20] Certainly a short article[21] by Kolbe from early 1856 implies acceptance of the law of saturation capacity, and gives a formula for acetic acid in which one methyl has replaced one oxygen of [dibasic] carbonic acid (C = 6, O = 8):

$$2HO.C_2O_4 \qquad HO.(C_2H_3)C_2O_3$$

carbonic acid $\qquad\qquad$ acetic acid

There then followed a paper[22] that Kolbe regarded as the foundation document for his 'carbonic acid' theory of organic compounds, particularly aldehydes, ketones and fatty acids. All are again derived from carbonic acid (C = 6, O = 8) and could be formulated thus:

$$2HO,C_2O_4 \quad HO,C_2(C_2H_3)O_3 \quad C_2(C_2H_3)HO_2 \quad C_2(C_2H_3)_2O_2$$

carbonic acid $\quad$ acetic acid $\qquad$ acetaldehyde $\qquad$ acetone

These offer a striking illustration of the law of saturation capacity formulated by Frankland but given here a powerful new application to organic chemistry by Kolbe. And therein lay the problem. Referring to his paper of 1855 to the British Association[23] Frankland informed Wurtz many years later:

> It was not until the following year that the law of atomicity was applied by Kolbe and myself to carbon compounds in a paper which, by mistake, appears in Kolbe's name only. ("Experimental Researches" p. 148). Throughout the paper, however, the views advocated are spoken of as ours jointly "wir sind der Ansicht" *etc*. This paper is dated Dec. 1856, about 6 months before Kekulé wrote his first word

> about atomicity (See Ann. Pharm. 1857, 104, 132) and nearly 2 years
> before the appearance of his celebrated paper on carbon compounds...[24]

The same complaint about 'inadvertent' omission of his name appears in Frankland's *Sketches*[25] and *Experimental Researches*[26]. The omission raises a number of problems that have been addressed by Alan Rocke; the following attempt at reconstruction, though differing in some respects, is much indebted to his account.[27]

Frankland's first reaction after seeing Kolbe's 1857 paper in print was one of great indignation, and for five years he *appears* to have ceased writing to Kolbe (though he did send him at least one off-print). Kolbe deeply regretted the silence of his English friend, as he indicated in two conciliatory letters[28] which began to heal the breach, early in 1862. A copy-letter[29] from Frankland to Kolbe in March 1862 has survived but not, apparently, the original. It is a useful source for Frankland's side of the story. A further letter from Kolbe on 24 November, 1863, was a long *apologia* for his conduct and effectively completed the reconciliation.[30]

In this last letter Kolbe reminded Frankland of his own insistence on the English reluctance to publish purely theoretical papers, so he decided to send his paper to *Annalen*. In deference to what he assumed Frankland would have wished, Kolbe had omitted his name as co-author. He cited a letter from Frankland, dated 18 January 1857 but now lost, generally agreeing with Kolbe's views and suggesting that 'something should appear from us'. From his copy-letter of 1862 it is possible to construe, with Rocke, that Frankland had intended his earlier remarks to mean a request for co-authorship. He wrote thus:

> Although I felt at the time that the memoir, if it appeared at all, ought
> to appear in both our names, yet, in order not to lose time by the
> rewriting of it, I acquiesced in your proposition, & I did this the more
> readily, as you had fully explained in the memoir that the views there
> were held by us jointly.

But this is also consistent with *single* authorship: 'it *ought* to appear in both our names, *yet* . . . I acquiesced in your proposition'. In any case if the paper contained this full explanation why have joint authorship? For a paper attributed in the title-line to two authors such explanation would be redundant. Moreover, Frankland went on to add that, with the subsequent appearance of Wanklyn's paper, which provided experimental corroboration of their theoretical views, 'my reluctance, which I might previously have felt as to the

Fig. 6.1 Hermann Kolbe (1818–84): Frankland's first chemical collaborator and lifelong friend. (Royal Society of Chemistry.)

publication of purely theoretical views, no longer existed'. But that reluctance could have been banished only after Wanklyn had published in early 1858, a year after Kolbe's paper had been in print. It was therefore irrelevant to the question as to whether in 1857 Frankland had clearly expressed a wish for co-authorship. Indeed the whole copy-letter is a curious document; not least because the original seems never to have reached its destination; it has not apparently survived and Kolbe questioned whether he had received it.[31] The copy-letter is not in Frankland's handwriting (which effectively eliminates it as a draft letter) and is manifestly not the original which for some reason Frankland might have failed to post. Had he decided not to send it after all there appears to have been little reason for retaining the copy, so the most likely scenario is either that Kolbe received and then forgot it, or, more probably, that it was lost in the post.

The sequence of events in 1856/7 is now tolerably clear. By early 1856 Kolbe had accepted Frankland's views and (according to his 1863 *apologia*) had urged upon Frankland the need for a joint theoretical paper but had met the reply outlined above: English journals do not publish that kind of article. Then, in October, a paper by the Italian chemist Rafaelle Piria appeared which was getting dangerously near to their own views.[32] In some alarm Kolbe put pen to paper and by December had completed the short article in question. A quick note to Frankland seeking permission to add his name as co-author brought the ambivalent reply referred to above. In fact, however, Kolbe had gone ahead with single authorship two weeks *before* that letter was written[33] It is now suggested that (a) he probably knew that his friend would not give unequivocal assent without having a hand in the actual text, (b) he was spurred to immediate action by the threat of being forestalled by Piria or someone else, and he could not wait for even a brief delay in the post from England, (c) that a generous acknowledgement to Frankland in the text would be as good as joint authorship, and (d) that when Frankland's reply did arrive it was so ambiguous that Kolbe saw no necessity to add the extra name at proof-stage.

If this analysis is correct it is likely that the paper which Kolbe had originally penned as from both authors had to be slightly adapted. Accordingly he set about changing the first person pronouns from plural to singular, but omitted to complete the task. This could have been due simply to the extreme pressure of time, or alternatively a crude device to confuse the issue of exact authorship. If so, he made one big mistake. He grossly underestimated the sensitivity to such questions likely to be felt by Edward Frankland.

One other question remains. On almost any scenario the omission of Frankland's name cannot have been accidental. Why, then, did Frankland always speak in later years of 'inadvertence' on the part of Kolbe? It could have been because this was the simplest way of short-circuiting an awkward question, one that he was now anxious to forget. But it is more likely that, later in life, he reciprocated his friend's generosity and did not wish to perpetuate an impression of double dealing or even carelessness. That would also explain why he omitted from his *Experimental Researches* the two papers written fairly soon after the controversy started. These were addresses to the Royal Institution[34] and the Chemical Society[35] on the carbonic acid theory; on the first occasion he was careful to assert joint ownership and in the second he omitted Kolbe's name altogether though he cited 'their' paper as though it were his own. Omission of such tendentious material was consistent with a later stance of reconciliation and friendship. However, given the amount of relevant documentation now missing, the ambiguities in some extant material, and the curious mix of pronouns in the controverted paper, it is probable that some critical facts of the case are lost beyond recall. We shall never know with certainty the true causes of this strange rift.

In the immediate aftermath of Kolbe's 1857 paper what goaded Frankland still more was the ascription to Kolbe alone as the author of such ideas, both in Carey Foster's report to the British Association[36] in 1859 and in Kolbe's own paper the next year.[37] Yet this is not sufficient to explain his pique for the preceding two years and one can only conclude that, despite being 'characteristically modest and extremely reticent about priority claims'[38], he thought a good deal more about these issues than about a friendship that had helped to establish his chemical career. Fortunately Kolbe was sufficiently large-hearted to offer a generous acknowledgement in 1865 when he admitted that 'Frankland had a large share, indeed much larger than is generally known, in the formulation of this hypothesis' and that it was due to Frankland that the idea of copulae was completely discredited.[39]

Thereafter a measure of harmony was restored to their relationship and their correspondence was resumed. In the letter to Playfair previously cited[40] Frankland builds on to his distinction between elements in states of *stability* and *saturation* an exposition of the nature of organic compounds, adding to his carbonic acid type another based upon carbonic oxide (CO). He argues that 'the double atom of carbon is biatomic in C_2Cl_2 & C_2O_2 & quadratomic in C_2Cl_4 & C_2O_4. The biatomic stage is the stage of stability, the quadratomic that of saturation' adding that 'all organic compounds of the C_nR_{n+1} & C_nR_{n-1} families are constructed upon these types'. Amongst his examples are:

$$C_2\begin{cases} C_4H_5 \\ Cl \\ Cl \\ Cl \end{cases} \quad C_2\begin{cases} C_4H_5 \\ Cl \end{cases} \quad C_2\begin{cases} C_4H_5 \\ H \\ Cl \\ Cl \end{cases} \quad C_2\begin{cases} C_4H_5 \\ H \\ H \\ HO_2 \end{cases} \quad C_2\begin{cases} C_4H_5 \\ HO_2 \\ HO_2 \\ HO_2 \end{cases}$$

| trichlorhydrin | chloride of allyl | bichloride of propylene | propylic alcohol | glycerin |

Frankland went on to add:

> You will see that this [is] only one mode of applying the above law of chemical combination to organic compounds. The application ought to be, & will be found, equally truthful in referring organic compounds containing hydrogen, oxygen or nitrogen to the inorganic typical compounds of these elements. . . [Thus] nitrogen compounds & their analogues whilst derived from typical inorganic compounds of nitrogen are also true derivatives from the carbon types. Thus trimethylamine is correctly written as a derivative of carbonic acid

$$C_2 \begin{cases} N(C_2H_3)_2 \\ H \\ H \\ H \end{cases}$$

In these ways Frankland sought to build an organic chemistry derived from his own concept of valency ('atomicity'), and virtually (though not explicitly) advanced the doctrine of tetravalent carbon.

As for the derived concept of catenation, or self-linking carbon atoms, it is fairly clear that Frankland was not involved in that extension of valency theory. This is hardly surprising as linking of electrochemically similar – indeed identical – atoms would seem an *a priori* improbability for a member of the radical school. It was rather the case, as T. E. Thorpe observed many years later, that the character of organic chemistry was determined, first by Frankland's discovery of fixed combining powers, and 'secondly the hypotheses of Kekulé and Couper, which graft themselves directly on Frankland's discovery, concerning certain peculiarities in the mode of combination of the element carbon, the organic element *par excellence*.'[41]

1.3 Chemical bonds

Linguistic innovations may not often be thought of as significantly advancing a science, but there are several examples in chemistry, of which the revolution in nomenclature under Lavoisier may have been the most obvious. By a most daring step of simplification Edward Frankland was responsible for another. In 1866 he introduced into chemistry a term that has been indispensable ever since, and was a further contribution to the general understanding of valency: the word 'bond'.

The word had, of course, a rather more general use in the past. James Parkinson wrote of a piece of fossilised wood where carbonate of lime was dissolved out with nitric acid leaving a flocculent sediment, 'the carbonate of lime having been so thoroughly diffused between these light particles, as to have been their only bond of connection'.[42] Writing of sulphuric and nitric acids Gay-Lussac and Thenard suggest that 'water appears to be the bond which unites their elements'.[43] But Frankland was now to give this ubiquitous word a highly specific meaning. His introduction is worth quoting in full:

> By the term *bond*, I intend merely to give a more concrete expression to what has received various names from different chemists, such as an atomicity, an atomic power, and an equivalence. A monad is represented as an element having one bond, a dyad as an element possessing two bonds, *etc*. It is scarcely necessary to remark that by this term I do not intend to convey the idea of any material connection between the elements of a compound, the bonds actually holding the atoms of a chemical compound being, in all probability, as regards their nature, much more like those which connect the members of the solar system.[44]

A slightly later version of this passage has a different conclusion:

> . . . the bonds which actually hold the constituents together being, as regards their nature, entirely unknown.[45]

In that spirit he named 'each unit of atom-fixing power a *bond* – a term which involves no hypothesis as to the nature of the connection'.[46]

In fairly exalted chemical circles the term was first used sparingly and with caution, or else was openly deprecated, especially by Williamson and others in the type tradition.[47] As will be seen (Chapter 10) the concept of valency was given even more powerful assistance by the visual representation of bonds, and this most effectively in the mass of popular literature that flooded England in the 1870s. Although Frankland did not invent such graphic formulae it was largely due to him that they spread so widely, even becoming known as 'Frankland's notation'. Today the term 'chemical bond' is in almost universal usage though its originator may be quite forgotten.

2 A prophet without honour?

The notion of saturation capacity that first emerged at Putney is, of course, only one component of a modern theory of valency, but it is nevertheless absolutely fundamental. Frankland was in no doubt as to what he had discovered. Writing in 1877 he made the following claim about his 1852 paper:

> It was evident that the atoms of zinc, tin, arsenic, antimony, *etc.* had
> only room, so to speak, for the attachment of a fixed and definite
> number of the atoms of other elements, or, as I should now express it,
> of the bonds of other elements. This hypothesis . . . constitutes the
> basis of what has been called the doctrine of atomicity or equivalence
> of the elements; and it was, so far as I am aware, the first
> announcement of that doctrine.[48]

One might therefore have expected a rapid transformation of the
whole of chemistry as this new and fundamental law of the science
became known and applied. The reality was very different. Far
from being hailed as the discoverer of atomicity (as he had so
recently been acclaimed as the discoverer of 'Ethyl') Frankland was
to savour, not for the first time in his life, the bitter experience of
rejection. This time, however, it had nothing to do with the social
stigma of illegitimacy but was concerned rather with his status as a
chemical discoverer. It probably hurt more. For the hard fact was
that the immediate impact of his new discovery was negligible.

Nor were things much better in the medium term. One of the
hardest blows must have come from Kolbe whose textbook of 1854
declared Frankland's views to involve an impossible assumption,[49]
though after correspondence (now lost) the two friends 'arrived at a
substantial agreement'.[50] Elsewhere the next few years saw plenty
of writing about valency, both in ordinary student textbooks and in
the occasional excursions made by chemists into the history of their
subject. In almost none of these is Frankland credited with
'valency', 'atomicity' or even 'saturation capacity'. His name does
not appear in the considerable historical introduction to the
Lehrbuch der organischen Chemie[51] by August Kekulé or in the
famous *Histoire des doctrines chimiques*[52] by Wurtz (or its English
translation[53]). When the latter author brought out a new book in
1880, *La théorie atomique*, Frankland was completely ignored with
respect to atomicity. As Frankland said years later 'This was the
more extraordinary because Wurtz and I were then and always
intimate friends'.[54] On receiving a copy from the author he wrote:

> Pray accept my best thanks for a copy of your excellent book 'Théorie
> Atomique', several chapters of which I have already read with much
> interest. It will doubtless prove a most useful help to students from the
> clearness with which even the most abstruse chemical laws are
> expounded. [In] your chapter on the definition & historical development
> of the doctrine of atomicity I think you have doubtless quite
> unintentionally done me some injustice. Whilst you described the part
> which Williamson, Odling, Kekulé and others have taken in this

all-important doctrine, no mention is made of what I have always thought was the first definite announcement of the law in a paper communicated by me to the Royal Society in May 1852 and printed in their Phil. Trans.[55]

This, however, was nothing new. Just as Frankland had once been inadvertently misrepresented over organometallic chemistry by Miller[56] so he was to suffer a similar fate over valency. Indeed Miller compounded his error in remarks made from the chair at the British Association meeting at Birmingham in 1865. Although Frankland was in the audience he would have listened in vain for any recognition from his fellow-countryman and former referee:

> Owing to the labours of many distinguished men, amongst whom the names of Williamson, Kekulé, Odling, Cannizzaro and Wurtz are prominent, a classification of the elements into families has been made; and that classification rests upon what is known as the *atomicity* of the elements.[57]

To be sure, Frankland had still to publish an extended treatment of classification of the elements[58] but he had more than touched on the subject in his paper of 1852. The fact had simply failed to register.

In his letter to Wurtz Frankland raised another vexed question of priority: who had first discovered the tetravalent carbon atom? The most generally held view seems to have been that advanced by Erlenmeyer in 1874. In 1857 (actually 1856) he thought that

> Kolbe had adopted Frankland's way of thinking and further extended it to a number of carbon compounds (Ann. Pharm. 101, 257), in that he derived the latter from carbonic acid anhydride (C_2O_4).[59]

Here, as elsewhere in the history of science, the reception of a new scientific doctrine is a complicated business, with a multiplicity of subtle effects coming into play. Some are obvious, some less so, but in the present case at least four reasons may be suggested by way of partial explanation for the neglect of Frankland's claims by his contemporaries.

2.1 The ambivalent rôle of organometallic chemistry

The first of these relates to the origins of valency theory in organometallic chemistry. It is surely at the very least a matter of some surprise that this elegant and universal doctrine emerged, not from a study of simple compounds like those encountered in elementary inorganic chemistry (water, ammonia, simple oxides *etc.*), but from a group of compounds so obscure and novel that it took the experimental genius of a Frankland to isolate and identify them. One is reminded of Kepler's recognition of elliptical planetary orbits from a study of minute and obscure variations established by Tycho Brahe, or of Fleming's discovery of penicillin from a stray spore of mould. From the obscurity of organometallic chemistry emerged a mainline chemical theory. In the correspondence of 1860 (cited above) between Frankland and Playfair, Frankland explicitly asserted that 'your arguments in favour of the truth of your theory are very strong & might probably have convinced me were I not accustomed to look at chemical compounds through the medium of organometallic bodies, which have led me to views differing somewhat, though perhaps not fundamentally, from your own'.

Another irony of the situation was more simple. It lay in the sheer fascination exerted by the organometallic compounds as heralding a wholly new branch of chemistry. The possibility that they might display a general theoretical law was lost sight of in their very novelty and reactivity. One did not often encounter organic materials that violently reacted with water and were spontaneously inflammable in air.

2.2 Radicals v. types

It is easy to conclude that, because Frankland was a representative of what was usually seen as the dying school of Berzelian dualism, his adherence to the radical theory would diminish his chances of recognition by followers of the type theory. This was certainly his own view when he suggested that others had overlooked the work of Kolbe and himself because 'it was the death blow to the types upon which so much of their reputation had been built. It was of course very disappointing for them to find, by the light of atomicity, that the belief in such types as $\left.{H \atop H}\right\}O$ [and] HCl betrayed a complete ignorance of the first principles of chemical combination'.[60]

However it must be remembered that few objections to Frankland's claims were made explicitly on these grounds. Frankland himself was far from being a thorough-going dualist of the old school and actually tried to bring type and radical theories together, and it is probably naïve to imagine that adherents of either school were quite so myopic as to fail to perceive any virtues in the work of the other. As F. R. Japp wrote in his Kekulé Memorial Lecture:

> A discovery, made by an adherent of the radical theory and correctly formulated by him in terms of the old equivalents, does not become the property of the first adherent of the type theory who happens to translate it into the new molecular weights.[61]

His reference to old atomic weights and equivalents suggests a further reason for Frankland's difficulties.

2.3 Atomic weights v. equivalents

A third explanation for the non-recognition of Frankland's contribution to valency may well lie in the atomic weight system he adopted. Previous examples have shown his consistent use of the old 'half' atomic weights as 8 for oxygen, 6 for carbon, 32.5 for zinc, *etc.* One of his great opponents (to whom we return shortly) was August Kekulé who in 1883 wrote a polemic 'History of valency' as part of his sustained campaign against Hermann Kolbe but who was persuaded by Volhard, the editor of *Annalen*, to withhold it from publication. It saw the light of day only in 1929 when the two-volume biography of Kekulé was published by Anschütz.[62] In this document Kekulé raised a number of general points and wrote:

> There is then the further consideration that most chemists in 1853 did not yet clearly distinguish the ideas of atom and equivalent, and that Frankland also did not make such a distinction, but spoke now of atoms, now of equivalents, whereas valency theory assumed a sharp difference between the two ideas.
>
> Frankland argued with a false atomic weight for oxygen, and had to be able to argue also for a false atomic weight for antimony, *etc.* His law does not even deal with atoms. As far as bodies are concerned, it denotes that they shall have a definite saturation capacity and deals with that relative quantity we regard as a radical . . . The law refers not to atoms but equivalents. It is right for equivalents but false for atoms, and therefore false in the form arrived at by Frankland.[63]

This criticism was echoed by others, including Kekulé's former student A. von Baeyer, who, in a paper on organic arsenic compounds, went out of his way to denigrate Frankland's view as 'purely superficial', for it 'vanishes immediately when the proper atomic weight for oxygen is adopted, and the formulae which contain an odd number of that element are doubled'.[64] Such objections have to be seen in the light of the widespread uncertainty in the 1850s as to what were the correct atomic weights (later spilling over into almost desperate battles to keep alive the belief in atoms at all[65]). Until there was consensus on this point a degree of anarchy pervaded the whole of chemistry, which was the main reason for calling the Karlsruhe Conference in 1860 (of which Kekulé was one convenor). The burden of Kekulé's objection lay not in the fact that Frankland was using incorrect atomic weights, but that *he was not using any at all*. Given the relationship that soon became recognised,

$$valency = atomic\ weight/equivalent,$$

and that *both* variables are therefore necessary to define valency, it is clear that Kekulé had a point. So, at least Frankland himself admitted (though he never saw the 'History' of his rival). Writing to Kolbe in 1881 he observed:

> In our application of the theory to carbon compounds, it was no doubt unfortunate for us that we were still hampered with the old atomic weights of C & O. It is owing to this circumstance alone, in my opinion, that our ideas were never properly recognised & appreciated; but this unfortunate circumstance obviously gives no right to others to appropriate our ideas, applying them to revised formulae and then re-issue them as their own & original inventions.[66]

Ten years later still he admitted that the wrong atomic weights did not merely alienate others but also inhibited themselves:

> One of the chief causes of the neglect with which the views of Kolbe and myself were treated was the blunder we made in adhering too long to the small atomic weights of C and O. This is the only reason why we did not develop the valency of C and *die Verkettung der Atome* years before they occurred to Kekulé.[67]

And it was this last-mentioned chemist who had much to do with the ostracism of Frankland, for Kekulé was the English chemist's chief rival in matters of valency and his implacable opponent in all disputes over priority.

2.4 *The opposition of Kekulé*

August Kekulé, famous in later years for his hexagonal structure for benzene, was a young German chemist who entered the laboratory of Liebig one year after the visit by Frankland. There followed a year in Paris, where Gerhardt's theories made a lasting impression, and a further spell in von Planta's private laboratory in

Switzerland. Then, late in 1853, he came to London to work under John Stenhouse at St Bartholomew's Hospital. He joined the small circle of leading chemists in the metropolis, including Hofmann and his students, but also Odling and Williamson, both admirers of the Parisian school of type theory. By late 1855 he was back in his home town of Darmstadt and shortly afterwards in his first settled position as a Privat-Dozent at Heidelberg. But his time in Paris and London had left its mark. He was a convinced upholder of the theory of types and, more important, author of a series of papers that were to lead from it to a recognition of 'polyatomic' (polyvalent) radicals, the concept of self-linking carbon atoms (catenation), a new 'carbon type' to add to those of Gerhardt and others

$$C \begin{cases} H \\ H \\ H \\ H \end{cases}$$

and an identification of the tetravalent carbon atom.[68] Such were some of the achievements of a man who remained bitterly opposed to any idea that Frankland, not he, should have been the founder of the theory of valency. The whole saga of a soured relationship between the two men is not merely a sad reflection on human nature, but also a problematic question for historians.[69] H. E. Armstrong, later Professor of Chemistry at the Central Technical College, London, was one of Frankland's most loyal disciples but also an associate of Kekulé. Though hardly impartial his words confirm the essentials of their failure to communicate:

> The problem first solved by Frankland was in the air – chemists everywhere had it in mind, especially in France. Kekulé was in London in 1854 and consorted with Williamson, who like himself was under the enthralling influence of Gerhardt. Had he consorted with Frankland, a man infinitely in advance of himself and most other chemists as a worker, his attitude would not have been so independent. I have not been able to discover that Kekulé made the least attempt to exchange views with Frankland, having joined another camp. We are in face of a psychological puzzle.[70]

Although this energetic defence of Frankland came 35 years after his death, and over 80 years since he and Kekulé might have met, subsequent evidence suggests it was not seriously in error. The thousands of items in the Frankland correspondence contain only two letters from Kekulé[71] (both about the affairs of the Chemical Society of which Frankland was then Foreign Secretary) and no copy-letters at all to him. The German chemist hardly appears in Frankland's autobiographical *Sketches*,[72] almost never in his published scientific papers[73] and rarely in the private papers (and then mainly in unflattering comments in correspondence with Kolbe).

If Frankland ignored Kekulé the compliment was returned with interest. With studied indifference Kekulé paid almost no attention to Frankland's views and (as we shall see) this went well beyond his rival's claims to priority in matters of valency. Acknowledging debts to other workers, but not to Frankland, Kekulé made the explicit claim:

> Unless I am mistaken, it was I who introduced into chemistry the idea of atomicity of the elements,[74]

which of course was exactly what he argued in his posthumous 'History of valency'.

Kekulé's opposition was therefore more important than that of anyone else because he was a claimant, not merely an observer, historian or reporter. And there was another reason. Writing to Kolbe, Frankland commented that 'Kekulé and Baeyer, however, stand upon a different footing: they have been aggressive and may therefore, with justice, be dealt with sharply'.[75] There can be little doubt that Kekulé's implacable opposition was a significant factor in diminishing Frankland and his work on valency in the eyes of the international community. The reasons are still rather problematic.

As already mentioned Kekulé opposed Frankland on the grounds of 'wrong' atomic weights and his departure from the orthodoxy of type theory. There were also other matters of theoretical organic chemistry on which they differed (Chapter 9). Related to much of this was the issue that Japp identified so clearly in his Kekulé Memorial Lecture:

> We are in a position, I think, to explain how it was that Kekulé came to ignore Frankland's work and to claim the theory for himself. To Kekulé, varying valency and, moreover, varying valency referred to equivalents instead of to atoms, was not valency at all, as he understood it. We must bear in mind the attitude of the two opposed schools in questions of chemical theory: how each seemed to labour under an absolute inability to place itself in the mental position of the other. I have no doubt that Kekulé paid very little attention to Frankland's theoretical views; that he evolved the doctrine of valency

in part independently and in part from the indications which he found
in the writings of Williamson, Odling, and Wurtz; and that afterwards,
perceiving that Frankland had put forward similar ideas, he came to
the conclusion that they had not been deduced by a legitimate process
and were not correctly stated.[76]

Other explanations may lie in the sheer problems of scientific
communication. Thus although the two men did meet in London
in the 1850s[77] it can only have been a rare encounter because
Frankland's commitments in Manchester, family and professional,
left little opportunity for visits to the metropolis. In any case it is
not clear why Frankland should have sought out a comparatively
obscure research student, four years his junior and only temporarily
in England. Yet one must not overstate such difficulties. We know
that in the later 1850s Frankland's works were circulating in
Heidelberg[78] and Kekulé must have been aware of them.

Perhaps the best solutions to the riddle are to be found in the
differing cultural values the two chemists espoused and the
different modes of life they experienced in the years ahead. Kekulé
was an organic chemist, dedicated to research and becoming world
famous for his benzene hexagon. He did little work outside organic
chemistry, had few diversions of a technical, financial or social
nature and remained the archetypal German professor. In almost
all these respects, as we shall see in due course, Frankland was
different. His later work in organic chemistry was largely confined
to a burst of activity from London in the 1860s, his scientific
interests ranged over a wide area of technical, physical and
biological chemistry, and many were the distractions experienced
by an upwardly mobile English chemist of which Kekulé would
know nothing.

Although these two chemical giants rarely did more than eye
each other over the fences of chemical doctrine that each had
erected, it is pleasant to record that at the end of their lives each
found generous and statesmanlike words to say about the other. At
Kekulé's famous *Benzolfest* in 1890 he uttered the following
much-quoted *résumé* of recent chemical history:

> Fifty years ago, the stream of chemical progress divided into two
> branches . . . At length, as the two branches had again approached
> much nearer to one another, they were separated by a thick growth of
> misunderstandings, so that those who were sailing along on the one
> side neither saw those on the other, nor understood their speech.
> Suddenly a loud shout of triumph resounded from the host of

adherents of the type theory. The others had arrived, Frankland at their head. Both sides saw that they had been striving towards the same goal, although by different routes. They exchanged experiences; each side profited by the conquests of the other; with united forces they sailed onward on the reunited stream. One or two held themselves apart and sulked; they thought that they alone held the true course, – the right fair-way, – but they followed the stream. . . My views also have grown out of those of my predecessors and are based on them. There is no such thing as absolute novelty on the matter.[79]

When Kekulé died, shortly after this event, eulogies poured in from all quarters. None was more brief, or more to the point, than that of Lord Lister, President of the Royal Society whose tribute was included in his next Annual Address:

> The death of August Kekulé will be felt as a severe loss to chemical science all over the world. Not only did his great activity in original research enrich organic chemistry with many new and interesting compounds, but his announcement of the tetradic valency of carbon, and, especially, his theoretical conception of the benzene ring, gave an impulse to the study of structural chemistry which has introduced order into the vast array of organic compounds, both of the alcoholic and aromatic types and has not yet fully expended itself.[80]

What is truly remarkable about this short speech is that, though the voice was that of the President, the words were those of another. In seeking information Lister had solicited, and received, that fair-minded and generous tribute from Kekulé's old-time rival, Edward Frankland.[81]

3 Vindication

Frankland had to wait over 20 years for anything approaching public recognition of his work on valency. An address given by Erlenmeyer in 1871 on 'The task of chemical education'[82] and reported in a textbook of 1877 by Naumann was an early example:

> Following the assumption by Williamson of monobasic and dibasic compound radicals in his 1851 paper on the constitution of salts, the regularity of combining-ratios of different elements was first recognised by Frankland in the year 1853 and explained in terms of their saturation capacity. (Frankland spoke of triatomic and pentatomic groups.) In 1856 both Odling and Kekulé formulated such saturation laws for compound and additionally for elementary radicals. In the year 1857 Kolbe had adopted Frankland's way of thinking and further extended it to a number of carbon compounds (Ann. Pharm. 101, 257),

in that he derived the latter from carbonic acid anhydride (C2O4). In the same year Kekulé (Ann. Pharm. 1857, 104, 132) carried the idea of Types, merely anticipated by Gerhardt, with fuller clarity into the way of representing the chemical combination of compounds, because he attributed Gerhardt's Types to mono-, di- and triatomic elementary atoms, and in a certain sense Kolbe's carbonic acid Type to a tetratomic carbon atom. In the above case, and still more in a paper appearing in 1858 (Ann. Pharm. 106, 129), Kekulé gives us to understand that he regarded it as necessary for representing the constitution of chemical compounds to bear in mind the constitution of radicals and to extend this even to the elements. Nevertheless he still held to fixed Types.[83]

In Frankland's letter to Wurtz of 1878, this paper was cited and its sentiments generally approved (though it was pointed out that the paper by Kolbe had in fact been a joint publication with himself and that several of the dates are late by one year). Frankland went on to say:

> In thus bringing under your notice the discovery of the essential principle of the law of atom-fixing power, *Sättigungscapacität*, or atomicity of elements, and, in conjunction with Kolbe, the first application of this law to compounds of carbon, I do not forget how much the present developments of this law owe to other chemists, especially to Kekulé and Cannizzaro. Indeed until the latter had placed the atomic weights of the metallic elements on their present consistent basis the satisfactory development of the doctrine was impossible. Whilst therefore I should not be surprised to hear different estimates put upon the value of the respective contributions of different chemists to the present condition of the law of atomicity, I do not think that my papers of 1852 and 1856 can be fairly altogether ignored in a history of this great chemical law.[84]

It was not, however, until the 1880s that recognition of Frankland as a founder of the theory of valency became at all general. Possibly his own *Experimental Researches* (1877) may have contributed to the new turn of events, for the volume robustly sets forth Frankland's own perception of the situation. But there must be some doubt as to the extent of its circulation except among close friends.[85] His written representations to Wurtz produced a positive response[86] and a year later the French chemist specifically declared Frankland as author of 'the idea of saturation-capacity of elementary atoms'.[87] It is probable that more substantial aid was forthcoming from the very man with whom Frankland had quarrelled over priority issues 25 years before, Hermann Kolbe. Spurred on by his vendetta with Kekulé and his fondness for literary controversy, Kolbe had determined to set the historical record straight with a series of articles '*meine Betheiligung an der Entwickelung der theoretischen Chemie*', which duly appeared in his own journal.[88] After receiving news of the project from its author[89] Frankland remarked:

> I fully appreciate the services which Kekulé rendered to chemistry by his subsequent developments of the doctrine [atomicity], but he has no claim whatever to the authorship of the theory or its application to carbon compounds and I am very glad to hear that you are about to put the matter in its real historical light.[90]

Kolbe's response was to send the first two sections of his *Betheiligung* to Frankland, accompanied with a letter sizzling with indignation against Kekulé (an uneducated man, incapable of thinking logically, author of the worst textbook of its kind *etc.*, *etc.*). By contrast Frankland was addressed as 'the principal founder of exact chemistry through your discovery of the saturation capacity of the elements'.[91] Frankland's reply to a further epistle[92] included the following statement:

> I never knew the former [Kekulé] had actually claimed the authorship of atomicity. That was *Frechheit* & *Unverschämtheit* [audacity and effrontery].[93]

This is truly remarkable in the light of Kekulé's clear announcement of 1864 in *Comptes rendus*.[94] Possibly Frankland had never read that paper (though it seems unlikely); perhaps he was suffering from selective amnesia; more probably he was attempting to calm down the infuriated Kolbe by implying that Kekulé's idiosyncrasies were not very well known and therefore not worthy of extended onslaughts.

The publication of Kolbe's version of chemical history must have brought more visibly into the public domain the accomplishments of his friend in the discovery of atomicity. A few years later Kolbe's son-in-law, the organic chemist Ernst von Meyer, brought out a new history of chemistry.[95] An English translation followed two years later.[96] Because Kekulé had responded angrily over the treatment of his part in establishing the theory of valency, Meyer sent a copy to Frankland with a request for his opinion.[97] In a long and cordial response[98] Frankland said:

> I am decidedly of opinion that you have treated the subject as far as possible in a perfectly impartial spirit and have done full justice to the contributions of Kekulé. There is no longer any doubt that Kekulé was anticipated by 5 years in respect of valency and by 6 months in respect of the application of the theory to carbon compounds.

This is almost a repetition of what he had written to Wurtz 12 years previously. [99]

Shortly after this Kekulé died, full of honour in his own country. A tribute was given, not merely by the President of the Royal Society (p.138), but also in a full Memorial Lecture at the Chemical Society. The lecturer, F. R. Japp, made plain his position: 'we must thus admit that it was with Frankland and not with Kekulé that the idea of the valency of elementary substances originated.'[100] He also added, in a private note to Frankland, 'of late years, German chemists have shown themselves more just than formerly, in discussing your claims'.[101] Examples included the revision by Anschütz of Richter's *Organische Chemie* which attributed to Frankland the application of valency 'for the elements of the nitrogen group'[102], an unjust limitation, perhaps, probably derived from Ladenburg's *Entwicklungsgeschichte der Chemie*, but 'an improvement on their former treatment'. It was probably to this time that we must date a somewhat self-satisfied but undated note in the Frankland papers which has, in his own hand, the words "*Frankland dem wir verdanken die Theorie der Valenz*" – Victor Meyer'.[103]

In his own country recognition first came from an unlikely source, the writer Herbert Spencer. To him Frankland wrote in 1880:

> You were the first to assert my claim to the discovery of the law of 'atomicity of elements' & I shall always owe you a deep debt of gratitude for this recognition of what I have always supposed to be my most considerable contribution to chemical science. I thought, however, that it was hardly fair to you to let the responsibility of the assertion rest upon you alone & so I afterwards asserted it for myself first in my Experimental Researches (page 145 – 146) and then in a letter to Sylvester, a copy of which I enclose.
>
> There appeared in August last in the 'Journal für praktische Chemie' (Kolbe's Journal) a very full acknowledgement of my claims to the authorship of the doctrine of atomicity, by a German Chemist Albrecht Rau. I did not know of this article until about a month ago, for the copy which I now send you lay all that time unopened on my study table amongst a heap of papers. I have marked the important passages.
>
> The article is obviously written in the interest of my friend Kolbe, & the writer attempts to show that the doctrine of atomicity was the natural outcome of that of conjugate compounds previously held both by Kolbe & myself; but, unless a doctrine can be said to be derived from its antithesis, I cannot for the life of me see how the doctrine of

conjugate compounds could lead to the law of atomicity for the first asserts that when A combines conjugately with B, the union is without any effect upon the combination of B with other elements (this was the essence of the doctrine of conjugate compounds). Whilst the law of atomicity asserts that B in combining with A thereby inevitably loses a definite amount of its capacity to unite with other elements.

Please return the blue pamphlet when you have done with it.[104]

Let the final word be from T. E. Thorpe, who had taught successively at Manchester, Glasgow, Leeds and London, and was now taking up a new post as Government Chemist. Acknowledging Frankland's congratulations he wrote:

> I am very much obliged to you for your kind letter. In my lectures & in my writings I have uniformly pointed out your relations to the conception of valency as the simplest act of justice to you – I do not know indeed that there should be any question about it. To my thinking your enunciation of the idea is perfectly clear & distinct. It is true it was like bread cast upon the waters but it came back after many days. Whatever your contemporaries here in England may have done or said, you may rest perfectly assured that posterity will set you right. Indeed I have good reason to know – for I may claim to know something of what the younger generation of chemists have been taught – both as a teacher myself and as an examiner. The matter is definitely set at rest without cavil or doubt. [105]

Notes

1 On the development of valency generally see W. G. Palmer, *A history of the concept of valency to 1930*, Cambridge University Press, 1965; and C. A. Russell, *The history of valency*, Leicester University Press, 1971.

2 W. Gregory, *Outlines of chemistry for the use of students*, 2nd ed. London, 1847.

3 *Idem, A handbook of organic chemistry*, London, 1852.

4 W. Gregory to Frankland, 10 January 1852 [RFA, OU mf 01.03.0222].

5 Copy-letter from Frankland to W. Gregory, 12 January 1852 [RFA, OU mf 01.06.0207].

6 L. Playfair to Frankland, 17 March 1860 [RFA, OU mf 01.04.0603].

7 Frankland to L. Playfair, 24 June 1860 [RFA, OU mf 01.04.0596].

8 E. Frankland, *J. Chem. Soc.* 1860, **13**, 177–235.

9 A. S. Couper, *Compt rend.*, 1857, **45**, 230–2; 1858, **46**, 1157–60; each paper appeared in several versions. On Couper see Russell (note 1), pp. 71–80, 133.

10 On these aspects of Kekulé's work see Russell (note 1), pp.61–71.

11 F. A. Kekulé, *Annalen der Chem.*, 1857, **101**, 200–13 (on mercuric fulminate), and 1858, **105**, 279–86 (fulminic acid).

12 F. A. Kekulé, *Annalen der Chem.*, 1857, **104**, 129–50.

13 F. A. Kekulé, *Annalen der Chem.*, 1858, **106**, 129–59.

14 W. Odling, *Proc. Roy. Inst.*, 1855, **2**, 63–6.

15 On Kolbe see A. J. Rocke, *The quiet revolution: Hermann Kolbe and the science of organic chemistry*, University of California Press, Berkeley, Los Angeles and London, 1993; on his work with Frankland on the valency of carbon, especially pp.181–209.

16 Frankland, *Phil. Trans.*, 1856, **147**, 59.

17 J. A. Wanklyn, *Proc. Roy. Soc.*, 1858, **9**, 341–5; Frankland never believed that zinc alkyls could 'break a bond between oxygen and carbon' [*Sketches*, 2nd. ed., p.191], though they can if specifically activated by catalysts as potassium *t*-butoxide.

18 Frankland to J. A. Wanklyn, 8 March 1858 [Deutsches Museum Archive, Munich, Urkundsammlung Nr. 3574].

19 H. Kolbe, *Ausführliches Lehrbuch der organischen Chemie*, vol. 1, Braunschweig, 1854.

20 Frankland, *Sketches*, 2nd. ed., p.192; the correspondence appears now to be lost.

21 H. Kolbe, article on 'Radicals and radical theory' in Liebig's *Handwörterbuch der reinen und angewandten Chemie*, vol. 6, Braunschweig, 1856, pp.802–7.

22 H. Kolbe, *Annalen der Chem.*, 1857, **101**, 256–65.

23 Frankland, *Rep. Brit. Assoc. Adv. Sci. (Trans. of Sections)*, 1855, **25**, 62.

24 Copy-letter from Frankland to A. Wurtz, 25 December 1878 [RFA, OU mf 01.04.0629].

25 Frankland, *Sketches*, 2nd. ed., p.192.

26 *Experimental Researches*, p.148.

27 Rocke, *Quiet revolution* (note 15), pp.184–7.

28 H. Kolbe to Frankland, 2 January 1862, and 5 February 1862 [RFA, OU mf 01.08.0567 and 01.03.0426].

29 Copy-letter from Frankland to H. Kolbe, 9 March 1862 [RFA, OU mf 01.03.0429].

30 H. Kolbe to Frankland, 24 November 1863 [RFA, OU mf 01.04.0073].

31 H. Kolbe to Frankland, October 1863 [RFA, OU mf 01.04.0069].

32 R. Piria, *Annalen der Chem.*, 1856, **100**, 104–6.

33 H. Kolbe to Vieweg, 4 January 1857 [Vieweg Verlag Archiv, Wiesbaden, no.122].

34 Frankland, *Proc. Roy. Inst.*, 1858, **2**, 538–44.

35 Frankland (note 8).

36 C. Foster, *Rep. Brit. Assoc. Adv. Sci.*, 1859, **29**, 1–22 (21); his statement that Frankland had advocated the same views in a Royal Institution lecture of 1858 is likely to have rubbed salt in the wound.

37 H. Kolbe, *Annalen der Chem.*, 1860, **113**, 293–332..

38 Alan Rocke's charitable phrase: Rocke, *Quiet revolution* (note 15), p.186.

39 H. Kolbe, *Das chemische Laboratorium der Universität Marburg und die seit 1859 darin ausgeführten chemischen Untersuchungen*, Braunschweig, 1866, pp.29–36.

40 Frankland to Playfair (note 7).

41 T. E. Thorpe, 'On the rise and development of synthetical chemistry', *Fortnightly Review*, May 1893, p.695.

42 J. Parkinson, *Organic remains of a former world*, vol. 1, London, 1804, p.386; I am indebted to Prof. W. H. Brock for this example.

43 J. L. Gay-Lussac and L. J. Thenard, *Extrait des Mémoires*, p.357 (27 February 1809).

44 E. Frankland, *J. Chem. Soc.*, 1866, **19**, 372–94 (377–8).

45 E. Frankland, *Lecture notes for chemical students*, 2nd ed., London, 1870, vol. 1, p.25.

46 *Ibid.*, p.18.

47 A. W. Williamson, *J. Chem. Soc.*, 1869, **22**, 202, in discussion of a paper by J. Wanklyn on the atomicity of sodium (pp.199–200).

48 *Experimental Researches*, p.145.

49 H. Kolbe, *Lehrbuch der organischen Chemie*, vol. 1, Braunschweig, 1854, p.23.

50 *Experimental Researches*, pp.147–8.

51 F. A. Kekulé, *Lehrbuch der organischen Chemie*, Erlangen, 1861.

52 A. Wurtz, *Histoire des doctrines chimiques, depuis Lavoisier jusqu'à nos jours*, Paris, 1869.

53 A. Wurtz, *A history of chemical theory from the age of Lavoisier to the present day*, trans. H. Watts, London, 1869.

54 Copy-letter from Frankland to E. von Meyer, 28 January 1892 [RFA, OU mf 01.01.0852].

55 Copy-letter from Frankland to Wurtz (note 24).

56 W. A. Miller, *Elements of chemistry, theoretical and practical*, London, vol. 3, 1st ed., 1857 [see Ch. 5 note 69].

57 W. A. Miller, *Rep. Brit. Assoc. Adv. Sci. (Trans. of Sections)*, 1865, **35**, 22–7 (24); Frankland was, however, cited for his work on organic acids.

58 Frankland, *Lecture Notes for chemical students*, London, 1866, p.32.

59 Copy-letter from Frankland to Wurtz (note 24).

60 H. Kolbe to Frankland, 19 May 1881 [Deutsches Museum Archive, Munich, Urkundsammlung Nr. 3572]

61 F. R. Japp, Kekulé Memorial Lecture, *J. Chem. Soc.*, 1898, **73**, 97–138.

62 F. A. Kekulé, 'A history of valency theory' in R. Anschütz, *August Kekulé*, Verlag Chemie, Berlin, 1929, vol. 1, p.554.

63 *Ibid.*, p.565.

64 A. von Baeyer, *Annalen der Chem.*, 1858, **105**, 265–76 (274).

65 W. H. Brock (ed.), *The atomic debates*, Leicester University Press, 1967.

66 Frankland to H. Kolbe, 6 March 1881 [Deutsches Museum Archive, Munich, Urkundsammlung Nr. 3570].

67 Frankland to von Meyer (note 54).

68 These and other contributions of Kekulé have been fully discussed in Palmer (note 1), pp.27–75; and in Russell (note 1), pp. 44–80, 93–105, 108–25, 134, 242–57, *etc.*

69 The matter has been discussed in more detail in C. A. Russell, 'Kekulé and Frankland', in J. H. Wotiz (ed.), *The Kekulé riddle*, Glenview Press, Carbondale, IL, 1993, pp.77–101.

70 H. E. Armstrong, Oration at the Lancastrian Frankland Society, reported in *Lancaster Guardian*, 26 January 1934.

71 F. A. Kekulé to Frankland, 18 January 1863, and 23 December 1863 [RFA, OU mf 01.03. 0081 and 0080].

72 Only once, in a letter to his wife, 1 May 1867, when Frankland mentions meeting Kekulé at an evening in Wurtz's house in Paris (reproduced in *Sketches*, 2nd ed., p.400).

73 The only exception I can find is a passing reference in a paper of 1862 relating to an apparently incorrect view held by Kekulé of a reaction between diethylzinc and triethoxyboron (*Experimental Researches*, pp. 283–4); in Russell (note 69) it is wrongly stated that Kekulé *accepted* Frankland's view of that reaction, the word 'not' being inadvertently omitted.

74 F. A. Kekulé, *Compt. rend.*, 1864, **58**, 510–4.

75 Frankland to H. Kolbe, 19 May 1881 [Deutsches Museum Archive, Munich, Urkundsammlung Nr. 3572].

76 Japp (note 61), pp.118–19.

77 Anschütz (note 62), p.39.

78 *Ibid.*, p.80.

79 F. A. Kekulé, *Berichte*, 1890, **23**, 1304–5.

80 J. Lister, Presidential Address to the Royal Society, 30 November 1896 [Royal College of Surgeons of England, London, Lister papers no. 70].

81 Frankland, undated note [RFA, OU mf 01.02.0505].

82 E. Erlenmeyer, *Ueber die Aufgabe des chemischen Unterrichts*, Munich, 1871, p. 26.

83 A. Naumann, *Handbuch der allgemeinen und physicalischen Chemie*, Heidelberg, 1877, pp.73–4.

84 Copy-letter Frankland to Wurtz (note 24).

85 Letters of acknowledgment for the volume exist from his absent son Fred, and also R. Biedermann, E. Brembridge, G. B. Buckton, J. Caddy, E. Dubois-Raymond, A. F. M. Duppa, D. G. Gilman, J. G. Greenwood, W. C. Henry, T. H. Huxley, J. Johnson, H. Kolbe, H. Kopp, J. Lubbock, J. C. Morton, J. H. Nicholson, W. Odling, L. Playfair, F. G. T. Reay, H. E. Roscoe, Lord Salisbury, W. L. Savory, C. Schorlemmer, B. Silliman, R. A. Smith, W. Spottiswoode, J. Stefan, H. R. Tedder, J. Tyndall, H. Weber and J. Young [RFA].

86 A. Wurtz to Frankland, 1 January 1879 [RFA, OU mf 01.04.0593].

87 A. Wurtz, letter of 29 December 1879 to the German Chemical Society, *Berichte*, 1880, **13**, 6–7.

88 H. Kolbe, *J. prakt. Chem.*, 1881, **23** (n.s.), 305–23, 353–79, 497–517; **24** (n.s.), 374–425.

89 H. Kolbe to Frankland, 11 February 1881 [RFA, OU mf 01.02.1447].

90 Frankland to H. Kolbe, 6 March 1881 [Deutsches Museum Archive, Munich, Urkundsammlung Nr. 3570].

91 H. Kolbe to Frankland, 20 March 1881 [RFA, OU mf 01.02.1439].

92 H. Kolbe to Frankland, 23 March [1881] [RFA, OU mf 01.02.1442].

93 Frankland to H. Kolbe, 29 March 1881[Deutsches Museum Archive, Munich, Urkundsammlung Nr. 3571].

94 Kekulé (note 74).

95 E. von Meyer, *Geschichte der Chemie, von der ältesten Zeiten bis zur Gegenwart; zugleich Einführung in das Studium der Chemie*, Leipzig, 1889.

96 *Idem, History of chemistry from the earliest times to the present day, being also an introduction to the study of the science*, trans. G. McGowan, London, 1891.

97 E. von Meyer to Frankland, 18 January 1892 [RFA, OU mf 01.07.0486].

98 Frankland to von Meyer (note 54).

99 Frankland to Wurtz (note 24).

100 Japp (note 61), 118.

101 F. R. Japp to Frankland, 3 March 1898 [RFA, OU mf 01.02.0675].

102 The error was still to be found in R. Anschütz and G. Schroeter (eds.), Richter's *Chemie der Kohlenstoffverbindungen*, Cohen, Bonn, 11th ed., 1909, vol. 1, p.24 (a book dedicated to August Kekulé), and in even later English and American editions.

103 Undated note [RFA, OU mf 01.01.0160].

104 Copy-letter from Frankland to H. Spencer, 12 April 1880 [RFA, OU mf 01.04.0607]. The letter to Sylvester (presumably the mathematician J. J. Sylvester) does not seem to have survived; the article referred to is by A. Rau in *J. prakt. Chem.*, 1879, **20**, 209–42. Frankland's son Fred, writing about his receipt of *Experimental Researches* remarks, somewhat enigmatically, 'It is too bad of Sylvester to rob you of the credit of *atomicity*. It is just like his wrong-headed blundering style' : Fred Frankland to Frankland, 27 September 1878 [RFA, OU mf 01.02.1064].

105 Letter from T. E. Thorpe to Frankland, 19 April 1894 [RFA, OU mf 01.01.0150].

CHAPTER 7

Manchester: 'The educational and commercial utility of chemistry'

1 Owens College

It is not an exaggeration to say that the Town Hall was seething with excitement. The burgeoning industrial city of Manchester ('Cottonopolis' to its detractors and admirers alike) was to take a great step forwards into the second half of the nineteenth century. On this night, 12 March 1851, the town of John Dalton, James Prescott Joule and many other scientific worthies, was to acquire its own university (or something very near one) giving special preference to applicants born within two miles of the Parliamentary Borough.[1] Manchester was to be the first provincial city in England to break the monopoly of Oxford and Cambridge, with a secular university having real professors and open access without reference to birth, rank or social status. Now the moment had come for its public inauguration, under the chairmanship of the city's chief citizen, the mayor. Unfortunately the Principal, a cleric of nervous disposition, was not well enough to attend so instead the large and enthusiastic audience was treated to an address by one of his staff (who rejoiced in the comprehensive title of 'Professor of the Language and Literature of Greece and Rome, and of Ancient and Modern History'). He was followed by the Professor of Mathematics.

At further meetings these learned discourses were followed by manifestos from the professors of natural history and chemistry. The final address was a spirited defence of his subject by the new incumbent of the chemical chair, who spoke on 'The educational and commercial utility of chemistry'.[2] His name was Edward Frankland.

As the speaker rose in his place the audience must surely have listened with more than usual attention. Not only was he very young – a mere 26 years – but was he not the first Englishman to receive a PhD from the University of Marburg? Someone might even have heard that he had discovered a whole new area of chemistry, which he called *organometallic*, though few would have had the faintest idea as to what exactly that was. But to most Mancunians chemistry was a matter of civic pride; they had turned out in their thousands to mourn the passing of John Dalton and were well aware of the international fame that he had brought to their city. In their Manchester Literary and Philosophical Society, the oldest 'Lit. & Phil.' in the land, chemical science had been prominent on the agenda. Thomas Henry, Secretary (1781) and President (1807) of the Society had been the head of a firm of chemical manufacturers bearing the family name and most famed for their magnesia, a sovereign remedy for indigestion. His son William was the discoverer of 'Henry's Law', connecting gas solubility and pressure, and other members of the clan also made their mark in chemistry. And there was a dawning recognition of the subject's importance to the new trades of calico-printing, dyeing, gas-making and so on. So Frankland started his lecture, and his new career, with the great advantage of relevance.

The college had been established from an endowment in the will of a Manchester merchant, John Owens (1790–1846) who, having neither wife nor children, left the whole of his residual estate, about £96 000, to a new Trust. The object was the establishment of 'an Institution for providing or aiding the means of instructing and improving young persons of the male sex (and being of an age not less than fourteen years) in such branches of learning and science as are now and may be hereafter usually taught in the English Universities'. Reflecting the nonconformist and liberal traditions of early Victorian Manchester there was to be no religious test for staff or students, and the teaching was to be devoid of all theological and religious references that might be deemed 'offensive'. Nevertheless some religious instruction was provided, which caused the (Unitarian) New College later to affiliate with University College, London, rather than with their near neighbour.

A search for suitable staff began early in 1850, and on 22 October

1850 the Trustees appointed the first Principal: the Rev. A. J. Scott, Professor of English Literature at University College, London.[3] Meanwhile Edward Frankland the young Professor of Chemistry at Putney College, London, had heard of a new opportunity for increasing his stipend to a level that would make marriage possible at last. A chair at Manchester would enable him to continue his research and teaching activities and was, after all, much nearer to his old home. As early as March 1850 he wrote to many of his chemical friends for a selection of references, all of which he duly printed as a small booklet.[4] Their fulsome praise denotes something of his international reputation, even allowing for the exaggeration customary on such occasions. Their efforts were successful and on 2 January 1851 Edward Frankland was appointed, marrying Sophie Fick in London just five weeks later. He was to receive a salary of £150, together with some income from students' fees. Manchester's first Professor of Chemistry was to take up office on 1 March.[5]

Other colleagues also made their appearance about this time. The above-mentioned Professor of Language and Literature of Greece *etc.* was a schoolmaster, Joseph Gouge Greenwood (1821–94); he was later to succeed Scott as Principal and eventually became Vice Chancellor of Manchester University.[6] The chair of Mathematics and Physics was filled by Archibald Sandeman (1822–93), many years later described by Frankland as 'a Cambridge wrangler, who, though an able mathematician, probably never made an experiment in his life'.[7] The Professor of Natural History was the Yorkshire-born surgeon and naturalist William Crawford Williamson (1816–95).[8] Some years after his own appointment Frankland was glad to welcome his former school-fellow James Helme Nicholson (1824–1901) who, in November 1853, was appointed clerk and librarian, and eventually became Registrar.[9]

The first home of the new Owens College was a house in Quay Street, rented at £200 per annum from Richard Cobden, MP, the Anti-Corn Law League campaigner. For Frankland there was the immediate problem of laboratory space. An appeal was launched, bringing in £9550.10s. mainly for that purpose. Temporary accommodation was made available to him in St John's Street as the new rooms would not be ready until autumn.[10] There were the usual problems of things getting lost in the move[11] but by November he was just about to recommence work on organometallic chemistry in his new laboratory.[12] Designed in consultation with himself it was one of the best chemistry suites in Britain, Frankland claiming its superiority over any in Britain in convenience for elementary and advanced students in lighting, warming and ventilation.[13] It had a lecture room for 150 students; a student laboratory measuring 51' × 21' with 42 places; cloakrooms, balance-rooms and store-rooms; two private laboratories, and a roof platform for 'experiments with gases'.[14] His experience in Giessen, Putney and Marburg was not wasted when it came to laboratory design.

Even before this desirable accommodation was available teaching had to commence. Frankland's pedagogic activities over the next few years must be a matter of considerable interest.

2 Education: professional academic

The College's first session, from March to July, opened with a class of 25 students. It was only a matter of days after the ceremonial opening that the chemical class found themselves in the lecture room of the new professor: tall and spare, full of intimidating nervous energy, very young but with a reassuring trace of familiar Lancashire accent. His teaching, had they but known it, was to inaugurate a new age of science education in England. He was one of the first of a new breed of professional academic scientists.

Many legendary teachers of science have their exploits viewed by posterity through the lenses of modern educational practice. Occasionally contemporary accounts survive but more often than not the imagination has to work in overdrive to form some impression as to what was actually taught, let alone the rationale or the manner of instruction. With Frankland we are extremely fortunate in that numerous sources survive from his own time. First, there are examination papers like the printed example[15] reproduced as Fig. 7.1.

The most obvious feature is the breadth of coverage and variety of emphasis, remarkable at any time in an introductory course and particularly at a time when chemical theory was far from settled. General principles abound (questions 1, 2, 3, 11, 14), but not at the expense of quite detailed knowledge (6, 9, 12, 13). As might have been expected from Frankland's own experience, analysis features

Owens College,

1851.

ELEMENTARY CHEMISTRY.

---

1. What do you mean by the term chemical affinity, and what influence have the forces of heat, light, electricity, and cohesion upon chemical combination?

2. By what three principal laws is the formation of chemical compounds governed?

3. Explain the meaning of the terms " equivalent number," " combining proportion," and " atomic weight."

4. What are the two principal views which are held by chemists relative to the constitution of salts?

5. What would be the equivalent number of a base contained in a neutral sulphate, consisting of 37.3 per cent of sulphuric acid and 62.7 per cent of the base?

6. Describe in equations the changes and decompositions which take place in manufacturing oil of vitriol from sulphur and nitrate of potash.

7. Suppose a manufacturer of nitric acid is offered two cargoes at the same price per ton, one of nitrate of soda and one of nitrate of potash, which should he take for the purposes of his manufacture, explaining your reasons for the selection, from your knowledge of equivalents?

8. Divide the metalloids into groups, and state your reasons for associating the various members of these groups.

9. What is the composition of bleaching powder? Explain in chemical language the processes of bleaching and of calico printing in the discharge style.

10. How would you endeavour to prevent infection in a sick room? Give your reasons for the adoption of the process you would employ.

11. Into how many groups or families can the metals be divided, and by what general re-agents can these groups be separated from each other?

12. Suppose a liquid suspected to contain arsenic were given you for analysis, how would you proceed to detect the poisonous ingredient?

13. How could you prove the presence or absence of potash, soda, lime, and iron, in a sample of manure?

14. Explain the general principles upon which the extraction of the metals from their ores depends.

E. FRANKLAND.

PROFESSOR.

Fig. 7.1 Examination paper, Owen's College, 1851. (R. Frankland Archives)

in several questions (11, 12, 13). It is also possible to discern a feel for controversy (4) and a recurrent urge to emphasise the applied or 'useful' aspects of the science (6, 7, 9, 10, 14).

It is fairly clear that the great emphasis in descriptive chemistry is on the common non-metals ('metalloids') though metals are not entirely neglected. This is repeated in a manuscript version of another paper[16] for Owens College (simply entitled 'Second examination'), the themes for which were

1. Physical and chemical properties of oxygen; equivalent number and specific gravity
2. Preparation of oxygen (with diagrams)
3. Preparation of hydrogen (with diagrams)
4. Physical and chemical properties of hydrogen; meaning of combustion
5. Combination of oxygen and hydrogen; states of water
6. Natural occurrence and extraction of nitrogen
7. Properties of nitrogen and chief compounds with H and O
8. Preparation and properties of 'protoxide of nitrogen', nitric acid and ammonia

Yet another draft ('Third examination: elementary chemistry') shows the same trend, with topics for examination being:

1. Carbon allotropes
2. Preparation of compounds of C and O, and rôle of one in animal and vegetable life
3. Sulphur: natural occurrence and manufacture
4. Preparation and properties of principal compounds of S and O
5. Preparation and properties of phosphorus and principal compounds with O and H
6. Use of silicates in agriculture
7. Detection of arsenic and distinction between arsine and stibine

Other MS papers exist in the same style and almost certainly from the same period though without a definitive inscription 'Owens College'. They include a 'First examination: Michaelmas Term' with the instruction 'Illustrate the answer to these questions by diagrams where necessary' and a major focus on silicon and the halogens,[17] together with a 'Second examination' of a more theoretical character.[18]

However it is in Frankland's own lecture notes that his intentions are most fully displayed. A course on 'Technological chemistry', delivered in 1853/4, discloses his great emphasis on processes of *combustion*, both for heat and for light.[19] The topics covered by these 17 lectures were as follows:

1. [missing]
2. Technology; heat; fuels
3. Specific thermal effects
4. Pyrometric thermal effect
5. Wood; peat; lignite
6. Coal: varieties
7. Coal: absolute thermal effect
8. Coal: thermal effects; charcoal
9. Coke; patent fuels; gaseous fuels
10. Light: gas lighting; illuminating gas; luminosity
11. Coal gas manufacture: retorts, condensers
12. Illuminating gas
13. Hydrocarbon process; distillation
14. Gas consumption
15. Estimation of gaseous constituents
16. Coal as a raw material; SO_2 from iron pyrites
17. Lead chamber process for H_2SO_4

It did not need to be like that. Alternative courses could have been given on chemistry and food, paint and colours, textile processing and many other topics. But – in these lectures at least – fuel combustion was the dominant theme. It is not unreasonable to infer the strong influence of the Admiralty coal investigations in London and of Bunsen's work on gas analysis in Marburg. Moreover, as will soon be seen, this subject of combustion was a dominant one in the consultancy work that Frankland was already beginning to develop. This course of lectures on technical chemistry was, he said, 'the first, so far as I am aware, ever delivered in this country'.[20]

Another surviving document is a notebook[21] containing, on the right hand pages, a course of eighteen lectures from the Manchester period (numbered 82 to 99). Let us call this Course A. Its contents were as follows:

82. Copper and sulphur; alloys; cadmium; bismuth
83. Lead, its oxides, alloys and sulphide
84. Palladium, rhodium, osmium and ruthenium
85. 'Salts of metals of 6th Group': halides of silver and mercury
86. Salts of copper and lead
87. 'Metals of 5th Group': platinum, iridium
88. Gold; tin
89. Tin
90. Molybdenum; tellurium; tungsten; vanadium
91. 'Salts of metals of 5th Group': Pt, Ir, Au, Sn
92. Organic chemistry; its nature; ultimate organic analysis
93. Empirical formulae from analysis.
94. Compound radicals
[95]. Homologous series
96. Aldehydes, acids, ketones, 'olefins', ammonium salts, amides, nitriles
97. Bases, ether salts
98. Aromatic analogues
99. Naphthalene and doctrine of substitution

At Lecture 92 the subject changes from inorganic to organic chemistry, and at that point the left hand pages (hitherto almost entirely blank) are filled with a different course of organic lectures, numbered 15 to 27 (Course B). These were obviously added after the Course A since after lecture 99 the few blank pages at the end of the book are occupied *on both sides* by the conclusion of Course B. Presumably A was given first, possibly for several years, and then was either replaced by Course B, or else amalgamated with it. Again it may be helpful to indicate the contents:

Course B
[14]. Object of organic chemistry; proximate organic analysis
15. Rational constitution of organic compounds
16. Homologous series: *I. Radical type*: electropositive radicals
17. Neutral radicals (ketones, haloid ethers); electronegative radicals
18: *II. Water type*: electropositive (alcohols, ethers); neutral
19. Electronegative (hydrated and anhydrous acids); *III. Hyponitrous acid type*
20. Electropositive (amines + P, As and Sb analogues)
21. Neutral (amides etc.); electronegative (imides, nitriles)

22. *IV. Nitric acid type*: electropositive (R_4N bases); neutral (R_4N salts); electronegative (Asand Sb acids)
23. [missing]
24. [missing]
25. Transformation of organic compounds by oxygen and halogens
26. Transformation by nitric acid and phosphorus [pent]oxide
27. Transformations by halogen acids, 'NO_3', reducing agents, water, alkalis, ammonia, PCl_5, ferments

Not a single lecture is dated but, by a most fortunate coincidence, a record[22] of Frankland's Manchester lectures by one of his students (Francis M. Spence) has recently come to light and these turn out to be the organic Course B. This 'worm's-eye view' is of great interest in complementing the notes of the Professor himself. It reveals, among many other things, a student's *ennui*, with doodles, sketches and other inconsequential jottings interspersed with frantic and highly detailed transcriptions as consciousness returned with a jolt. On the whole, however, it is not hard to discern the lecturer's argument and even recapture some of his distinctive phrases and (in one case) a most unusual but deliberate mis-spelling: *olifenes*. Since each lecture is dated by Spence it becomes possible to identify the provenance of these lecture notes of Frankland as Manchester, 1854/5. It appears from Spence's notes that syllabus B was followed fairly closely but that material from A was incorporated from time to time.

Course A was obviously the final stage of a journey through the whole of chemistry. Its inorganic part is remarkable chiefly for the thoroughness of treatment, awareness of recent work and constant reference to applications. However it is in the organic lectures that Frankland delivered at Manchester that the chief interest appears to lie. In addition to the rapid survey in Course A there was the much more detailed and theoretically inclined Course B, and a further course entitled 'General principles of organic chemistry' which appears to date from 1857.[23]

The courses have much in common. All used the old atomic weights, all were aware of recent developments (theories of Laurent and Gerhardt, Hofmann's work on amines *etc.*), and all maintained what was to become the traditional view of Wöhler's famous synthesis of urea in 1828. This is the view that the synthesis eliminated once for all any vital forces from organic chemistry and thus dissolved the barrier separating it from the rest of chemical science. More recent scholarship has thrown considerable doubt on such claims and insisted that vitalism survived for a long time, only gradually disappearing as more and more laboratory syntheses were accomplished.[24] The myth of the overthrow of vitalism in 1828 has been attributed to the obituary[25] of Wöhler by Hofmann in 1882. However, revisionist authors (including the present one) may find some embarrassment in the following notes by Frankland in early 1855:

> Vital Force thought to form alone organic bodies. It was even thought that organic compounds could not be artificially made until organs of plants and animals were artificially produced. Hence this afforded convenient line of distinction between inorganic and organic compounds. The artificial formation of Urea by Wöhler in 1828 therefore revolutionised our ideas & destroyed this natural boundary between inorganic and organic bodies [Frankland's underlining].[26]

A few years later almost identical words are used ('The artificial formation of Urea by Wöhler in 1828 however broke down this distinction between inorganic and organic compounds'), but new support is found from the researches of Kolbe and Berthelot.[27] Admittedly Frankland does not say in so many words that 'vitalism has been eliminated from organic chemistry' but the sense is surely there. It is entirely consistent with a lecture he was shortly to give at the Royal Institution 'on the production of organic bodies without the agency of vitality', claiming that the syntheses by Wöhler and Kolbe 'completely broke down the barrier between the so-called "organic" and "in-organic" bodies'.[28] Thus by the mid-1850s at the latest the traditional view of the urea synthesis was being strongly advocated by at least one leading chemist in Britain.

Frankland was of course writing and lecturing as a *chemist*. He was not a biologist and in fact seems to have left room for vital forces to operate within physiology. Thus he wrote 'the chemist . . . studies the changes which chem. compounds suffer by action of chemical reagents, – whilst the physiologist, using chemical results, studies the metamorphoses which such bodies suffer under influence of Vital Force'.[29] Neither was he a philosopher. It is extremely unlikely that he would have been bothered by the ambiguities so clearly exposed by John Brooke; Frankland was a practical chemist for whom such subtleties were largely irrelevant. Alan Rocke has observed that the Wöhler legend 'was created in the immediate aftermath of the event itself' and as early as 1843 was enshrined in Kopp's influential *Geschichte der Chemie*. 'By the early 1850s, if not before, the myth of a definitive refutation had become ensconced in the German textbook literature'.[30] Here, then is clear evidence that Frankland stood within that German tradition, one that harks right back to Kopp himself. It is hardly surprising in the light of his experience at Marburg and Giessen, and especially in view of his own collaboration in organic syntheses with Hermann Kolbe (p. 28).

There may be further reasons for this promotion of the Wöhler legend in Manchester. Though logically unnecessary there often was in the nineteenth century a connection between an anti-vitalist stand in chemistry and the adoption of a generally materialist world view. Conceivably Frankland had begun to respond to a hitherto unconscious urge to espouse the philosophy of materialism, in which case the overthrow of vitalism in any context would only be good news. That may or may not have been the case in the 1850s. What is much more certain is that Frankland recognised the pedagogic value of a fully reductionist approach to organic chemistry where all explanations in terms of vitalist or other occult forces must be rigorously excluded. He had to undermine all arguments that could prevent students from analysing chemical phenomena in strictly chemical terms. Vitalist explanations might seem to 'save the phenomena' now and again, but they would never lead to the fuller understanding of constitution and reactivity that chemists so fervently sought. As those who have taught organic chemistry ever since will have to admit, in this respect he was entirely right.

One other aspect of Frankland's teaching of organic chemistry at Manchester may be discerned in all the available documents, though developing with the passage of time. This reflects a feature of his thinking already met in connection with his early research: a recognition of the value of a type classification but combined with a dedication to the electrochemical ideas of Berzelius. Again he was in a minority but this (as usual) troubled him not at all. He took four main types, based on what he considered to be H, HO, NO_3 and NO_5; within each were examples of electropositive, electronegative and neutral substances. The neutral members were regarded as derived from electropositive and electronegative parts; thus ketones were [in our language] combinations of alkyl and acyl radicals,

Table 7.1. *Frankland's scheme of organic classification*

Types	Electropositive	Neutral	Electronegative
Hydrogen (or *Radical*)	radicals [as Me] hydrides of radicals [as Me.H] double radicals [as Me.Et] 'olifenes'	ketones haloid ethers [alkyl halides]	radicals [acyl] hydrides [aldehydes] haloids [acyl halides]
Water	alcohols	compound ethers or ethereal salts [esters]	hydrated acids [carboxylic acids]
	ethers		anhydrous acids [acid anhydrides]
Hyponitrous acid	cmpd. N ammonias [amines]	imides	amides
	cmpd. P ammonias	nitriles	cmpd. ammonias [with subst. alkyl gp.]
	cmpd. As ammonias cmpd. Sb ammonias protoxide bases [AsR$_2$O]		haloid cmpds. of protoxide bases
Nitric acid	caustic N bases	haloid salts of bases	cmpd. As acids
	caustic P bases	haloid salts of binoxide bases	cmpd. Sb acids
	caustic As bases caustic Sb bases binoxide of As bases binoxide of Sb bases		

haloid ethers combinations of alkyl radicals and halogen, esters as combinations of an alkyl and acyl radical with oxygen. This gave him an explanatory scheme within which he fitted all his teaching about homologous series. It accomplished little else and was to be discarded completely after the advent of structure theory. But that was well after his years at Manchester. Table 7.1 summarises his scheme in its final form.

On Frankland's teaching strategy it remains to comment briefly on the part given to laboratory work. Not much evidence from Frankland himself has survived and the best source is probably Frank Spence, though the material is sparse enough. From his notebooks we learn of a system of Group Separation of the metals in which the precipitating reagents were successively:

$$HCl$$
$$H_2S$$
$$NH_4Cl + NH_4OH$$
$$NH_4HS$$
$$NaOH$$

which has a remarkably familiar ring to it. Other specific tests included the famous one involving colorations of a borax bead. Gas analysis was also performed by students and gas densities were determined by weighing and by diffusion. That much is certain from Spence's notebooks. An open testimonial for a Mr Bertrand reports that over a two year period he had 'made himself practically acquainted with the qualitative and quantitative analysis of inorganic bodies as well as many of the methods used for the commercial valuation of chemical substances used in the arts and manufactures'.[31]

It is improbable that organic preparations were neglected though decisive evidence is not yet available. Attendance at practical classes was a pre-requisite to admission to the course on technical chemistry.[32] The large student laboratory can hardly have been left idle for long. Also, as in the German laboratories, visitors were welcomed for short periods of intensive practical training. An example was J. F. W. Johnston who, in the spring of 1855, spent two months in Frankland's laboratory being instructed in the techniques of gas analysis.[33] A similar period was spent by E. K. Muspratt who sought to learn metal assaying by the dry method. However his experience was less happy for he claimed many years later that 'Frankland could teach me nothing, so I got a text-book on assaying, and worked by myself in the basement, where the assay apparatus were placed'.[34]

3 Commerce: research consultant

From what is known of the teaching strategy of Edward Frankland at Owens College one might have expected a gradual build-up of student numbers, an accelerating programme of academic research, a rising reputation of both department and college, and for Frankland himself a growing sense of confidence and well-being. In fact none of these things was true, and the picture that emerges is rather of a bright young chemist struggling for survival and battling against disillusionment. There were two reasons for this sad state of affairs. The first lay in the nature of the teaching task, allied to the poor abilities of many students and the low expectations of the community. Hard work was simply not enough and we now have evidence to suggest that, for all the meritorious content of the lectures, the manner of their delivery may well have been less than inspiring. Unquestionably student numbers declined. Secondly, he received a relatively low salary of £150 p.a. (which took into account the clause in his contract that Frankland was 'not required to devote his time and attention exclusively to the business of the Professorship' though he was required to provide laboratory instruction at least four days a week in term time). These two disabilities remained for all his time at Manchester and, towards the end of that period, became even more visible, with seemingly disastrous consequences. Possibly sensing difficult times ahead Frankland did not wait long to look outside teaching for stimulus and support. Within days of his appointment he had begun to exploit 'the commercial utility of chemistry' and within a few years had established himself as one of Britain's leading academic consultants to industry. Of the hundreds of individual documents surviving from his Manchester period many deal with the subjects of industrial consultancy, particularly several Letter Books each with several hundred copy-letters between their battered covers.

Before examining Frankland's strategy in detail a single topic for consultancy will be sketched as a not untypical example of the whole.

3.1 The case of 'hydrocarbon gas': analysis of coal and coal-gas

One of the first industrial enterprises to catch Frankland's eye in Manchester was concerned with improvements to gas-lighting. Coal-gas by itself gave poor illumination before the invention of the gas-mantle. It was early realised that ethylene was the only effective illuminant in coal-gas, though present in very small quantities. It was also believed that the effect 'is in direct proportion to the quantity of carbon contained in a given volume' of the gas.[35] Many devices were invented to achieve this, such as the 'naphthalising' process of George Lowe,[36] whereby naphtha was introduced into the raw coal-gas (a process employed for some time in Buckingham Palace) or the use in Paris of water-gas flames impinging on a platinum gauze.[37] At exactly this time Frankland wrote to Liebig that he had recently been investigating a new process of gas-making, for 'a company of commercial men'.[38] This was the Hydrocarbon Gas Company, set up to exploit a patent of Stephen White whereby water-gas was passed through the bed of hot coal being carbonised, to the great enhancement of illuminating power. The theory of the process was simple. During the production of ordinary coal-gas many potential illuminants (including ethylene) were destroyed by pyrolysis on the retort walls. This process could be reduced by rapidly sweeping out the newly-formed gases, and also by diluting them with another combustible gas. White's choice was water-gas, prepared in an adjacent retort by passing steam over heated coke or charcoal. This had the additional advantage of 'dissolving' other less volatile carbon-rich compounds from the tar so increasing further the luminosity of the flame. The retorts were of conventional design, though having a horizontal tray for most of their length to increase their mechanical strength. It was claimed that the process economised on materials, labour and wear and tear, as well as producing a gas with much higher illuminating power.

In the summer of 1851, using different types of coals, Frankland put White's process to the test. His experiments were performed at a gas-works attached to the mill of George Clarke & Co. at Ancoats. His conclusions were unambiguous, above all for the famous Boghead Cannel from Scotland, 'the richest gas material known'[39], yet useless for domestic lighting on account of smoke and smell. With the water-gas process there was spectacular improvement, well over 200% in some cases.[40] Frankland became an ardent advocate of the 'Hydrocarbon Process' (as it was curiously known). He strongly recommended it to enquirers,[41] lectured on White's 'hydrocarbon gas' at both the Manchester Lit. & Phil.[42] and the Royal Institution, and reported on it to the Chemical Society.[43] He was, however, severely taken to task by the *Journal of Gas Lighting*. After the journal accepted his paper on Hydrocarbon Gas and praised him as a 'talented author' it launched a vitriolic attack on the whole process and accused Frankland of 'long-winded reports on bungling experiments', suggesting that he would be better employed remedying the numerous failures that had come to light.[44]

Meanwhile Andrew Fyfe, Professor of Chemistry at Aberdeen, had undertaken his own analyses of Boghead Cannel and concluded, not only that Frankland's analyses were wrong, but that White's Process led to a *loss* of illuminating power.[45] By the time his results were published Frankland was away on a continental holiday, but when Fyfe returned to the subject[46] in March 1854 they could no longer be ignored and Frankland also resumed his study, comparing the illuminating powers of gases produced by the old and new processes. This time he worked at a gas-works at Westhoughton; his results on Boghead coal confirmed his previous experiments, and he determined to reply to Fyfe in the *Journal of Gas Lighting*.[47] Frankland went on to issue a printed *Report* on White's resin gas retorts installed at the Ancoats mill.[48]

Despite the attack by the *Journal of Gas Lighting* one cannot but be impressed by the care and thoroughness of Frankland's experiments which were, as a distinguished contemporary remarked, 'conducted with scrupulous accuracy'[49]. All gases were collected over mercury, ethylene being estimated by absorption in fuming sulphuric acid, carbon dioxide with potash, and the rest by combustion with excess oxygen.[50] Oxygen itself was estimated by Liebig's method using alkaline pyrogallol.[51] Indeed a partnership to undertake gas analyses between Frankland and John Leigh (the director of Manchester gas-works) had been advertised in the *Journal of Gas Lighting*, though it does not seem to have been carried into effect.[52] The lessons in analytical chemistry learnt in the Marburg laboratory were not wasted, and it was Bunsen's photometer that was often used to determine illuminating power.

It is sometimes asserted that water-gas admixed with coal-gas did not appear in Britain till the 1890s[53], but it was certainly being used in Manchester 40 years earlier. It even attracted a visit from that pioneer in coal-gas manufacture, Samuel Clegg.[54] However it did not succeed on anything like a large scale for a variety of reasons. The water-gas retorts have to be of iron, not earthenware, and tend to crack under operating conditions (which involve alternate heating and cooling)[55]; the process is more likely to suffer from carelessness by workmen[56]; much tar is consumed, which mattered little at first but the days soon passed when it could be asserted that 'tar is of very little value'. Frankland's work for the Hydrocarbon Gas Co. was a landmark in gas analysis, being at the time the only thorough investigation into the chemical composition of the gases evolved from the different kinds of gas-coal.[57]

Frankland's encounter with the Hydrocarbon Process led to investigations of the coals themselves, and of their suitability for gas-production. Each such analysis brought a fee of 20gn.[58] He advised that resin was better for domestic and small works, but cannel for large works (as much cheaper).[59] On the other hand Trinidad bitumen was quite unfit for gas manufacture, Brunswick coal being much to be preferred.[60] Not all coal samples were examined for their gas-making potential, and complete analyses were performed on a wide range of coals from a specimen containing 12% iron pyrites[61] to high quality anthracites.[62] No coal however can have haunted his dreams as persistently as that product of the Scottish Lowlands, Torbanite, otherwise and ominously known as Boghead Cannel.

The saga began in 1850 when the owners of the Torbanehill estate near Bathgate, Mr and Mrs William Gillespie, entered into a contract with a firm of Falkirk iron masters, James Russell & Co., giving them rights to certain minerals including coal (with a royalty at 6d per ton) but excluding all others.[63] Realising the unexpected value of one of the minerals, known appropriately as Torbanite, *alias* Boghead Cannel, the owners sought to exclude it from their agreement with Russells on the grounds that *it was not truly coal* but a bituminous schist. Frankland was asked by Russells to adjudicate, and in April 1852 wrote to them with the opinion that Boghead Cannel was indeed a true coal and not a bituminous shale (largely on the basis of a very small percentage of sulphur and of the action of organic solvents).[64] A subsequent letter revealed that the Gillespies had tried to get Liebig to say it was not coal.[65] The latter immediately wrote to Frankland for more information[66] and was quickly advised by Hofmann to settle a fee first.[67] Litigation was inevitable and the case was heard in Edinburgh from 29 July to 4 August 1853. In addition to Frankland, witnesses for the defence (Russells) included Graham, Stenhouse and Hofmann. Judgement was given for the defendants.

Even as these proceedings were in train Frankland was preparing for further litigation in a related area. Within a year of the previous case two Manchester entrepreneurs, Edward Binney and James

Young, had set up their works at Bathgate in order to extract
paraffin by low temperature pyrolysis of coal from that estate.
There then arose a legal dispute between Young and Stephen
White, and a further case was heard in the Queen's Bench Division
in the next year. Young regarded the manufacture of 'Hydrocarbon
gas' as an infringement of his patent for obtaining paraffin from
coal. Frankland, acting for his 'Hydrocarbon' friends, was to argue
that people had known paraffin could be obtained from coals long
before Young took out his patent. There were plenty of precedents,
most notably that of Reichenbach[68] who in 1833 had reported that
paraffin was extractable from 'certain oleaginous matters termed tar
derived from the destructive distillation of wood and its derivatives
lignite and coal', and Frankland had used Reichenbach's methods
to obtain paraffin from Salford Gas Works tar.[69] Regarding the
attendance of William Brande as a witness Frankland wrote: 'I have
no doubt that a skilful but <u>mild and gentlemanly</u> cross-examination
will make him one of our best witnesses as was the case in the
Boghead action in which he very fairly and candidly admitted
everything we wanted'.[70] This time, however, Frankland was 'alone
on one side of a case with nearly every chemist in England and
Scotland against me on the other'.[71] Not surprisingly he lost, but
not before acquiring a lifelong taste for the profitable business of
being an expert witness in law.

3.2 *Forming an industrial strategy*

From the abundant data available it is difficult to believe that
Frankland did not consciously adopt a strategy designed to
maximise both his income and his reputation in the local community
and the wider world of industry. Whether deliberate or not his
strategy was successful and long outlasted his few years in
Manchester. Several elements may be identified.

First, he sought to *establish a web of local contacts*. Almost as soon
as he had arrived in Manchester Frankland was able to meet the
Mayor,[72] and shortly afterwards that dignitary invited him to serve
on the books sub-committee of the Free Library.[73] He joined the
Manchester Literary and Philosophical Society on 29 April, 1851,
the same day as Peter Spence and his academic colleagues
Sandeman and Williamson.[74] This was still an important forum for
scientific discussion. He served on their Council in 1856/7. The

Manchester Photographic Society included many local chemists and found a ready recruit in the new chemical professor; he became Vice President in 1855.[75] Other local institutions with which he was associated included the National Public School Association (founded in 1847 to advocate 'a general system of secular education')[76] and the Manchester Working Men's College (probably the one at Ancoats).[77]

His institutional connections were of course supplemented by contacts with important individuals. In 1853 he obtained signatures from all three Manchester Fellows of the Royal Society (Joule, Fairbairn and Schunck)[78] in support of his own candidature for FRS (which he was awarded in that year[79]). He became a correspondent of many other Manchester notables including Edward Binney (geologist and President of the 'Lit. & Phil.')[80], the chemists E. A. Parnell,[81] Crace Calvert and Angus Smith.[82] He developed a lifelong friendship with the engineer James Nasmyth. Out of their discussions emerged designs for digesters capable of conducting chemical reactions under considerable pressure. These later proved to be of great value to other chemists, including Hofmann[83] and Williamson. Apologising for a delayed delivery-date to the latter Nasmyth expressed the hope to Frankland that he would 'keep a keen look out to commercialize . . . and turn them to profitable results', for then, 'with such powerful tools as money can furnish', he would be able 'to set out as pioneer of that glorious band who are soon in the van of discovery'.[84] Such advice is unlikely to have fallen on deaf ears.

Nor was local industry neglected. From the Westhoughton gas works to Nasmyth's Bridgewater Foundry at Patricroft he was frequently to be seen on site, relating his chemistry to ever wider fields of technology. On the pressing new problem of pollution he advised on the manufacture of black and white ash by Muspratt at Newton Heath. Though admitting that the by-products were doubtless 'injurious to vegetable life' Frankland was prepared to certify that 'no trouble or expense seems to have been spared'.[85] On a similar theme the firm of Peter Spence was also to benefit from Frankland's advice, as it did in the matter of coal economy by using superheated steam for an engine and exhaust steam for evaporation and other processes.[86] It would doubtless have been feasible to continue with a multitude of local problems but Frankland had bigger fish to fry.

His second aim was therefore to *build up a number of major industrial contacts*. The Hydrocarbon Gas Co., already mentioned, was one of these. Modestly acknowledging the impropriety of requesting a large fee Frankland suggested 150 gn p.a., to include all experiments but no consumables or travel expenses![87] Clearly this was a firm whose custom had to be cultivated, even if his demands bordered on the outrageous. Less ambitious were his requests to the Royal North Lancashire Agricultural Society, one of many such bodies influenced by the agricultural chemistry of Liebig and now much interested in the composition of fertilisers. As early as September 1851 Frankland was analysing guano for them and, after the Agricultural Meeting at Lancaster that month,[88] was to receive the plaudits of his uncle for his part in the public discussion.[89] A letter of March 1852 confirms his willingness to act as their Consulting Chemist, though does not mention terms.[90] A year later his appointment does not appear to have been confirmed.[91] He regretted that he had been consulted only twice in the previous year and had been expected to attend the two-day Annual Meeting without fee.[92] However after another year had passed he indicated a new willingness to be reappointed, at an annual fee of 50 gn. However analyses for members would be charged for (at half-price) and travelling expenses would have to be added.[93]

His consultancies were diversified chemically as well as geographically. From his days at Putney had come a useful contact with the famous brewing firm of Watney & Co., the brothers James and Daniel owning the Stag Brewery at Pimlico (closed 1959) and Daniel a distillery at Battersea. Now, in Manchester, Frankland is engaged in an obscure dialogue with them on the purification of spirits. He sends a bottle of 'prepared liquid for experiments upon spirits' of which 20–800 drops have to be added to a gallon of the spirit 'according to the degree of impurity'. Symbolically as well as literally this potion has to be kept in the dark.[94] What it contained is hard to guess. A few days later he sends to Daniel Watney some samples he has himself purified[95] though later admits that they contain ether due to 'an overdose of the purifying agent'.[96] If the term 'ether' is used in its modern sense one can only envisage some powerful dehydrating agent like sulphuric acid, though if it means 'ester' (as it generally did in the brewing trade) one might have a carboxylic acid in, or formed by, the mysterious liquid. His previous proposals of rectification with alkalis, digestion with charcoal *etc.* are now recognised as 'too inconvenient and costly', which leads him to the use of chlorine to remove the unwanted and odiferous oils. Conceivably an aqueous solution of chlorine might have been used – which would explain the need to exclude light and the possible production of esters from acids produced by oxidation of alcohols. He does refer to that reagent in a later letter[97] and it is at least plausible that odours due to dissolved sulphides could be removed by chlorine and esters be produced by an acid catalysed reaction given a high enough alcohol content (40% in 'spirits' and 6–8% in beers of that period).[98]

Frankland was a perennial optimist and within three months of this debacle had persuaded the firm to appoint him as consulting chemist for a period of ten years on the (unspecified) terms that they had suggested.[99] Two years later he was still in touch with them regarding the use of solidified mash residues as a manure; he wrote strongly in its favour.[100]

A final example of a large-scale consultancy must be that with the firm of Cox, Score & Co. This was a partnership between E. Score and Stephen Cox who set up in 1852 as 'alkali and vitriol manufacturers' on the site of an earlier works at Netham on the banks of the Severn just outside Bristol.[101] Now, like others in the same trade, they were faced with a new and massive problem of pollution. That problem may be understood in the light of the complex of processes summed up in Table 7.2, where major products are in bold type, including three noxious by-products. The by-product from the first stage of the soda process, hydrogen chloride, is a gas which avidly dissolves in water to make hydrochloric acid. If allowed into the atmosphere it causes acid rain that can wreak havoc on vegetation, iron railings, and stonework, to say nothing of its effects on human beings in the vicinity. The second by-product, carbon monoxide, can be burned as a fuel, but the third, calcium sulphide, is a gelatinous solid that yields the poisonous gas hydrogen sulphide on exposure to acid in the atmosphere. Tips of this substance, over a century old, still emit the characteristic smell of rotten eggs. Thus the atmosphere is polluted (by hydrogen chloride) and the land and water (by calcium sulphide).

In a situation of this kind Cox and Score were worried men. Somehow they had heard of an entrepreneurial academic in distant Manchester who showed more than a slight interest in problems of

Table 7.2. *The soda industry and its by-products*

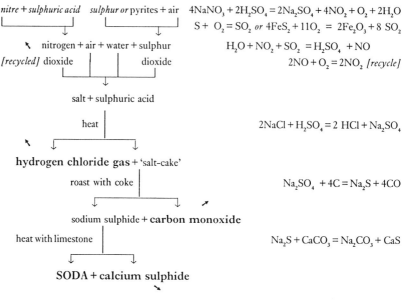

nitre + sulphuric acid   *sulphur or* pyrites + air $4NaNO_3 + 2H_2SO_4 = 2Na_2SO_4 + 4NO_2 + O_2 + 2H_2O$

$S + O_2 = SO_2$ or $4FeS_2 + 11O_2 = 2Fe_2O_3 + 8\ SO_2$

[recycled] nitrogen + air + water + sulphur
dioxide   dioxide

$H_2O + NO_2 + SO_2 = H_2SO_4 + NO$

$2NO + O_2 = 2NO_2$ *[recycle]*

salt + sulphuric acid

heat

$2NaCl + H_2SO_4 = 2\ HCl + Na_2SO_4$

hydrogen chloride gas + 'salt-cake'

roast with coke

$Na_2SO_4 + 4C = Na_2S + 4CO$

sodium sulphide + **carbon monoxide**

heat with limestone

$Na_2S + CaCO_3 = Na_2CO_3 + CaS$

SODA + calcium sulphide

industry. So, just in case he could help, they wrote to Frankland.

Frankland replied promptly, offering to visit the site for a fee of 5 gn per day for the first two days and 3 gn thereafter. He believed that the experience of the Muspratt plant nearby had shown that hydrogen chloride could be effectively washed out of the effluent gases by a stream of water. His next comment disclosed a political awareness that was to manifest itself with remarkable frequency in his later life. He wrote:

> Another point however of scarcely less importance is to make the surrounding neighbourhood believe you have perfect condensation, for without this faith stupid farmers will detect muriatic acid in every diseased ear of wheat or decayed branch of quickthorn, and old fish women will be quite certain that their oysters and mussels are in the last agonies of death from the same cause.[102]

The Bristol public were up in arms and four days later Frankland suggests that the firm is facing a powerful conspiracy. However 'I think we shall have no difficulty in disappointing all the expectations of nuisance'. Then, turning to the other pollution problem, he makes the remarkable assertion that hydrogen sulphide is not dangerous, though it does bring unwelcome notoriety. He follows this with the even more astonishing suggestion: 'Is it possible to take the waste in flats [boats] and throw it into the Severn?'[103]

In subsequent correspondence Frankland offers his own design for HCl condensers,[104] regrets his inability to comment on nitre consumption in the manufacture of sulphuric acid, and comments that northern manufacturers are 'not at all particular' about steam pressures.[105] Through 1853 he sent a stream of letters about a new plan for making bleaching powder,[106] about the analysis of black ash samples,[107] about lead chamber design and about the estimation of bleaching-powder.[108] Having announced his own method for economising in the use of nitre in the lead chamber process ('which for the present I request you to keep perfectly secret') he suggests that he become a consulting chemist to the firm at a fee of 100 gn (an arrangement which works well 'for a London distillery').[109] Whether the firm jumped at this generous offer is not clear, but Frankland was still writing to them (about the manufacture of ammonium sulphate) in the summer of 1855.[110] Score soon left the partnership (moving to Runcorn to make fertilisers[111]) and Cox went bankrupt in 1859. In 1855 Frankland wrote to Tyndall mentioning a 10-day stay at Bristol.[112] A subsequent letter revealed that 'the Bristol affair is still in abeyance' but he would only consider making a change in view of worsening conditions at Owens College.[113] Presumably the firm was offering alternative employment; if so he must have rejected it. However, while it lasted, Frankland's consultancy with the firm must have been lucrative and challenging. It seems to have been one of his major *coups* but such large industrial contacts must be seen in the light of another aspect of his policy.

This was to *accept a wide range of smaller individual assignments*. One of the best examples of these would be the large number of water analyses that came his way from all kinds of sources. Thus domestic water from the Preston to Lancaster Canal was shown to be soft enough but too rich in vegetable matter, needing a reservoir for oxidation.[114] On the other hand a sample of well water was too polluted and unfit to drink.[115] To the chairman of Owens College Trustees he advised that water with considerable organic matter should be boiled and filtered before use.[116] Frankland seems to have been one of the first chemists to offer their services to the growing number of railways.[117] By 1855 he was analysing samples of water for use in locomotives.[118] Though adequate supplies could be taken from the Canal it would be better to draw on the supply from the water works; possibly he was considering the Lancashire and Yorkshire Railway with its headquarters at nearby Horwich. As for the mode of water supply he investigated a lead-poisoning case associated with the use of lead pipes at Manchester waterworks,[119] and expressed to one correspondent his preference for lining lead pipes with zinc rather than tin.[120] On the related question of sewage he gave an opinion on the efficacy of Herapath's patent for its purification,[121] and (for the Patent Solid Sewage Manure Co.) a view on Wicksteed's process for removing 'ingredients which are most valuable for manure'.[122] Here, in Manchester, foundations were being laid for the work that would occupy much of his later life and bring him the most immediate fame: the chemical analysis of urban water supplies.

Analysis was the order of the day in Frankland's laboratory, not only of coal and water but of all manner of curious substances including linseed oil,[123] gypsum,[124] gold mud[125] and potassium chloride.[126] Each brought its own modest financial reward: 20 gn for analysing coal for gasification purposes,[127] 3 gn for well water[128] or 2 gn for partial water analysis.[129] And over and above all these activities was the growing business of being an expert chemical witness at a court of law.

When the railway company objected to his charges for water analysis (5 gn per sample) he was assured by Crace Calvert that this was 'a very moderate charge for a complete analysis'[130] and reminded by Angus Smith that 'we cannot do these by machinery'.[131] Another objection arose over some work on the bleaching of silk (for which he had suggested ozone). His report[132] received no acknowledgement for over a year, so, as he had asked for 'a mere nominal fee' of 15 gn, he successfully threatened legal action.[133] On acknowledging the subsequent cheque he expressed resentment at the opinion that 'my charge was too high', adding that the value placed upon chemists' services was 'anything but flattering to the cultivators of our science'.[134]

To assess whether Frankland made reasonable demands upon his clients is not easy and requires a knowledge of payments made elsewhere. The opinions of Calvert and Smith suggest perhaps a closing of the ranks but also a genuine assent to the level of fees. As Smith said later:

> No man in Manchester has ever made a decent living, or indeed any
> living at all, by analytical chemistry, though four or five had made their
> way by consulting in addition.[135]

There was probably a considerable discrepancy between fees
demanded in London and the provinces. Thus Thomas Graham
told Frankland that fees varied widely and that, for court cases in
London, the daily figure was 3 to 5 gn, with 20 gn as maximum for
longer periods. Water analysis would cost 10 gn.[136] It therefore
seems that Frankland's charges for individual analyses were not
unreasonable, though whether that applied to his retainer fees for
larger contracts is another matter.

With all this variety of external consultant work it is not
surprising to note that Frankland made every effort to *transfer
information from one situation to another*. In an atmosphere of
industrial secrecy this was not always easy. Writing to Cox, Score &
Co. in Bristol about collection of tar and ammoniacal liquor from
gas-making plant he noted that a company at Vauxhall were doing
this effectively but added that they are 'very jealous of the
inspection of their coking apparatus, but *it would perhaps not be
impossible to get a sight of it*' [my italics].[137] Such suggestions of
industrial espionage were quite in keeping with general attitudes in
the chemical industry at the time. To the same firm he wrote about
the now demolished alkali works of Muspratts, proposing to offer
them information on the basis of what he had seen:

> Since my last visit to Bristol I have been present at the pulling down of
> the condensing apparatus at Newton-le-Willows and had thus the best
> opportunity of seeing the internal construction and taking the
> dimensions of the troughs &c.[138]

He even claimed that he had written to Muspratts for details of
their pitch/fireclay mixture but had had no reply.[139] In another
letter to Watney he refers to the latter's quest for a process to
convert waste starch into sugar. He writes:

> In conversation with a chemical manufacturer the other day a patent
> process for effecting this desirable result accidentally came under my
> notice, and the description which he gave me of the process impressed
> me so much with its importance that I requested him to present me
> with a copy of the specification. I have read this specification carefully
> over and now enclose it for your inspection . . .[140]

A similar situation is recorded in the Laboratory Book of Peter
Spence, alum manufacturer at Pendleton, for whom Frankland was

a consultant; probably the firm mentioned was Pochin & Co. who had previously been manufacturing crude alum cake from china clay and who did acquire a licence from Spence.[141]

> I have just got into new correspondence through the kindness of Dr Frankland with a firm at Bristol who are commencing the Alum manufacture by clay and are not succeeding at all and are advised by him to apply for a licence to work under my patent.[142]

Frankland was merely being a 'middle-man' and suggested quite properly that a licence be acquired. He was not always to be so circumspect. The most spectacular case of misuse of *privileged* information came with the famous prosecution in August 1857 of Peter Spence for creating a 'public nuisance' by his Pendleton Alum works.[143] Many expert witnesses were called, amongst them Edward Frankland. He had long been employed by Spence to advise on this very problem and had been helped by him in other ways. The court was shocked to learn that this friend and advisor of the plaintiff had been persuaded to act as witness for the prosecution. This led to bitter recriminations from Calvert and Smith (who appeared for the defence) and from others whose sense of fair play was outraged. As the *Manchester Guardian* observed:

> Dr. Frankland had become a deserter, and after partaking of the bounty of Mr. Spence and taking advantage of his works for the instruction of his pupils he had been induced to join the ranks of the enemy.[144]

Frankland's ability to make enemies was not always on questions of finance, and another bitter feud can serve to illustrate a further aspect of his strategy as a consultant. This was a tendency to *retain consultancies for himself alone*. Frankland was a great soloist. Partly this must have reflected his determination to maintain the highest standards of analytical work. Even his rivals had to admit to an extraordinary aptitude for sustained accuracy in experimental investigations. But there were other considerations. An instructive example is the case of Sheridan Muspratt.

Sheridan Muspratt[145], as well as being a son of the alkali manufacturer James Muspratt, was a former pupil of Liebig, founder of the Liverpool College of Chemistry and author of the famous dictionary *Chemistry, theoretical, practical and analytical*.[146] In the early 1850s, however, he had the misfortune to cross swords with Frankland on an issue where he simply could not win. The question was the identity of an ester ('ether') apparently obtained from some experiments on wine or beer. It is possible that these may have been connected with the purification process that Frankland had devised on behalf of Watney & Co. On the basis of boiling temperature and of Muspratt's own analyses Frankland concluded that the product was probably 'oeanthylic or caprylic ether' (*i.e.* ethyl heptanoate or octanoate).[147]

Muspratt was clearly dissatisfied and evidently asked Frankland to do some more work, drawing the uncharacteristically tart rebuttal: 'I am exceedingly busy with several commercial investigations which render it impossible for me just now to collect the information you require'.[148] Nevertheless he appears to have returned to the topic and concluded that the boiling range of the liquid implied a fairly impure mixture but after fractionation and hydrolysis he obtained pelargonic [nonanoic] acid. On this basis the 'ether' would have been ethyl nonanoate. He communicated this information to Muspratt in another letter (no longer extant). Muspratt replied aggressively: 'I believe that when I stated it to be in the valeryle series I was nearer the mark than you were. Time will show. These investigations are most difficult or else chemists would be able to tell at once what an ether was'.[149]

Frankland expressed amazement at his correspondent's inability to see the issue as settled, for he had by now obtained 'a large and beautiful specimen of pelargonic acid'. Frankland went on to comment on another chemical problem, in some way related to tannin. Muspratt had evidently got into difficulties with this also and had sought Frankland's help. Once again the request was declined, on the grounds that Muspratt had offered him 'a fee which after the arrangement we had made I could not prevail upon myself to take'. Whatever that 'arrangement' may have been the fee was manifestly too low and the work involved too much. Even Frankland's concluding attempt to be conciliatory had a sting in its tail:

> I do not intend to imply that your results were either incomplete or imperfect. I only know that the results of commercial analysis are very frequently so when low fees are charged & the character of our science is thus greatly damaged.

In the same letter he made a remark that effectively terminated any partnership between them, and not merely on the grounds of chemical disagreement: 'I fear that the wide differences in our professional charges will prevent us working together in these matters as we had contemplated'. Apparently Muspratt charged 50 gn for a journey to London but only 4 gn for an investigation 'for which any extensive Whiskey Dealer in the kingdom would gladly pay 40 or 50 guineas'.[150]

The Muspratt saga rumbled on for another few months, other persons in the international chemical community becoming involved. Liebig learned of the row and wrote to Hofmann with the opinion that his former pupil was mad, and his publication on carmufellic [carophyllic?] acid the work of a schoolboy. In any case Frankland had done much work for him; 'it really does one no honour to be connected with such an impudent man'.[151] A week later Frankland was reinforcing Liebig's jaundiced view of his former pupil with the remark that Muspratt was a 'monomaniac' with the 'unheard of vanity' to use Liebig's name 'to puff himself before the public'.[152]

Writing to another of Liebig's former pupils, William Gregory, Frankland reported:

> I have had a great deal of trouble with Muspratt lately & really believe he is going mad. I enclose you for your own perusal a letter received from Liebig a short time ago on the subject & which will show you the extent of Muspratt's hallucination. After my reply to that letter Liebig wrote M. stating that he should decline all further communication with him. Muspratt is in a tremendous rage at me for stating in answer to L.'s letter that I made the analysis of P[elargonic] ether. Soon after sending me that analysis I declined making any more for him and have consequently had none from him since.[153]

In reply Gregory disclosed that it was he who alerted Liebig to the pretensions of Muspratt, at which news Liebig 'expressed himself in terms which would astonish that simple minded and modest youth not a little. I never saw anything so bitter from the pen of our friend, but of course it is private. . . If M. does not take care he will bring an avalanche about his ears from Giessen'.[154] This letter must have arrived at almost the same time as two further missives from Muspratt, dated 12 and 15 January, blaming Frankland for the rift between himself and Liebig and accusing him of breaching confidence in the matter of the pelargonic ester. One of them included the remarkable sentiment 'neither you nor Liebig himself can teach me anything in science'.[155] After several false starts Frankland composed a stern rebuttal of all charges, remarking that 'one of the most eminent of our British chemists' had joined him in those researches and had presumably been the source of Liebig's information.[156] It appears that this must have been William Gregory.

It is pleasant to record that less than two years later Frankland was writing a friendly letter to Muspratt, generously commending his *Dictionary* which was just completed.[157] The storm had passed but its fury had indicated as little else could do the strength and complexity of Frankland's conviction as to the hazards as well as the value of industrial consultancy.

A further aspect of his policy of personal control was the taking out of numerous patents in his own name, for example a method for making ammonium and potassium sulphates for artificial manures.[158] This was a widespread practice in Victorian England and frequently featured in Frankland's Manchester Letter Books.

Only in one respect did Frankland permit himself any deviation from the principle of sole control. That was when he allowed assistants to conduct some of the more routine activities of the laboratory. At Manchester he had at least four. First was W. J. Russell, whom Frankland brought from Putney and whose work on coal analysis was acknowledged in his 1852 paper on hydrocarbon gas.[159] Following Russell's departure in 1853 to work with Bunsen in Heidelberg, the next assistant was C. J. Tufnell, whose conditions of work included 'general surveillance' of the laboratories, acting as lecture assistant and, 'as far as time will allow to assist the Professor in his experiments and researches when required'.[160] Then came Wilhelm Gerland (1854) and Frederick Guthrie (1855), former students of Kolbe.[161] To Gerland Frankland evidently delegated some lecturing as well, for his name appears frequently in the notebooks of Frank Spence. Both Gerland and Guthrie are referred to in a letter of 1855 by Sophie Frankland to her family in Germany.[162] A letter to Tyndall reveals that Russell is now back, engaged in flame research and capable of doing 'a year's work in a couple of months'[163]; he was therefore staying the winter in Manchester instead of returning to Heidelberg. Also present was a former student James Wanklyn, working chiefly in organometallic rather than industrial research. Expected in the following week was E. K. Muspratt, the student of assaying who was nevertheless regarded by Frankland as a member of the team. It is clear that the borderline between 'assistant', research student and general dogsbody was a fairly imprecise one.

Given the Herculean efforts put into industrial and other consultancies one inevitably wonders how far Frankland's lifestyle in Manchester was driven simply by financial considerations.

Given his poor salary it is clear that a certain amount of commercial work was a sheer necessity. Nevertheless by 1855 he was enquiring about the possibility of investing £1800 at $4\frac{1}{2}$% with Stockport Corporation.[164] Whether his financial targets were too high is a matter for individual judgement. There were at least two occasions when he yielded to other priorities. One was in his early and extensive dealings with the Duke of Argyll whom he advised on the matter of nickel extraction from deposits on his estates. At one point Frankland wrote to him:

> My position at the Owens College is not so good as to make me independent of professional fees yet I feel a delicacy in making a charge in the present instance inasmuch as I volunteered to make the investigation when Mr Nasmyth related to me your Grace's fortunate discovery and expressed his desire to ascertain the merits of the ore. If my experiments and observations have been of real service to your Grace it is a source of great satisfaction to me'.[165]

Not unnaturally for the times, approval from so exalted a figure would have been its own reward. Four years later, in the letter to Tyndall just cited, Frankland admits 'I am pitching commerce to the devil as much as I can afford & working hard at my researches'.[166] Here, in an established reputation as a researcher, lay another reward, more satisfying and enduring than an inflated bank account. So many an academic has thought before and since. Evidently Frankland felt the same.

4 Values, human and otherwise

For all his preoccupation with the educational and commercial utility of chemistry, and with the rewards obtainable by its pursuit, financial and otherwise, Frankland was far from being a soulless automaton bent only on grinding out scientific facts. His Manchester years witnessed development of numerous other values that had little to do with chemistry or commerce.

4.1 Family

Ever since that memorable picnic in July 1849 where Edward Frankland and Sophie Fick entered on their long engagement the weary months at Putney must have seemed interminable. Marriage in those days would have been unthinkable given a low salary

Fig. 7.2 Edward Frankland in studious pose. (R. Frankland Archives)

derived from an institution whose future was far from certain. At least that would have been the view of a German academic family and also of the rising British middle classes to which Frankland so earnestly aspired. But with a new post at Owens College in sight it suddenly became a practicable proposition. Accordingly plans were laid for marriage a week before he took up his duties in Manchester.

The bride travelled with her brother Heinrich[167] to Ostend where they were met by Edward for the sea-crossing to London. They took a house at Adelphi Terrace near the Strand from which it was but a short drive to the church, the famous St. Martin-in-the-Fields in Trafalgar Square.[168] Sophie entered on the arm of her uncle Heinrich, who had travelled separately. Apart from her brother and uncle the only other family present were Edward's uncle John Frankland[169] and his wife Mary, residents of Lambeth.

The very choice of the wedding location conceals some important 'family values'. Why was the marriage in England and not Germany, and why in London and not Lancashire where most of Edward's relatives lived, including his mother and step-father? The reason Frankland gives for the first choice is rather curious: 'our marriage in Cassel was absolutely impossible, because, according to the Hessian law, no such marriage could be permitted, unless the guardians of the poor of my parish guaranteed that my wife should never become chargeable to the Gemeinde of Cassel', and no such guarantee was possible.[170] It is also likely that Frankland would refuse to be married in a Catholic church (to which the Ficks belonged and which would have been inevitable in Germany). As for the famous church in London its very distance from Lancashire and nearly all known relatives was a positive advantage. At weddings questions of ancestry always arise, if only because the legal documents require identification of parents. This, as we have seen, was absolutely ruled out in the case of Edward's natural father and in fact even the marriage register gave him a false name ('Edward Frankland').[171] It was a big risk to take, for the penalties for misrepresentation were as for perjury. But then Frankland was playing for high stakes, and in the heart of London the chances of anyone knowing the truth were small, save only the couple from Lambeth, and presumably they were sworn to secrecy.

If Victorian family and financial values determined the church to be used, the next few days of honeymoon were not wholly free from those of chemistry. The four days at Windsor and subsequent two in London included trips for Edward to inspect the architecture of chemical laboratories in London and Oxford. Fortunately the ubiquitous younger Heinrich was able to care for his sister at such times, as well as accompanying them both to the Lyceum Theatre one evening. All too soon the express train had to be boarded for Manchester, with Heinrich still in tow. Awaiting them was their

Fig. 7.3 Sophie, née Fick, first wife of Edward Frankland (1821–74).
(J. Bucknall Archives.)

new home and the warmest of welcomes from Edward's mother
and step-father, Margaret and William Helm. Sophie's first
impressions are best conveyed in a letter to her own parents:

> The closer we came to Manchester the graver became my spirits, that
> as I entered the town I could hardly hold back my tears. When I

climbed down from the coach and when Edward led me into the house I was excited and apprehensive. There I met his parents who received me with such love and joy. It was a most moving moment that I shall never forget. I can tell you that we were right about his parents and they have exceeded all my expectations and every moment I spend with them I love and value them more. A true description of their personalities would appear to be superfluous, since Heinrich can describe them better orally than I could in writing.

The house they rented was 18 St George's Terrace, Dawson Street, in a poor area near Regent Bridge over the River Irwell. However Sophie confessed that 'the elegance and comfort of it surpass all my expectations'.[172] There they remained until Christmas Day 1852,[173] after which they moved to the southern edge of the city, at Longsight. Their new home in Richmond Grove was christened *Hanse Haus* after the Marburg hostelry in which they became engaged. In those days Longsight was almost open country, and we find Frankland complaining to his landlord about the lack of promised improvements, including an iron fence 'for the protection of my garden from the inroads of cattle', for they were almost nightly visited by cows and horses 'to the great destruction and damage of my property.'[174] By 1857 *Hanse Haus* was also dignified as 'no. 43' since it had many near neighbours.[175] By a remarkable coincidence it remains, sole survivor from Victorian days, on the west side of the street where all else is modern.

The early years of Frankland's marriage seem to have been happy, dogged only by the relentless pressure of work and a measure of ill-health for his wife. An invitation to London was accepted for himself but not for his wife: 'As Mrs. Frankland is in a rather delicate state of health I scarcely think she will venture from home this winter.'[176] She was in fact pregnant but other ailments were not far away. She contracted a severe cold and sore throat during a visit to Edinburgh in 1853. No less a physician than James Young Simpson was called in to advise, and he suspected lung disease.[177] This was not confirmed by a specialist in such matters who prescribed further attention and an immediate return to Manchester. After two months' convalescence at Southport she was pronounced well. In fact she was in the early stages of a second pregnancy.

The Franklands' first child, a daughter, Margaret Nannie ('Maggie'), was born on 25 March 1853 and a son, Frederick

Fig. 7.4 The Frankland's home for most of their time in Manchester: *Hanse Haus* named after the hostelry in Marburg, and situated in Richmond Grove, Longsight. (Author.)

William, 18 April 1854. A third child, Sophie, arrived on 3 November 1855. Such was the improvement in their financial position that, after the second baby, they were able to engage 'an excellent nurse'. Illness struck again when Fred developed acute diarrhoea and failed to respond to medical treatment. Neither of the babies could be breast-fed and no alternatives seemed successful in the case of Fred, reduced to 'little more than a bag of bones'.[178] His father concluded that what was necessary was a fluid whose composition approximated to that of human milk. Given that cow's milk contains more casein and less lactose ('milk sugar') he devised a method for curdling de-creamed milk with rennet, filtering out residual casein and adding lactose to the whey. To this was added milk enriched with cream.[179] The result was spectacular, Fred recovered and Frankland reported his 'artificial human milk' to the Lit. & Phil. as a tailpiece to a talk on Liebig's extract of meat.[180] Many years later, in the 1880s, it was to become the basis of a commercial enterprise.

One other aspect of life in Victorian times was that the family holiday was rapidly becoming an institution. In their first summer Edward and Sophie had a delayed 'honeymoon' in Tenby, S. Wales, though the real purpose of the trip was a technical investigation of local coal (for which Frankland was paid 'a considerable fee'). In 1852 the couple took an extended break in Cassel[181] and in the following year they made the abortive trip to Edinburgh (Maggie being in the care of her grandmother). With two young children they were more circumscribed, but in 1854 they took lodgings on the Esplanade at Blackpool, 'a mere village then'. Here again they made a premature return on account of illness, though this time it was an outbreak of scarlet fever in the lodging. It may have reminded Frankland of the flight from Salford of himself and his mother to avoid the plague of cholera in 1832.[182] However it was a blessing in disguise as Edward and Sophie were then able to spend several weeks in the Derbyshire Peak District, leaving the children in the care of their grandmother and nurse. Thereafter they had better fortune when Frankland purchased a three-bedroom cottage in the village of Windermere to which they regularly repaired, complete with children, nurse and grandparents. Here they (or at least Frankland himself) would walk, or sail on Lake Windermere.[183] He had also installed 'facilities there for performing any chemical operation of moderate character', displayed as an incentive for his friend Tyndall to make a visit.[184]

Frankland's mother and step-father, Margaret and William Helm, played a larger rôle than that of welcoming parents or dutiful grandparents. Edward was devoted to his mother and, perhaps more surprisingly, to the man she married when he was almost six years old: William Helm, cabinet-maker. Some time after Frankland left Lancaster they moved to Disley in Cheshire, not far from Stockport.[185] They settled into 'Dryhurst Lodge', a substantial detached house, built about 1832, along the new turnpike road to Buxton.[186] Frankland first visited their new home in the summer of 1850, and it was while staying there in January 1851 that he attended the interview at Owens College.[187] Their nearness to Manchester meant much to Edward and there is no doubt that he saw them often. The same cannot be said of his natural father, Edward Lowndes (as he now was), for contact between the two was almost entirely forbidden. All that there was consisted of a few terse letters touching upon the financial settlement that was, for Lowndes, the price of total secrecy. For Frankland that arrangement may have eased some of the burdens of supporting a growing family; for the Helms it must have added considerably to the comfort of the many years still left to them. By an ironic twist the very negation of traditional values embodied in Lowndes led to their enrichment in the families of Helm and Frankland.

4.2 Religion

When Frankland completed his autobiographical *Sketches* he created the clear impression of a man whose early religious enthusiasm had been entirely replaced by a sturdy agnosticism. How far this was strictly true, and why Frankland should have wished to portray himself thus, will be matters for discussion later. Meanwhile it is pertinent to ask how far the Manchester years saw a perceptible change in his religious values.

An important indicator, not mentioned in his *Sketches* and fairly inaccessible today, is his Inaugural Lecture of 1851. It is a remarkable document, shot through with a conviction not merely that chemistry is wonderful but also that it offers superb illustrations of design in nature. Frankland was, of course, not the first chemist to feel that way and there was a long tradition from Boyle to Prout of seeing the science in terms of natural theology. Two paragraphs are worth quoting in full:

> The position which chemistry now occupies amongst the sciences is a most elevated and honourable one, whether we regard it as exemplifying the wisdom and goodness of the great Creator in the transcendently-beautiful, perfect, and harmonious laws which he has impressed upon matter, as being the medium of elevating and refining the intellect of man, or as furnishing him with increased resources, comforts, and conveniences.
>
> It is remarkable that so few illustrations have been drawn from this science by the writers on natural theology, a circumstance which can only be accounted for by the almost total neglect of chemistry as an essential branch of a general liberal education. The astronomer is

struck with a feeling of awe and solemn dread when he surveys the grand and overwhelming scale of the universe, and finds his puny ideas of magnitude and distance utterly confounded. In like manner, the geologist and the natural historian trace with admiration the finger of God in the wonder-exciting changes that have occurred in the strata of our globe and the races of animals and vegetables which have inhabited it since order and harmony were called out of chaos. But the chemist experiences a peculiar delight and inexpressible feeling of love to the beneficent Author of creation, as he sees gradually unfolded in his experiments the incomparable beauty, the surpassing skill, the kind care, and the wonderful adaptation exemplified as well in inorganic as in organic nature. Wonderful as are many of the phenomena which meet our view in the study of inorganic chemistry, they are still surpassed by the admirable and skilful arrangements which we encounter at every step in our examination of the organic division of the science; here Omnipotence seems to have exhausted itself in endless variety, evolved from the simplest materials, – the great bulk of the now almost innumerable organic compounds being produced by the grouping of three or four elementary substances, that, according to their position in the groups of which they are the constituents, form the most widely different classes of bodies, and strikingly exhibit the simple means which the All-wise Creator employs for accomplishing the most important ends. By a slight alteration in the grouping of precisely the same elements, the chemist sees the most virulent organic poisons metamorphosed into the common articles of our daily food, and by the substitution of one element for another in the decaying groups which constitute contagion and miasma, beholds these fearful compounds transformed into harmless substances, utterly incapable of the further propagation of disease.[188]

Of course this can be written off as conventional pandering to the crowd, of telling the audience what it wanted to hear. However Frankland makes the specific point that what he had to say was largely new, and therefore *not* to be expected. Moreover natural theology had been a particularly Anglican propensity and the audience at Owens College was likely to be full of Dissenters who were not noted for this kind of theological thinking. Again, these paragraphs occupy a central part of his address and were integral to his theme. He went on to recommend the science 'for the bright glimpses of the Deity which it discloses at every step'. Finally, if he were merely playing to the gallery he would be guilty of gross insincerity, and for all the faults one may discern in his character this was not one of them. Of a future agnostic or even atheist there is not a trace in Frankland's Inaugural Lecture.

That however is by no means to say that formal religion played a large or even visible part in his life at Manchester. His reply to a letter from the Rev. James Fleming, minister of Lancaster Congregational Church (to which he belonged as a youth), reveals more. He gives a series of reasons (or excuses) for not regularly attending a church of the Independent [Congregational] persuasion, basically because there is not one sufficiently near. He and Sophie had, however, taken a seat at the local parish church, St. George's. He asserts 'it is my earnest desire to have again the privilege' of hearing Independent preaching again. Yet he adds significantly that 'my views however on some points of our faith and discipline are somewhat altered' so that he might have difficulty in making 'a confession of faith as is requisite for a member'. He gives no further details, saying he hopes to discuss matters with Fleming face to face.[189] This clearly indicates a growing doubt concerning confessional statements, coupled with an appreciation of much that was distinctive about Independency, not least the high quality of intellectual discourse. What repelled him even at this stage was evidently what he saw as mindless dogmatism. The unexpurgated edition of his *Sketches* contains several derogatory references from this period to Roman Catholicism with which such dogmatism was particularly associated in his mind. How far this was a reaction to the conservative religious traditions of his wife, or a reflection of the anti-Catholicism of his Ulster friend Tyndall, it is impossible to say. Nor can we be sure that he wavered between faith and doubt during those years at Manchester. That the convictions of his youth lingered on can hardly be questioned, however, and one laconic sentence from Tyndall's *Journal* for 3 May 1857 has a suggestive note: 'A cold, cold morning: Frankland went to hear Spurgeon'.[190] He had been in London to lecture at the Royal Institution, and now went to hear the 23-year old Baptist whose congregation filled the vast Surrey Music Hall. He was rewarded with a sermon on 'Regeneration'.[191] Possibly it was mere curiosity that impelled him to hear the preacher who was taking London by storm;[192] conceivably it was something more.

4.3 *The arts*

For a man who was to value the arts highly in later years the Manchester period was relatively barren. With one exception there is little evidence that Frankland found sufficient time to indulge in musical or other artistic activities. Perhaps the city itself was less conducive than others in that respect. It was only in the year of Frankland's departure (1857) that the Halle Orchestra gave its first concert. At almost the same time Prince Albert opened in Manchester the great Exhibition of Art Treasures of the United Kingdom. So when Frankland gave a series of lectures at the Royal Institution 'on the relations of chemistry to graphic and plastic art' the emphasis was firmly on the manufacture and preservation of the objects described.[193]

The one area of the arts that cast its spell over Frankland during his Manchester years was photography, at that time the almost exclusive preserve of chemists. Here he had made the acquaintance of the chemical manufacturer John Dale, of Roberts, Dale & Co.,[194] and had learned from him the new process of collodion photography. This was a technique to use glass as the material for supporting the photosensitive silver iodide in negatives. Potassium iodide was mixed with collodion, an ethereal solution of nitro–cellulose. This was applied to a glass plate which was allowed to dry but 'activated' just before exposure by immersion in a solution of silver nitrate. It was thus a 'wet' process, using cumbersome apparatus but faster than its predecessors (the Daguerreotype or Fox Talbot's calotype). Frankland acquired the necessary apparatus and used it at home and in Cassel, where his pictures created a local sensation. Some have still survived, though most were ruined as a varnish film placed over the collodion peeled off.[195] It was about this time that Frankland was exploiting the properties of collodion in the manufacture of balloons (for demonstrating properties of gases).[196]

4.4 Away from it all

A favourite scheme for those Victorians able to take advantage of it was to throw off the effects of overwork by travel. As early as 1851 Sophie and Edward Frankland were able to spend a month in London, chiefly to enable them to view the Great Exhibition. They travelled via Wakefield and the Great Northern Railway, for by that route (he said) first class tickets could be obtained for five shillings each. Renting the house of Samuel Clegg (the celebrated gas engineer and then a professor at Putney College), they took a bus 'every day' during August to inspect the wonders of the Crystal Palace at Hyde Park. Sophie's brother Ludwig joined them for a fortnight. He savoured the different kinds of snuff on display with great delight, but (as emerged years later) was chiefly there to see his sister for the last time. As we have seen (p. 89) he had opposed her marriage because he thought she would not survive a year, which caused Frankland to remark wryly 'I have ever since placed little reliance upon doctors' opinions'.[197] That pleasant month in London set the tone for many future trips away from Manchester.

'Getting away from it all' is no modern luxury. In Frankland's case he managed to escape from the cares of Manchester in two ways. One was by undertaking lectures elsewhere, especially at the Royal Institution in London. These had the additional advantage of increasing his income still further. A course of ten lectures on 'Technological chemistry' on Thursdays in 1853 brought him 55 guineas, while the seven lectures on chemistry and the arts, given on successive Saturdays in 1857, attracted a fee of 50 guineas.[198] On some occasions at least he would take the train from Manchester on the Wednesday morning and return home on Friday at 4 a.m.[199] However such continual activity made its own demands. Writing to Tyndall he warned him that Royal Institution lectures 'keep you continually on the rack for six months'. His own course of 1853 led to indigestion for the whole year.[200] It was probably that experience that led him later in the same year to decline invitations to lecture at the Chorlton-on-Medlock Mechanics' Institution (and three others in the neighbourhood) 'on account of my very arduous duties at the Owens College'.[201] The decision was sound though the reason slightly disingenuous. Even so, he managed to deliver a couple of informal Friday evening lectures at the Royal Institution, including one in June 1854 'On the dependence of the chemical properties of compounds upon the electrical character of their constituents'. Here, in the birthplace of electrochemical dualism, Frankland showed (in effect) how it had developed from Davy to himself, concluding that in organometallic chemistry was the clearest demonstration of his thesis.[202]

Much more therapeutic than lectures were the outdoor activities and travels that were to become such a prominent feature of his later life. Some of these were at Windermere, presumably with his family, where two boats were available 'for fishing, shooting & general rusticating & brain-clearing purposes'.[203] In 1853 a few weeks' break in Ireland were planned.[204] For in those early days his lifelong friend John Tyndall was fellow-conspirator in planning mountaineering trips abroad. At least one of these plans was justifiably vetoed by Sophie who, with young children and indifferent health herself, was precluded from such adventures and understandably desired her husband's company. This is a rare documented example of Sophie successfully exerting her own rights. In serio-comic vein Tyndall's expostulation also reveals something of his friend's general state of health at the time:

> Mrs. Frankland . . . does not know how I would have taken care of you, and prevented your lank carcase from being squashed in the crevasses of the ice, she does not know how I would have fed you on the choice mountain milk, and kept everything dyspeptic out of your stomach's way'. If she knew all this – if she should contemplate the prospect of your returning to her a compact mass of bone and muscle weighing at least thirteen stone, with vigour in your heart sufficient to carry you triumphantly through the mist and soot of ugly Manchester during the coming winter – she would I am sure forgive the effort I made to confer all these benefits upon you and her.[205]

Evidently a compromise was reached in the end, for Frankland was able to spend a few days with his friend in early September 1856, meeting Tyndall at the Finstermuntz Pass in the Tyrolese Alps.[206] Little did he know what further adventures of that kind awaited them in the years ahead after 'smoky Manchester had' exerted its final toll.

5 The end of the road

That all was not well with the new College did not become apparent for several years, despite Frankland's recognition of the 'arduous duties' of his office and consequent ill-health as early as 1853. In a moment of rare frankness he made to Tyndall the classic complaint of beleaguered academics the world over: the poor quality of the student body:

> To be eternally dragging mere children through the 5 groups and no further is mere mechanical work, scarcely superior to breaking stones, and I feel it to be for me a waste of time.[207]

Writing to Bunsen in 1856 Frankland contrasted the quality of students at Heidelberg and Manchester; the former would include many capable of original research, but at Owens College 'there is

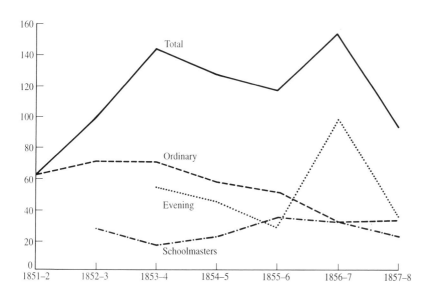

Fig. 7.5 Student attendances at Owens College, Manchester. (Author.)

rarely any further desire for knowledge other than the testing of "soda-ash" and "bleaching powder"'.[208]

In the case of the Professor of Chemistry there was the additional irritant that his stipend was only half that of his professorial colleagues. This would not have been alleviated by a letter from Alfred Neild, an alderman and former mayor, enquiring if the Trustees could not be relieved of all expenditure for practical classes. Frankland's robust response[209] included the suggestion that this might be possible if Trustees paid all other expenses, up to £10 p.a. for new instruments and £40 to the Professor for consumables used in lectures, with new college fees to be divided between Professor and Trustees in a ratio of 1:2. The worthy alderman's reactions have not been recorded.

Within two years matters became much more serious. The gathering storm clouds were so ominous that the future of John Owens' enterprise was seriously in doubt. The years 1857–8 were the darkest ever for the college, epitomised by the widely circulated joke involving Roscoe (Frankland's successor)[210]:

> MAN (*to Roscoe, standing at College entrance*): Maister, is this th'Neet Asylum?
> ROSCOE: Not yet, my man, but if you come in 6 months' time I fear it will be!

The threat came, not primarily from financial difficulties or even teacher tedium, but rather from diminishing student numbers and a plummeting morale among the staff. The decline in numbers is summarised in Fig. 7.5:

The curve causing most concern was of course that of the full-time 'ordinary' students. By 1857 there were 34 of these, of whom only 15 worked in the chemical laboratory. The drastic decline in public confidence that these numbers signified stemmed from a variety of causes. There was no shortage of suggested explanations: the absence of any preparatory training; lack of residential accommodation; alarm at entrance examinations; the unsectarian nature of the college, and so on. The matter was addressed in staff reports requested by the Trustees.[211] The Principal, in a rather anodyne report, conceded that professional students would not go to Manchester for training so Owens must concentrate on non-professionals. Two Professors (Sandeman and Williamson) put the blame on local Mancunian culture: it simply did not appreciate the value of higher education (particularly for wealth creation). This was important because the College had no reputation at a distance and therefore had to depend on local support. The point was amplified by Frankland, who added that the training was too exclusively classical. He later observed 'science, as represented by chemistry and natural history, was kept in the background'.[212] Perhaps Roscoe put it as well as any:

> The institution at that time had not gained in any degree the confidence of the public, and it had for many years to fight a stiff battle, and it was only gradually that the idea could be generally grasped that Science could be made an efficient instrument of education, and that such an education was not only compatible with, but absolutely necessary for, a successful manufacturing and industrial career.[213]

The other question of staff confidence clearly reflected adversely on the conduct of the Principal, A. J. Scott. Not unreasonably Frankland's report concluded with a request for periodical meetings of the teaching staff to discuss internal management of the college. As the Trustees appeared to do nothing about the reports, action by the staff was the only recourse. It was Frankland who held a meeting for the professoriate at his house where it was resolved to invite the Principal to appoint a deputy who could be readily accessible to those staff who wished to see him. As Williamson remarked 'The meaning of this procedure on the part of the staff was doubtless understood by the Principal, who on May 28, 1857, resigned his office'.[214] He was shortly afterwards replaced by J. G. Greenwood. By now, however, it was too late for Frankland, and shortly afterwards he removed to a new phase of life in London. In so doing he believed that he had very seriously offended the Trustees of Owens College,[215] as may be judged by the need to employ his old friend Thomas Ransome to value his stock of apparatus and chemicals before he could leave.[216] Within one week he took up his appointment at St Bartholomew's Hospital, London.

Assessments of Frankland's work at Manchester must depend on many considerations. It is, perhaps, hardest to agree with Lapworth and Myers in asserting that the work Frankland did from 1851 to 1857 'is not surpassed by any other that has come from the Chemical School'[217] for the simple reason that almost all his organometallic work, together with his formulation of the idea of valency, really belong to his earlier period. There is a difference between performance and publication. Frankland's academic research that genuinely originated in Manchester is of a very minor character. Indeed some would go so far as to say that his departure was 'a blessing in disguise'[218] for it left his successor, Roscoe, superbly equipped to bring the Department out of its doldrums. Like Frankland he was a former student of Bunsen, but unlike him he was not so beguiled by invitations from industry. In fact, however, Frankland had made a major contribution to applied science, had persuaded multitudes of industrialists of the importance of chemistry, had pioneered effective chemical education and had set on its feet the first English university department of chemistry outside Oxford, Cambridge or London. If indeed it is correct to go so far as to identify 'a specific London–Manchester school that extended from Edward Frankland to Christopher Ingold',[219] the founder of that school must have a rôle of quite exceptional importance in the rise of British organic chemistry. However it was to be in London, not Manchester, that that rôle was to find its fulfilment.

Notes

1 On the early history of Owens College see J. Thompson, *The Owens College: its foundation and growth*, Cornish, Manchester, 1886; E. Fiddes, *Chapters in the history of Owens College and of Manchester University, 1851–1914*,

Manchester University Press, Manchester, 1937; P. J. Hartog (ed.), *The Owens College, Manchester*, Cornish, Manchester, 1900; reminiscences in *Owens College Magazine*, April 1886.

2 Although Thompson (note 1), p.135, implies that all four professors addressed the same public meeting, the prologue to their published lectures (*Introductory lectures on the opening of Owens College, Manchester*, 2nd ed., Longman, London, 1852, p.xiv) states that only the mathematics Professor spoke on 13 March, the other two speakers following on 14 and 20 March respectively.

3 On Scott (1805–1886) see Thompson (note 1), pp.164–93.

4 The referees were Graham, Playfair, Liebig, Bunsen, Phillips, Gregory, Hofmann, Daubeny, Williamson, Wöhler, Andrews, Blyth, Ronalds, Miller, Brande, Winkelblech, Kopp, Brodie, Rammelsberg, Zwenger, Leigh, Hunt, Varrentrapp, Rose, Laurent, Gerhardt, Fremy, and Bromeis, together with his former employers Cowie and Edmondson: printed booklet *Testimonials in favour of Edward Frankland , Ph.D., F.C.S.*, London, 1850 [RFA, OU mf 01.03.0900].

5 Agreement between Owens College and Edward Frankland, 2 January 1851 [JBA, OU mf 02.02.1679]. The other candidates were Crace Calvert, R. Angus Smith and John Stenhouse, the last two being also pupils of Liebig who, embarrassed by requests to support three of his former students, nevertheless did so, though considered Frankland the strongest candidate (Liebig to Hofmann, 12 April 1850, letter 57 in W. H. Brock (ed.) *Justus von Liebig und August Wilhelm Hofmann in ihren Briefen (1841–1873)*, Verlag Chemie, Weinheim, 1984). Hofmann, however, preferred Stenhouse (Hofmann to Liebig, 22 April 1850, letter 58, *ibid.*).

6 On Greenwood see *DNB*.

7 Frankland, obituary of Tyndall, *Proc. Roy. Soc.*, 1894, 55, xviii (xix); Frankland complained that Tyndall was unable to get a post in the early 1850s, so presumably he too had been a candidate at Manchester. On Sandeman see Boase, and Venn, *Alumni Cantabrigienses*.

8 On W. C. Williamson see *DNB* and his own *Reminiscences of a Yorkshire naturalist*, Redway, London, 1896. He was Frankland's doctor during the Manchester period and remained on cordial terms for many years later; see his letter of congratulation to Frankland on his second marriage: 1 June 1875 [RFA, OU mf 01.01.0547].

9 On J. H. Nicholson see T. C. Hughes, *The literary associations of the county town of Lancaster and its surrounding districts*, Lancaster Guardian, Lancaster, 1929, p.89.

10 Copy-letter, Frankland to Royal Society, September 1851 [RFA, OU mf 01.06.0110].

11 Copy-letters, Frankland to G. Hunt, 25 September 1851 and 12 November 1851 [RFA, OU mf 01.06.0111 and 0120]. The missing item was a sample of guano for analysis.

12 Copy-letter, Frankland to J. Liebig, 22 November 1851 [RFA, OU mf 01.06.0127].

13 A. Lapworth and J. E. Myers, 'Chemistry and Manchester University', in W. H. Brindley (ed.), *The soul of Manchester*, EP Publications, Wakefield, 1974, p 93.

14 R. H. Kargon, *Science in Victorian Manchester: enterprise and expertise*, Johns Hopkins University Press, Baltimore, 1977, p.160. Details of the laboratory are in *Manchester Guardian*, 22 March 1851, p.8, 13 August 1851, p.5, and 7 July 1852, p.9.

15 Printed examination paper, 1851 [RFA, OU mf 01.02.1891].

16 MS examination paper [RFA, OU mf 01.02.1885].

17 MS examination paper; the topics were: 1. Natural occurrence of B and Si; preparation of pure boric and silicic acids; 2. 'Allotropic compositions' of silicic acid; large-scale preparation of it for agricultural purposes; 3. Soluble silicates in architecture; 4. Preparation and properties of chlorine; 5. Bleaching powder and its uses; 6. Analysis of 'chloride of lime'; 7. Effects of chlorine on 'contagion and miasma'; 8. Manufacture of HCl; tests for it; 9. Preparation, properties of bromine and iodine; 10. Occurrence of fluorine and preparation of HF [RFA, OU mf 01.02.1882].

18 MS examination paper; the topics were: 1. Chemical affinity: differences from gravity, cohesion and heat; 2. Chemical affinity: magnification by cohesion and heat; 3. Given 3 acids and 3 bases in same solution, how many salts? 4. Most favourable state for exertion of chemical affinity? 5. Chemical equivalents; law of multiple proportions; 6. Calculation of equivalents; 7. Chemical names for simple compounds; 8. Halides; 9. 'What is the atomic theory?' 10. Salts from -ic and -ous acids. 'Do these acids unite with metallic oxides?' [RFA, OU mf 01.02.1888].

19 Frankland, Lecture notes 'Lectures on technological chemistry 1853/4' [RFA, OU mf 01.06.0367–0418].

20 *Sketches*, 2nd ed., p.129.

21 Frankland, untitled notebook [RFA, OU mf 01.06.0650–0772].

22 Two notebooks of Francis M. Spence [Peter Spence Archives, Laporte Industries Ltd., Widnes]; Francis ['Frank', d.1909], was son of the chemical manufacturer Peter Spence at Pendleton and his eventual successor.

23 Frankland, Lecture notes 'General principles of organic chemistry' [RFA, OU mf 01.06.0454–0502].

24 Particularly relevant studies include D. McKie, 'Wöhler's synthetic urea and the fate of vitalism: a chemical legend', *Nature*, 1944, **153**, 608–9, and J. H. Brooke, 'Wöhler's urea, and its vital force? – a verdict from the chemists', *Ambix*, 1968, **15**, 84–114.

25 A. W. Hofmann, *Berichte*, 1882, **15**, 3152–3.

26 Frankland, Lecture entitled 'Object of organic chemistry' (note 21) [RFA, OU mf 01.06.0714].

27 Frankland, Lecture 1 (note 23) [RFA, OU mf 01.06.0455].

28 Frankland, 'On the production of organic bodies without the agency of vitality', *J. Roy. Inst.*, 1858, **2**, 538–44.

29 Frankland, Lecture entitled 'Organic chemistry' (note 21) [RFA, OU mf 01.06.0716].

30 A. Rocke, *The quiet revolution: Hermann Kolbe and the science of organic chemistry*, University of California Press, Berkeley, 1993, pp.240–1.

31 Frankland, open testimonial, 30 June 1855 [RFA, OU mf 01.04.2061].

32 Frankland, Lecture notes (note 19).

33 M. W. Rossiter, *The emergence of agricultural science: Justus Liebig and the Americans, 1840–1880*, Yale University Press, New Haven, 1975, p.136.

34 E. K. Muspratt, *My life and work*, Bodley Head, London, 1917, p.91.

35 S. Clegg, *A practical treatise on the manufacture and distribution of coal-gas*, 2nd ed., London, 1853, p.42.

36 E. A. Parnell, *Applied chemistry, in manufactures, arts and domestic economy*, London, 1844, pp.145–50; S. Everard, *The history of the Gas Light and Coke Company, 1812–1949*, Benn, London, 1949, p.107.

37 *Illustrated London News*, Supplement, 17 May 1851.

38 Copy-letter, Frankland to Liebig, 22 November 1851 (note 12).

39 Clegg (note 35), p.156.

40 Frankland, *Mem. Manchester Lit. &. Phil. Soc.*, 1852, 10, 71–120; *Experimental Researches*, pp.490–535. The account is based on his article in *J. Gas Lighting*, 1851–2, 2, 123, 247, 262 and 285.

41 *E.g.*, Copy-letter, Frankland to J. Barker, 12 July 1852 [RFA, OU mf 01.06.0263].

42 Frankland (note 40).

43 A manuscript report was read by Frankland to the Chemical Society on 15 December 1851 (*J. Chem. Soc.*, 1852, 4, 344). According to 'a Correspondent' it attracted 'much attention', though it is hard to believe that it was 'the most interesting and instructive document on the chemistry of gases which has ever appeared' (*Manchester Guardian*, 24 December 1851, p.7).

44 *J. Gas Lighting*, 1851–2, 2, 123, 262.

45 A. Fyfe, *J. Gas Lighting*, 1851–2, 2, 369–73 .

46 A. Fyfe, *J. Gas Lighting*, 1853–4, 3, 359–61.

47 Frankland, *J. Gas Lighting*, 1853–4, 3, 422–4 (May), and copy-letter to the Hydrocarbon Gas Co., 13 April 1854 [RFA, OU mf 01.08.0106]; their diverging analyses appear to have arisen largely through differences of scale.

48 Frankland, page proofs for printed *Report* (n.d.) [RFA, OU mf 01.04.1435].

49 Clegg (note 35), p.162.

50 *Report* (note 48).

51 Copy-letter, Frankland to Liebig, 22 November 1851 (note 12).

52 Copy-letter, Frankland to W. G. Ginty, 1 July 1852 [RFA, OU mf 01.06.0257]; Ginty was Stephen White's manager at Ancoats.

53 *E.g.*, T. I. Williams, *A history of the British gas industry*, Oxford University Press, Oxford, 1981, pp.60–1; see also a series of articles on 'History of the manufacture of water-gas' in *J. Gas Lighting*, 1851–2, 2, 244, 264, 282 and 333.

54 Clegg (note 35), p.155.

55 *Experimental Researches*, p.486.

56 J. S. Muspratt, *Chemistry, theoretical, practical and analytical*, Mackenzie, London, n.d. [1853], vol. ii, p.156.

57 References to this work may also be found in Clegg (note 35) and Muspratt (note 56).

58 Copy-letter, Frankland to W. Etchells, 4 July 1852 [RFA, OU mf 01.06.0258].

59 Frankland to C. Johnson, 1 February 1852 [Lancaster Public Library, MS 7144].

60 Copy-letter, Frankland to – de Castro, 26 January 1852 [RFA, OU mf 01.06.0149].

61 Copy-letter, Frankland to J. Myers, 26 March 1853 [RFA, OU mf 01.06.0305].

62 Copy-letter, Frankland to Pembrokeshire Iron & Steel Co., 24 February 1855 [RFA, OU mf 01.08.0243].

63 See J. A. Hassan, 'Relationships between coal, gas, and oil production: a nineteenth-century case study', *Ind. Archaeology Rev.*, 1978, **2**, 277–89; the Russells' consulting chemist was Frederick Penny, former student of Liebig.

64 Copy-letter, Frankland to J. Russell & Sons, 13 April 1852 [RFA, OU mf 01.06.0228].

65 Liebig to Hofmann, 20 June 1853, in W. H. Brock (note 5), letter no. 117.

66 Liebig to Frankland, 20 June 1853 [JBA, OU mf 02.02.1641].

67 Hofmann to Liebig, 26 June 1853 (note 5), letter no.118.

68 C. Reichenbach, many German papers from 1830 and *Edinburgh New Phil. J.*, 1834, **16**, 376–84.

69 Copy-letter, Frankland to the Hydrocarbon Gas Co., 7 May 1853 [RFA, OU mf 01.06.0311].

70 Copy-letter, Frankland to W. G. Ginty, 16 August 1853 [RFA, OU mf 01.06.0355]; at one stage, however, Brande pronounced Boghead Cannel as 'a new peculiar material': Muspratt (note 56), p.83.

71 Frankland to Liebig, 27 June 1853 (note 5, letter 119).

72 Sophie Frankland to her parents, 10 March 1851 [JBA, OU mf 02.01.0155].

73 E. Edwards to Frankland, 4 July 1851 [RFA, OU mf 01.06.0196].

74 *Mem. Manchester Lit. & Phil.*, 1852, **10**, 226–30.

75 Papers of Manchester Photographic Society [Manchester Record Office, M33/1–2].

76 Frankland to R. W. Smiles (Secretary), 10 January 1854 [Manchester Record Office, National Public School Association papers, M136/2/3/1113]; this was formerly the *Lancashire* Public School Association .

77 Frankland unsuccessfully attempted to persuade Joule to lecture there: J. P. Joule to Frankland, 20 July 1857 [RFA, OU mf 01.04.0610].

78 Frankland to Playfair, 25 January 1853 [Imperial College Archives, Playfair Collection, no. 279].

79 C. R. Weld to Frankland, 3 January 1853 [RFA, OU mf 01.01.0215].

80 Copy-letter, Frankland to E. W. Binney, 15 January 1852 and 28 June 1858 [RFA, OU mf 01.06.0177 and 01.07.0152]; Binney to Frankland, 18 June 1858 [*ibid.*, 01.07.0150].

81 E. A. Parnell to Frankland, 27 September 1852 [RFA, OU mf 01.02.0245].

82 Copy-letter, Frankland to F. C. Calvert and R. A. Smith, 5 June 1855 [RFA, OU mf 01.07.0173]; F. C. Calvert to Frankland, 7 June 1855 [*ibid.*, 01.07.0181]; R. A. Smith to Frankland, 28 November 1852, and 8 June 1855

[*ibid.*, 01.02.0256 and 01.07.0175]. On Calvert (1819–73) see *J. Chem. Soc.*, 1874, **27**, 1198–9; on R. A. Smith (1817–84) see *J. Chem. Soc.*, 1885, **47**, 335, *DNB* and A. Gibson and W. V. Farrar, 'Robert Angus Smith, F.R.S., and Sanitary science', *Notes & Records Roy. Soc.*, 1973–4, **28**, 241–62.

83 Hofmann to Frankland, 28 April 1855 [RFA, OU mf 01.03.0274].

84 Nasmyth to Frankland, 22 January 1856 [RFA, OU mf 01.02.0279]; on Nasmyth (1808–90) see S. Smiles (ed.), *J. Nasmyth, An autobiography*, London, 1885, and *DNB*.

85 Copy-letters, Frankland to Muspratt & Co., 2 and 22 February 1852 [RFA, OU mf 01.06.0172 and 0164].

86 T. L. Elliott, D. Thomas and J. H. Harwood, 'History of Peter Spence and Sons Ltd.', *Spence Magazine*, 1960–1, p.14.

87 Copy-letter, Frankland to – de Castro (of Hydrocarbon Gas Co.), 25 April 1854 [RFA, OU mf 01.08.0115].

88 Copy-letters, Frankland to Hunt, 25 September and 12 November 1851 (note 11).

89 Richard Frankland to Edward Frankland, 13 December 1851 [RFA, OU mf 01.03.0286].

90 Copy-letter, Frankland to G. Hunt, 10 March 1852 [RFA, OU mf 01.06.0166].

91 Copy-letter, Frankland to [illegible], 9 April 1853 [RFA, OU mf 01.06.0308].

92 Copy-letter, Frankland to Royal North Lancashire Agricultural Society, 28 May 1853 [RFA, OU mf 01.06.0315].

93 Copy-letter, Frankland to Royal North Lancashire Agricultural Society, 28 March 1854 [RFA, OU mf 01.08.0094].

94 Copy-letter, Frankland to D. Watney, 6 November 1852 [RFA, OU mf 01.06.0286]; Daniel Watney, younger brother of James, was part-owner of the Stag Brewery, Pimlico, and (by 1853) owner of the Battersea distillery: London *Directories*; W. P. Serocold, *The story of Watneys*, Watney Combe Reid & Co. Ltd., London, 1949; and J. Watney, *Beer is best: a history of brewing*, Peter Owen, London, 1974, pp.106–7.

95 Copy-letter, Frankland to D. Watney, 10 November 1852 [RFA, OU mf 01.06.0281].

96 Copy-letter, Frankland to D. Watney, 24 November 1852 [RFA, OU mf 01.06.0283].

97 Copy-letter, Frankland to D. Watney, 29 August 1854 [RFA, OU mf 01.08.0141].

98 Personal communication from Dr Ray Anderson of Allied Breweries Ltd.

99 Copy-letter, Frankland to D. Watney, 17 February 1853 [RFA, OU mf 01.06.0302].

100 Copy-letter, Frankland to D. Watney, 20 February 1855 [RFA, OU mf 01.08.237].

101 R. Holland, 'Alkali production at Netham', *J. Brit. Industrial Archaeological Soc.*, 1983, **16**, 28–30.

102 Copy-letter, Frankland to G. S. Stott and S. Cox, 13 May 1852 [RFA, OU mf 01.06.0233].

103 Copy-letter, Frankland to S. Cox, 17 May 1852 [RFA, OU mf 01.06.0235].

104 Copy-letter, Frankland to S. Cox, 27 May 1852 [RFA, OU mf 01.06.0241].

105 Copy-letter, Frankland to Cox, Score & Co., 2 October 1852 [RFA, OU mf 01.06.0269].

106 Copy-letter, Frankland to Cox, Score & Co., 16 February 1853 [RFA, OU mf 01.06.0295].

107 Copy-letter, Frankland to Cox, Score & Co., 26 February 1853 [RFA, OU mf 01.06.0303].

108 Copy-letter, Frankland to Cox, Score & Co., 23 July 1853 [RFA, OU mf 01.06.0349].

109 Copy-letter, Frankland to Cox, Score & Co., 5 October 1853 [RFA, OU mf 01.06.0359].

110 Copy-letter, Frankland to S. Cox & Co., 2 June 1855 [RFA, OU mf 01.08.0260].

111 Presumably the Runcorn Bone Works. In this enterprise he was directly encouraged by Frankland, who offered his services should the move take place and quickly followed this proposal with a long letter explaining in detail the manufacture of superphosphate and nitrogenous fertilisers: Frankland to W. Score, 18 November and 1 December 1854 [RFA, OU mf 01.08.0171 and 0177].

112 Frankland to Tyndall, 22 October 1855 [Royal Institution Archives, Tyndall papers, 9/E2.5].

113 Frankland to Tyndall, 13 November 1855 [Royal Institution Archives, Tyndall papers, 9/E2.6].

114 Copy-letter, Frankland to B. P. Gregson, 25 October and 8 November 1851 [RFA, OU mf 01.06.0113 and 0118].

115 Copy-letter, Frankland to G. Carruthers, 30 September 1852 [RFA, OU mf 01.06.0267].

116 Copy-letter, Frankland to J. F. Foster, 1 November 1852 [RFA, OU mf 01.06.0278].

117 The earliest reference to chemical analysis in the service of the railways that I have otherwise been able to find is a letter from William Bouch of 24 October 1859 relating to analysis of River Eden water by the Newcastle chemist Thomas Richardson [Newcastle-upon-Tyne City Library, Letters and cuttings from the early history of railways, f.61]. The first railway chemistry laboratory was set up at Crewe in 1864.

118 Frankland to J. B. Worthington, 23 December 1854, 29 January, 20 and 24 February 1855, and 30 October 1856 [RFA, OU mf 01.08.0215, 0227, 0233, 0239 and 0278].

119 W. A. Miller to Frankland, 1 July 1855 [RFA, OU mf 01.03.0275].

120 Copy-letter, Frankland to Sir John Potter, 26 January 1852 [RFA, OU mf 01.06.0153].

121 Copy-letter, Frankland to Cox, Score & Co., 18 January 1854 [RFA, OU mf 01.08.0076].

122 Copy-letter, Frankland to J. R. Ralph, 6 January 1852 [RFA, OU mf 01.06.0209].

123 Copy-letter, Frankland to Binney (note 80).

124 Copy-letter, Frankland to L. P. Aston, 14 February 1852 [RFA, OU mf 01.06.0202].
125 Copy-letter, Frankland to Cox, Score & Co., 12 October 1852 [RFA, OU mf 01.06.0273].
126 Copy-letter, Frankland to W. Dentith, 21 and 23 July 1853 [RFA, OU mf 01.06.0342 and 0347]; Dentith was a drysalter of 5 Lime Grove, Greenhays.
127 Copy-letter, Frankland to W. Etchells 4 July 1852 [RFA, OU mf 01.06.0258].
128 Copy-letter, Frankland to Carruthers (note 115).
129 Copy-letter, Frankland to Foster (note 116).
130 Calvert to Frankland (note 82).
131 R. A. Smith to Frankland, 6 June 1855 [RFA, OU mf 01.07.0175].
132 Copy-letter, Frankland to H. Gregson, 18 May 1852 [RFA, OU mf 01.06.0238].
133 Copy-letter, Frankland to H. Gregson, 24 June 1853 [RFA, OU mf 01.06.0316].
134 Copy-letter, Frankland to H. Gregson, 29 June 1853 [RFA, OU mf 01.06.0318].
135 R. A. Smith, in Kargon (note 14), p.144.
136 T. Graham to Frankland, 15 August 1851 [RFA, OU mf 01.07.0185].
137 Copy-letter, Frankland to Cox, Score & Co., 2 October 1852 (note 105).
138 Copy-letter, Frankland to Cox, Score & Co., 9 February 1853 [RFA, OU mf 01.06.0290].
139 Copy-letter, Frankland to S. Cox, 12 June 1852 [RFA, OU mf 01.06.0248]. No trace of Frankland's letter to Muspratts on this matter has been discovered.
140 Copy-letter, Frankland to D. Watney, 28 October 1853 [RFA, OU mf 01.08.0048].
141 Elliott, Thomas and Harwood (note 86).
142 Laboratory Book of Peter Spence, 28 October 1853 [Laporte Industries Ltd. Archives, Widnes].
143 J. Fenwick Allen, *Some founders of the chemical industry*, Sherratt & Hughes, London, 1906, pp.278–80; the firm was found guilty of nuisance but not of being injurious to health.
144 *Manchester Guardian*, 24 August 1857, p.4.
145 On J. S. Muspratt see M. D. Stephens and G. W. Roderick, 'The Muspratts of Liverpool', *Ann. Sci.,* 1972, **29**, 287–311 (293–300).
146 Muspratt (note 56).
147 Copy-letter, Frankland to J. S. Muspratt, 9 May 1851 [RFA, OU mf 01.06.0197]; it is extremely unlikely that either of these esters (or of the nonanoate) would be present naturally in concentrations exceeding 0.5 mg/l, though they could have arisen through some acid-catalysed esterification of the free acids.
148 Copy-letter, Frankland to J. S. Muspratt, 19 July 1851 [RFA, OU mf 01.06.0210].
149 Reported in a copy-letter from Frankland to Liebig, 22 November 1851 (note 12).
150 Copy-letter, Frankland to J. S. Muspratt, 7 August 1851 [RFA, OU mf 01.06.0205].

151 Liebig to Hofmann, 15 November 1851 (note 5, letter no. 78).

152 Copy-letter, Frankland to Liebig, 22 November 1851 (note 12).

153 Copy-letter, Frankland to W. Gregory, 1 January 1852 [RFA, OU mf 01.06.0207].

154 Gregory to Frankland, 10 January 1852 [RFA, OU mf 01.03.0222].

155 Cited in copy-letter from Frankland to Liebig 16 February 1852 [RFA, OU mf 01.06.0162].

156 Copy-letter, Frankland to J. S. Muspratt, 16 January 1852 [RFA, OU mf 01.06.0142].

157 Copy-letter, Frankland to J. S. Muspratt, 9 November 1853 [RFA, OU mf 01.08.0034].

158 English Patent 938, 25 April 1855: Kargon (note 14).

159 *Mem. Manchester Lit. & Phil. Soc.*, 1852, **10**, 71; *Experimental Researches*, p.516; on Russell (1830–1909) see *DNB*; G. C. F[oster], *Proc Roy. Soc.*, 1910–11, **A84**, xxx–xxxi; W. A. T[ilden], *J. Chem. Soc.*, 1918, **113**, 339–50.

160 Copy-letter, Frankland to C. J. Tufnell, 21 July 1853 [RFA, OU mf 01.06.0339].

161 Rocke (note 30), pp.212–13; Gerland was strongly recommended: Kolbe to Frankland, 18 August 1854 [RFA, OU mf 01.07.0158].

162 Sophie Frankland to family, 18 November 1855 [PCA, OU mf 03.01.0073].

163 Frankland to Tyndall, 13 November 1855 (note 113).

164 Copy-letter, Frankland to L. Vaughan, 10(?) January 1855 [RFA, OU mf 01.08.0219]; however at about the same time he was negotiating with his natural father for payment of a large sum of money, presumably part of the secrecy deal negotiated over his paternity: Copy–letter, Frankland to E. C. Lowndes, 15 January 1855 [RFA, OU mf 01.08.0221].

165 Copy-letter, Frankland to Duke of Argyll, 30 December 1851 [RFA, OU mf 01.06.0134].

166 Frankland to Tyndall, 13 November 1855 (note 113).

167 Heinrich's visit to England included the usual sights of London (the Zoo, the Stock Exchange, shops and riverside), followed by the rain and smoke of Manchester. He remarked that 'the satisfaction of my curiosity' was the only real enjoyment from the visit: H. Fick, *Ein Lebensbild. Nach sienen eigenen Aufzeichnungen dargestellt und erganzt*, von Hélène Fick geb. Ihlee, J. Schabelitz, Zurich, 1907.

168 They had to have their 'usual abode' within the parish, so their Marriage Allegation refers to 'Edward Frankland of the Parish of St Martin-in-the Fields' [Marriage Allegation, 3 February 1851, Vicar General's Office, Lambeth Palace].

169 John Frankland (b. 1808) was Margaret Frankland's youngest surviving brother; he was a piano-tuner by trade. His wife Mary was a witness to the marriage.

170 *Sketches*, 2nd. ed., pp.125–6.

171 Marriage register, St-Martin-in-the-Fields, 1851, entry 492.

172 Sophie Frankland to parents, 10 March 1851 [JBA, 02.01.0155]; also in *Sketches*, 1st ed., pp.292–305. The road, long since swept away, was near where the present A56 crosses the Mancunian Way.

173 Copy-letter, Frankland to W. Booth, 22 June 1852 [RFA, OU mf 01.06.0251].

174 Copy-letter, Frankland to F. Darragh, 24 October 1853 [RFA, OU mf 01.08.0042]; Darragh was a joiner, of Salford.

175 Sophie Frankland to parents, 23 June 1857 [PCA, 03.01.0103].

176 Copy-letter, Frankland to Watney, 6 November 1852 (note 94).

177 *Sketches*, 1st ed., p.310. Simpson was the famous pioneer in chloroform anaesthesia; he later wrote to Frankland seeking information about procurement of 'hydruret of amyl' [pentane] for possible use as an anaesthetic: J. Y. Simpson to Frankland, 30 January 1856 [RFA, OU mf 01.02.0230].

178 *Sketches*, 1st ed., pp.312–13. The passage was removed in the 2nd edition, possibly out of deference to Fred's feelings.

179 Typed recipe dated 1854 [RFA, OU mf 01.03.0587]. This work seems to have escaped the attention of previous historians, including V. A. Fildes who deals with little after 1800: 'The early history of artificial feeding', *Maternal & Child Health*, 1986, **11**, 222–7, 259–63, 'The early history of the infant feeding-bottle', *Nursing Times*, 1981, **77**, 128–9; 168–70, and *Breasts, bottles and babies: a history of infant feeding*, Edinburgh University Press, 1986.

180 *Manchester Guardian*, 2 December 1854; transcript [RFA, OU mf 01.04.1152]; *Experimental researches*, pp.843–4..

181 *Sketches*, 2nd ed., p.277–80.

182 *Lancastrian Chemist*, p.47.

183 *Sketches*, 1st ed., pp. 310–14.

184 Frankland to Tyndall, 22 October 1855 (note 112).

185 Frankland was writing letters to Disley (from Germany) by mid-1849: Frankland, *Diary*, 24 June 1849 [JBA, OU mf 02.02.1304].

186 1851 Census Return [PRO, HO 107/2153].

187 *Sketches*, 2nd ed., pp. 276 and 119.

188 *Introductory lectures* (note 2), pp.112–33 (118–20).

189 Copy-letter, Frankland to J. Fleming, 31 December 1851 [RFA, OU mf 01.06.0139].

190 J. Tyndall, *Journal* for 3 May 1857 [Royal Institution Archives].

191 C. H. Spurgeon, *The New Park Street pulpit*, vol. III, repr. Banner of Truth Trust, London, 1964, p.185.

192 Spurgeon and his preaching had recently attracted the attention of *The Times* leader-writer: *The Times*, 13 April 1857, p.6.

193 *Sketches*, 2nd ed., pp.133–4.

194 A local firm of dye manufacturers whose employees were to include such notables as H. Caro, C. A. Martius and C. Schorlemmer. Their Bismarck Brown was otherwise known as 'Manchester Brown'.

195 *Sketches*, 2nd ed., pp.279–80.

196 As in his correspondence with Tyndall, *e.g.* letter dated 'Sunday afternoon' [1855] [Royal Institution Archives, Tyndall papers, 9/E2.2].

197 *Sketches*, 1st ed., p.307.

198 Royal Institution Managers' Minutes, 15 November 1852 and 2 February

1857 (facsimile reproduction, ed. F. Greenaway, by Scholar Press, London, 1976).

199 Frankland to Fick family, 10 April 1853 [PCA, 03.01.0047]; similarly he declined work at Manchester gasworks on the grounds that the RI lectures kept him away for two days a week: Copy-letter, Frankland to Gibb, de Castro & Co., 4 April 1853 [RFA, OU mf 01.06.0307].

200 Frankland to Tyndall, 21 April 1855 [Royal Institution Archives, Tyndall papers, 9/E2.4].

201 Copy-letter, Frankland to W. Thynne, 5 November 1853 [RFA, OU mf 01.08.0050].

202 Royal Institution, *Notices of the meetings*, 2 June 1854; *Sketches*, 2nd ed., p.134.

203 Frankland to Tyndall, 21 April 1855 (note 200).

204 Copy-letter, Frankland to J. Russell & Sons, 4 July 1853 [RFA, OU mf 01.06.0320].

205 Tyndall to Frankland, 4 August 1856 [Royal Institution Archives, Tyndall papers, 9/E7.8].

206 J. Tyndall, *The glaciers of the Alps*, Everyman Ed., J. M. Dent, London, 1906, pp.28–30.

207 Frankland to Tyndall, 12 November 1855 [Royal Institution Archives, Tyndall papers, 9/E2.6].

208 Frankland to Bunsen, 5 March 1856 [Heidelberg University, Bunsen papers; cited by Kargon (note 14), p.164].

209 Copy-letter, Frankland to A. Neild, 20 February 1854 [RFA, OU mf 01.08.0080].

210 *Record of Jubilee Celebrations at Owens College Manchester*, Manchester, 1902, p.5.

211 Thompson (note 1), pp.153–6.

212 *Sketches*, 2nd ed., p.155.

213 H. E. Roscoe, *Record of work done in the Chemistry Department of the Owens College 1857–1887*, Macmillan, London, 1887, pp.1–2.

214 W. C. Williamson, *Reminiscences of a Yorkshire naturalist*, Redway, London, 1896, pp.139–40; at a college meeting the professors, surely tongue-in-cheek, regretted Scott's resignation and astonishingly claimed it was 'wholly unexpected' by them: Thompson (note 1), pp.185–6.

215 *Sketches*, 2nd. ed., p.135.

216 Legal Agreement between Frankland and Owens College, 23 September 1857 [RFA, OU mf 01.07.0022].

217 Lapworth and Myers (note 13).

218 D. S. L. Cardwell, *James Joule: a biography*, Manchester University Press, Manchester, 1989, p.109.

219 M. J. Nye, *From chemical philosophy to theoretical chemistry*, University of California Press, Berkeley, CA, 1993, p.194.

CHAPTER 8

Return to the Metropolis

1 'Accomplished chemical pluralist'

Frankland's move to London was to some extent born out of desperation. The situation at Manchester had deteriorated so badly that it was beyond his capacity to improve it, given his lifestyle as a consultant, his personal priorities and the simple fact that he had been there too long. New blood was needed and, as we have seen, it was supplied by Henry Enfield Roscoe who came to the rescue just in time. But the move was also an outcome of a burning ambition to be in the centre of things, to be fully recognised as a leading chemist, to be socially accepted by the chemical community, in short to prove to himself that he had overcome his inherited disabilities. In no way could Manchester pretend to offer the advantages of London for these purposes and so he alerted his friends in the metropolis to his need of a post and one of them provided the answer.

Henry Bence Jones (1814–73) was a physician at St. George's Hospital, London, and a former student of Liebig and Graham. He was also a Manager (member of the governing body) for the Royal Institution and a friend of Tyndall. Conceivably he first met Frankland at the RI Lectures in 1853. However, by 1856 their friendship was established and, together with Tyndall, Bence Jones made it his business to look out for suitable vacancies. He did not have long to wait. A post soon became available at St. Bartholomew's Hospital, London.

1.1 St. Bartholomew's Hospital

This ancient Hospital has survived from its foundation in 1173 until its threatened destruction by the Department of Health in the 1990s. With new buildings in the mid-eighteenth century, but still on its original site, it advertised a course 'on the application of chymistry to medicine' in 1788. By the 1850s it was one of four London medical schools teaching practical chemistry (the others being Guy's Hospital, Aldersgate Dispensary and Sydenham College).[1] This facility remained until 1981. William Brande served as 'professor' of chemistry from 1836 to 1841, and in February 1851 the appointment was announced of John Stenhouse, a pupil of Liebig.[2] He soon acquired a demonstrator, F. A. Abel, who left a year later to begin a career of distinguished service at Woolwich Arsenal; he was to receive commendation for his work at Barts from Frankland himself.[3] From 1853 to 1855 Stenhouse's assistant was the youthful F. A. Kekulé, who began to develop with Odling, Williamson and others the germs of structure theory (and had his famous 'dream' on the Clapham omnibus). It was Frankland's misfortune to have been absent from London for those crucial years in the history of chemistry, a fact that may go some small way to explaining his subsequent failure to relate to Kekulé's chemical ideas. If only he had known this in 1855 his efforts to escape from Manchester might have been made with redoubled energies. However the days of Stenhouse were numbered and in July 1857 he resigned 'owing to the present very feeble state of my health',[4] apparently a serious but not fatal form of paralysis. His misfortune was Frankland's opportunity.

Once he heard the news (presumably from Bence Jones) Frankland wrote promptly to him and to James Paget, then Professor of Anatomy at Barts. As a well-known physician, and acquaintance of many influential people who frequented the lectures at the Royal Institution, Bence Jones was in a position to pull a few strings. By July 11 he was advising Frankland that

> I have in consequence told him [Paget] that you are a candidate, that I will write a note for you to the committee stating that you are desirous of the appointment, & Mr. Paget will get his colleagues to recommend the committee to elect you on the 20th of this month lecturer at St. Bartholomew's.[5]

Thus Frankland's appointment was clearly 'fixed' by his friend. Bence Jones went on to indicate the conditions of the appointment. Between October and March there were to be three weekly lectures, at 10 a.m. on Mondays, Wednesdays and Fridays (with a fortnight's holiday at Christmas). For each week of lectures he earned five guineas, bringing him about £130 for half a year, but there were no other commitments during that period. From May to July, however, there were four morning classes of practical chemistry, each student paying for apparatus as well as a fee of two guineas. As 60 students were expected that meant another £130 or so. He could also take as many private laboratory pupils as he could get and as many assistants as he could afford. An undated letter from Paget to Bence Jones indicated also that coal and gas were supplied by the Hospital and that Stenhouse would be willing to sell to his successor any apparatus required ('at a valuation'), and informing him that 'very little can be made by laboratory pupils against the competition of the [Royal] College of Chemistry'. In an almost illegible comment Paget appears to quote Stenhouse to the effect that Frankland could expect a clear £350 p.a.[6]

The appointment was duly made by a committee of governors; Frankland was to be 'Lecturer on chemistry'.[7] On 22 July Bence Jones wrote to Frankland wishing him every success. His lectures really should start on Friday 3 October but the following Monday would be acceptable. He was to work out a prospectus and take it to Paget, who would then tell him everything. Thanks to Jones' efforts the committee had been manipulated, Frankland was coming to London, and 'it was all most satisfactory'.[8] From his new home in Kent the Manchester engineer James Nasmyth also rejoiced that his old friend had come south 'for good' and that he would be 'so near and so often in contact with first rate men in other departments of science'.[9]

A measure of Frankland's determination to leave Manchester lies in his willingness to accept apparent demotion, from a 'professorship' to a 'lectureship' (even though such terms did not always convey the distinction we recognise today). So there was no Inaugural Address to give. Few, if any, of his lecture notes from this period have survived and it is likely that he recycled the abundant material from Manchester days. One extant but undated document is a manuscript 'Introductory Address'.[10] It might have been expected to have come from his first years at the hospital but the opening paragraph refers to the death of the institution's 'Dr. Baly' in the past year. Assuming this was the famous William Baly, Physician to the Queen, the date of the lecture must be 1861 or 1862. It was certainly an introduction to chemistry aimed at medical students. If not exactly a 'hard sell' for chemistry it came quite near it. Frankland made a repeated appeal to the 'dignity' of the medical profession. Once having a virtual monopoly of scientific training it was now in danger of losing its pre-eminence to civil and military departments of the state. On questions of natural phenomena 'the doctor is constantly appealed to as the highest available authority' and the profession cannot afford to lose such confidence. To modern minds it seems a curious argument, appealing more to vanity than anything else, but it was reinforced by more mundane considerations. Natural science was to be commended 'as a means of mental culture and also on account of its many applications to medicine'. Here, rather than in the rarefied regions of natural theology, Frankland hoped to strike a chord with his student audience. Accordingly he plunged into a discussion of the atmosphere and its rôle in respiration. He began by referring to a current controversy about ozone in the atmosphere. Since the early 1850s use of potassium iodide test papers had convinced many observers that, especially in rural and coastal districts, there were small, but measurable, amounts of ozone in the atmosphere. These were believed to contribute to the salubrity of air away from towns and cities. In fact, however, the estimates were far too high since, as we now know, other substances can lead to positive readings, especially oxides of nitrogen. So Frankland's doubts have a strangely modern ring about them:

> Dr. Frankland referred in the following terms to the supposed influence of ozone in purifying the air, and to the fallacy of the test which is now so generally used in meteorological observatories for the detection of this substance in the atmosphere:
>
>> Without waiting to enquire into the grounds upon which the claims of ozone as a constituent of the air rested, numerous observers began to gauge the atmosphere for this purifying and health giving body, and the indications of ozone test papers were, and still are, believed by many to afford valuable information respecting the sanitary condition of the atmosphere. Contemplating the industry with which such observations have been made it is painful to have to come to the conclusion that up to the present moment we have no reliable evidence whatever that ozone is present amongst the constituents of our atmosphere.

He then alluded to the deleterious effects of organic matter in the air and to its estimation by Dr. Angus Smith's test which had proved that the air of Manchester contains on an average more organic matter than that of London in a proportion of 6:4, the comparative rate of mortality in the two cities being as 31:24. This test had also proved that the air on the high grounds in the northern suburbs of London contains only about $\frac{1}{3}$rd as much organic matter as the air in the city; that there are human dwellings in our large towns which are surrounded by an atmosphere fouler as regards organic matter than that of a piggery; and that the amount of organic matter in the average London atmosphere is about 10 times as great as that contained in the pure sea-breeze. On the subject of respiration he said:

> We here come at once into contact with phenomena which may be said to be at the very foundation of all philosophical views of vitality, and which phenomena can receive an explanation from chemistry and physics alone. It is here that those forces originally stored up and conserved in vegetable organisms find their outlet. By this act of respiration is brought into activity that dynamic power which manifests itself in the aggregation of phenomena which we call life.

By this well-chosen example he had touched upon several matters of public concern about health, he had commented on a live issue in chemistry, and had come perilously near a reductionist view of the life process. The students cannot have found it dull! As if to emphasise his insistence on a materialist explanation of organic phenomena he cited recent work in laboratory syntheses 'without any assistance from vitality'. On this occasion he wisely omitted Wöhler's preparation of urea. The Address concluded with a ringing call for his audience 'to furnish themselves with all available knowledge as to be able efficiently to cope with disease, prolong life and relieve human suffering.'

From his predecessor Frankland acquired a demonstrator whom Stenhouse had appointed in September 1854. This was John Attfield who, for several years, was daily with him as lecture assistant and then research assistant. He obviously learned quickly from Frankland that spare time can be spent to advantage, for while at Barts he wrote many chemical articles for Brande's *Dictionary* and for the *English Cyclopaedia*, as well as revising the chemical portions of Clegg's standard work on coal-gas manufacture.[11] After Frankland's death Attfield was to write: 'my first quarter of an hour with him revealed his great powers of perception and his transparent simplicity and nobility of character'.[12] This loyal assistant left in 1862 to become Professor in the Pharmaceutical Society, where he remained until 1896. Meanwhile further help was to come. Frankland writes:

> Shortly before leaving St. Bartholomew's, I became acquainted with Mr. Frank Baldwin Duppa who had gone through a course of chemical training under Hofmann at the Royal College of Chemistry, and who was anxious to learn gas analysis in my laboratory. I soon found Duppa to be a most enthusiastic young chemist, and after he had acquired facility in gas analysis, I invited him to join me in a research, already commenced, on the action of zinc ethyl upon boracic ether; but he had only worked for a short time upon this subject before his health gave way, and he was ordered by his doctor to spend the winter abroad.[13]

For reasons that will soon be apparent St. Bartholomew's Hospital was not to prove a great place for chemical research, though as the above quotation indicates, some work on organometallic chemistry was undertaken. Neither were short-term industrial consultancies as available in the south as they had been in Manchester. However he maintained an interest in the affairs of 'Paraffin' Young. A letter from his Mancunian friend Edward Binney enquired whether Frankland (and Warington) intended to serve Young in his forthcoming legal action against the Clydesdale Chemical Co.[14] Frankland replied that they did not feel constrained 'to take any active part against you on behalf of strangers – so consider ourselves retained by you'.[15] He conducted some experiments on 'Gillard's water gas'. This was a product obtained from the action on charcoal at a bright red heat of superheated steam under high pressure, injected through small orifices. Together with a Mr Wright, Frankland conducted experiments on rabbits to determine its toxicity. The gas was burnt in a cylinder of platinum gauze to emit a bright and steady light; Frankland concluded that more of this 'Hydroplatinum gas' would be needed than coal-gas to produce equivalent illumination, but it had the great environmental advantage of not producing sulphur dioxide.[16] He wrote to the Gluten Food and Starch Company with analyses of their bread, cited Liebig on flour and commended gluten as an ingredient of cattle food and a source of 'yellow prussiate of potash'.[17]

Not all his work was carried out in the laboratories of St. Bartholomew's Hospital. He was often seen at the Royal College of Chemistry. At the request of the Metropolitan Board of Works he and Hofmann worked on deodorisation of the Thames, suggesting iron(III) chloride as a remedy[18] (in which they were opposed by

Odling and Letheby). In 1861 Frankland was working with Hofmann on protecting various kinds of stone from degradation by acid rain.[19] Two years later he was collaborating with Tyndall at the Royal Institution on a closely related problem: measuring the degrees of pollution in London air.[20]

At about the same time Frankland was being consulted by C. W. Curtis, of the explosives manufacturers Curtis & Harvey, on a variety of questions relating to gun-cotton.[21] He was now displaying a new interest in such matters as explosives. This would probably have been encouraged by another opportunity that arose for supplementing his income, even though it was to militate against further sustained research for the next few years. The new opening came from an old-established institution known as Addiscombe College.

1.2 Addiscombe College

There was a curious sense in which the fate of the Lancastrian-bred chemist Edward Frankland was on occasion bound up in a rather improbable partnership, with one of the great institutions of Britain's imperial history, the Honourable East India Company.[22] Not for the first time was he to become involved in activity on its behalf.

The East India Company was not merely a trading organisation. It required engineers to build and maintain bridges and railways, for example. The need for these persons to be properly trained seems to have been recognised rather late in the Company's history, but by 1848 the idea was conceived of using an existing establishment, the Putney College of Civil Engineering. In the following year the intention was announced of 'improving the College system by the addition of classes in metallurgy and general engineering for Indian officers'.[23] An unusual spectacle at the previous annual prize distribution had been the presence of no fewer than four directors of the East India Company.[24] This was only a few months before Frankland's return in 1849, so he doubtless encountered something of the East India mentality in the subsequent year.

The Company also had other responsibilities of an awesome magnitude. To maintain civil order so necessary for commerce it ran an army that by the 1820s had exceeded a quarter of a million, larger than that of any European power. By the beginning of the century the need for training in England was urgently apparent. Just as the Company had founded its college in 1807 at Haileybury[25] for educating its new breed of civil servants, so two years later it set up a military college,[26] to provide its army with trained personnel. In 1809 it acquired Addiscombe House, an early eighteenth century mansion on the outskirts of Croydon, Surrey, and formerly the property of Lord Liverpool. The building was greatly extended, with chapel, parade ground, dormitories and a huge enclosure where fortifications could be modelled in sand. The fees were £30 p.a., students being admitted for a two year course prior to posting with Engineers, Artillery or Infantry. Its head was always an army officer (usually retired). Cadets awoke each day to Reveille at 6 a.m., followed by Parade and Chapel, and then embarked on a regime of study and other activity appropriate for their final vocation. Only for two hours each day could they do as they liked.

To this unlikely establishment came Edward Frankland when, in 1859, he took up an additional post as lecturer in chemistry and physics at the Royal Indian Military College at Addiscombe.

With Frankland's total lack of military connection or interest it is pertinent to ask what possessed him to take on this additional duty, so remote from his other responsibilities at the hospital. The educational task before him does not offer a ready explanation. The austere regime was coupled with a correspondingly military curriculum in which science was far from prominent. In 1843, for example, the subjects taught, with their relative weightings, were: mathematics (150), fortification (100), military drawing (40), military surveying (40), civil drawing (40), Hindustani (80), French (40), Latin (20).[27] A little later chemistry, physics and geology appear as subjects, but with very minor ratings; they clearly played a 'service' rôle to more important concerns. This impression is borne out by consideration of the staff. For the main subjects there was a staff of 'professors' who were all, at least at one stage, Wranglers from Oxford or Cambridge. A list from 1850 includes all their names, and is followed by a 'Lecturer in geology' and 'Lecturer in chemistry' at the tail of the list and in the company of 'Instructor in Fencing &c.' and 'Steward and Purveyor'. More important even than lack of social kudos was the evident lack of laboratory facilities.[28]

Yet the strange thing about unpromising Addiscombe was that

its parent body had long had more than a dallying interest in science as well as engineering. To further its own ends the East India Company had sought from at least the turn of the century to promote studies in botany, chemistry and other branches of natural knowledge that could help to exploit and manage the vast material resources of India.[29] For some years the Royal Institution itself played an important part in this strategy.[30] However Frankland was not merely to locate himself within a long-standing if unexpected tradition of science, but also to find his place within a network of part-time academic workers that already included several well-known people. The 'Lecturer in geology' in 1850 was David Thomas Ansted, Fellow of Jesus' College, Cambridge and former associate of Adam Sedgwick. He held the additional post of Professor of Geology at King's College, London, and had been from 1844 to 1847 Assistant Secretary of the Geological Society.[31] He had also been one of four King's men teaching part-time at Putney College from at least 1845 to 1848.[32] The 'Lecturer in chemistry' at the same time was Edward Solly, who worked on the border between chemistry, electricity and horticulture, and was an associate of Faraday, having lectured at the Royal Institution in 1841. Solly had also been Honorary Professor at the Royal Horticultural Society in 1845–6 but from 1849 spent most of his energies as director of the Gresham Life Assurance Society and as an amateur antiquarian.[33] It was said that Frankland was his direct successor.[34] Thus Putney College and the Royal Institution already had links with Addiscombe. To their number could be added the third metropolitan institution with which Frankland had been connected, the Museum of Economic Geology. A letter from Lord Lincoln to Lyon Playfair in 1845 had assured that ambitious young man that there could be no objection to his going to Addiscombe; it would be a useful addition to his income and would not interfere with his duties at the Museum.[35] It can be seen that Frankland had good precedent for augmenting his income by part-time teaching and plenty of reason for choosing Addiscombe College for that purpose.

His lectures would usually be at 7 p.m., during one of the periods labelled 'Study'.[36] It is not at all clear exactly what he was teaching, but probable that it was on a somewhat *ad hoc* basis. For example he wrote to the French chemist Cahours explaining that a planned visit to Paris *en route* for Germany in September 1860 had to be cancelled on account of a sudden demand for a new course of lectures at the college.[37] It happened that in 1849 his predecessor Solly had published what purported to be a *Syllabus* of his own chemistry lecture courses at Addiscombe, though it amounts rather to a long series of condensed lecture notes.[38] Using the old atomic weight system ($O = 8$) the lectures appear to have been little more than a dreary trudge through the elements and their compounds in a purely descriptive manner, with little attempt to spell out underlying principles. What is more surprising, perhaps, is that the gentlemen of the E. India Academy were given extremely little on those parts of chemistry that (despite the book's title) they might have been expected to find specially relevant, such as explosives or metallurgy. Possibly that may have been one reason for replacing Solly, but in any event Frankland was given a golden opportunity to inject new interest and relevance into the course. Just as some of his teaching at Manchester was devolved upon his assistant Gerland, so now we find Attfield coming over from St. Bartholomew's Hospital to help out by 'lecturing at Addiscombe'.[39] How far Frankland saw this as a convenient device for finding more time for himself, or as a genuine attempt to give assistants training 'on the job', is difficult to say.

While at Addiscombe Frankland took the opportunity to observe shell-firing practices and commented on these in a letter to Stokes about some work he had recently been doing on the influence of atmospheric pressure on the burning of time-fuses.[40] As physics was included on his agenda he probably taught those aspects of it relating to gunnery (excluding ballistics) and fortification. He is likely to have had much to say about the chemistry of explosives, particularly as he continued to be in touch with his friend Abel at Woolwich Arsenal. Indeed there were rumours in the air that the whole establishment might be moved to Woolwich. In early 1861 Frankland was telling his friend Dean Dawes that Addiscombe had developed itself into something more extensive than he had expected. The Dean replied with the hope that, if the move to Woolwich did take place, Frankland would go with it as full Professor.[41]

How far Frankland's oscillation between central London and Croydon might have continued in the absence of external interference it is impossible to say. In fact matters were taken out of his hands. In 1857 the London newspapers were reporting a dreadful uprising within the Bengal army, ostensibly a result of incompetence by

British officers but doubtless connected to general unease at the pace of westernization in the country. After the fall of Delhi, Cawnpore and Lucknow the Indian Mutiny (as it became universally known) was suppressed. But things in India could never be the same again, the most spectacular consequence being the downfall of the East India Company, with sweeping political and administrative changes in India itself. In the following year the government of India was transferred from the Company to the Crown, as in due course was the Company's private army. Almost overnight the rationale for Addiscombe College had disappeared and closure was inevitable. On 6 June 1861 the Lieutenant Governor of the College conveyed this sad information to his staff. In almost the only reference to Frankland in Vibart's history of the college the reception of that news is described. It is in an appropriately obsequious letter from the mathematics professor, Jonathan Cape (who had served for nearly 40 years), to the aforesaid Lieutenant Governor, Sir Frederick Abbott:

> All the professors assembled at my house on Friday after the parade, together with Dr. Frankland, Mr. Sdetky, and 3 or 4 other gentlemen; and after the collation I read to them your communication of the 6th inst. They were all exceedingly obliged to you for the kind feelings which were expressed . . .[42]

So, in 1861, Addiscombe College ceased to exist. It could be said to have served India well.[43] For Frankland it had been another opportunity to increase his income, to see a wholly new facet of British life and to broaden his experience of teaching. And a further opening was soon to arise, this time at a scientific establishment basking in several decades of national prestige, the Royal Institution.

1.3 The Royal Institution

The Royal Institution of Great Britain was – and still is – situated in Albemarle Street, off Piccadilly. Founded in 1799 at the instigation of the American inventor Benjamin Rumford and others, the RI set out to diffuse and promote knowledge of science among the people, especially the poor. Many of the original Managers were landowners on the grand scale, and it has been argued that the early emphasis on philanthropy was partly an attempt to divert the working class from emulating their counterparts in the French Revolution. The point is very debatable but the fact remains that, within a very few years, the Royal Institution's public activities were directed more to the *élite* of London's society than to the poor and uneducated. The young professor of chemistry Humphry Davy (1778–1829) had lectured with such vitality and flair that crowds flocked to his lectures, blocking the narrow road with their carriages and causing a prohibition of south-bound traffic, thus creating London's first one way street.[44] Certainly science was being popularised but, ironically, not among those for whom its delights were originally intended. Like many another institution ostensibly created for the lowest orders of society the Royal Institution became more and more successful as its *clientèle* moved up the social scale. These lectures launched the tradition that continues today and now includes not only the Friday evening Discourses but lectures for schools and television. Unlike St. Bartholomew's Hospital, and certainly unlike Addiscombe Academy, the Royal Institution seems likely to survive into the twenty-first century.

Another intention in the minds of its founders may have been the promotion of agricultural science from which all landowners stood to profit. Yet, by another strange irony of fate, the research for which the Royal Institution was to achieve lasting fame turned out to be in other areas. From the fields of chemistry and electricity Davy founded the new science of electrochemistry and in so doing discovered half a dozen new elements including sodium and potassium. His successor Michael Faraday laid the foundation of modern electrical science, working on electromagnetism, as well as electrochemistry, and in effect discovered the dynamo.

This famous institution must often have occupied Frankland's thoughts during the relatively barren years in Manchester. The experience of lecturing there in 1853 and 1857 must have opened his eyes to a whole new world of prestige lectures, established scientific tradition and membership of a small but growing band of chemical *élite*. Who knows how far his ambitions may have been extended by encouragement from friends like the Manchester engineer James Nasmyth? Writing about Frankland's 1853 Lectures at the Royal Institution Nasmyth observed with characteristic generosity (and not a little prophetic insight):

> 'I doubt not the result will be that you become one of their chief men and some future day worthily occupy the place of the great men who have been before you there . . . and I look forward with confidence to

your admirable qualities and profound knowledge placing you at the
very head of chemical science in England'.

He went on to say 'I am rejoiced to know from my worthy friend
Warren De La Rue that you are fully appreciated'.[45]

Such encouragement is one thing, an actual opening is quite
another. In the event two quite separate happenings coincided to
provide Frankland with his first real opportunity since 1857 to
display his talents at the Royal Institution. One was the appointment
as Secretary to that body of the hyperactive Henry Bence Jones, the
man who had pulled the strings so successfully for Frankland in
connection with St. Bartholomew's Hospital. He it was who,
against considerable odds, sought to make the RI once again a great
centre for pure scientific research, as in the days of Davy and the
young Michael Faraday. This comes out clearly in his biography of
Faraday and his history of the RI, both published in the early 1870s
and both harking back to a rather romanticised view of the old
heroic days.[46] But what Bence Jones lacked in historiographical
sophistication was more than compensated for in his genuine
devotion to research and his determination to promote it at all costs
within the Royal Institution. Moreover he had an uncannily good
eye for spotting potential greatness in a researcher. As early as 1853
he had helped to engineer the appointment of young John Tyndall
as Professor of Physics, bringing him directly from the obscurity of
his teaching post at Queenwood. Now another former employee of
that Hampshire academy was to be lured to the Royal Institution.

The second event which opened the way for Frankland was the
retirement of Michael Faraday in 1861 from his lectureship, owing
to the progression of a long and debilitating illness. He remained,
however, as Superintendent of the House. New chemical lectures
urgently needed to be provided and the chemical lecturer at St.
Bartholomew's Hospital was approached. Frankland consented to
give ten lectures before Easter on inorganic chemistry, at a fee of
ten guineas each. Later that year he followed in Faraday's footsteps
in a series of six Juvenile Lectures in the Christmas holidays, on 'air
and water'.[47] They were followed by a further ten, for adult
audiences, on chemical affinity. For these he was paid £100 and
given a card to secure admission for 'your boy or anyone else'.[48]
There followed a correspondence with Bence Jones in which the
latter expressed a forlorn hope: 'I wish I could retain you always as
Professor of Chemistry but that alas is impossible for want of funds,

THE ROYAL INSTITUTION.

Fig. 8.1 The Royal Institution, Albemarle Street, London. (Author.)

but as much as I can I should like to attach you to the R.I.' [49]An invitation to give further lectures next year was, however, declined in order to make time for more original work. This would have been a bitter pill for the research-minded Bence Jones to swallow: the very reason he wanted to hire Frankland for the RI was to bring in a new research dynamic, but research was given as the reason for refusal. Writing from St. Bartholomew's Hospital Frankland indicated that acceptance would mean that he would have to 'resign the practical class portion of my duties here in favour of my senior assistant'.[50]

However the eager Secretary was not to be deterred. Within days he was writing:

> 'Tyndall wants a rest from lectures. Could you if we make you Professor of Chemistry give me six juvenile lectures & a course of 10 lectures on Thursdays before easter?'

For these services he offered £50 and £150, and in future £200 ('or something less or more if I cannot get this, but not less than £150 for 12 lectures yearly'). He would also have 'half the laboratory to work in and all I can get to help you'.[51]

In a document that seems to have been a reply to this letter Frankland conceded that he would be happy to be Professor of Chemistry on the terms suggested, though was careful to stipulate a future minimum of £200 for 12 lectures annually, 'to be increased to £300 as soon as practicable'. That half of the laboratory that was suitable for chemical operations should be designated for himself (though he understood that Tyndall wanted to retain it). He also expected 'one laboratory man or his equivalent' to be available for lecture preparation and laboratory work.[52]

At their meeting on 3 November 1862 the Managers of the Royal Institution recommended Frankland for election as Professor of Chemistry from May 1863, a proposal duly – and unanimously – ratified by the Members.[53] He had already written to the Governors of St. Bartholomew's Hospital, seeking permission to retain his post there. He pointed out that there would be no overlap of lecturing times, since the Royal Institution Lectures (12) were due on Tuesday, Thursday and Saturday afternoons (January-Easter) while those at St. Bartholomew's Hospital were scheduled for Monday, Wednesday and Friday mornings.[54] This was agreed and he continued to hold the Hospital appointment.

Congratulations were quick to arrive, including a letter from Maxwell Simpson[55] and a remarkably prescient one from Odling written a fortnight *before* the Mangers' formal decision. Given the assurance from Bence Jones that the post was permanent, 'no other appointment of a successor to Davy could have given such general satisfaction in chemical circles'.[56] A letter some months later from Hermann Fehling expressed the hope that Frankland would not follow Faraday in turning from chemistry to physics![57]

By March 1863 he had finished his lecture course at St. Bartholomew's Hospital and proposed to start work on lactic acids in the RI laboratory.[58] Later that year he wrote to Faraday justifying the almost exclusive use of a laboratory assistant for chemistry alone. He requested one trained and educated assistant at £60 *p.a.* and 'one uneducated laboratory man at £1 per week', recommending his present senior assistant, John Broughton, for the new assistant's post.[59] This was duly granted by the Managers. When Broughton resigned (November 1866) his duties were divided, part of them going to his replacement Herbert McLeod, whom Bence Jones described as 'a first rate assistant'.[60] In 1864 Frankland recommended a Mr Tingle as non-salaried assistant in succession to Mr Bowdler.[61]

Table 8.1. *Frankland's Lecture Courses at The Royal Institution up to 1867*

Year	Period	Days	No. of lectures	Lecture Titles
1853	April–June	Thursdays	10	Technological chem.
1854–6	–	–	–	–
1857	April–June	Saturdays	7	Chemistry and art
1858–60	–	–	–	–
1861	Jan.–April	Saturdays	10	Inorganic chemistry
1862–3	Dec.–Jan.	various	6	Air and water
1863	Jan.–March	?	10	Chemical affinity
1864	Feb.–May	Saturdays	12	Metallic elements
1864–5	Dec.–Jan.	various	6	Chemistry of a coal
1865	April–June	Tues./Thurs.	12	Organic chemistry
1866	Feb.–March	Tues./Thurs.		Non-metallic elements
	April	Tues./Thurs.	8	'Chemistry'
1866–7	Dec.–Jan.	various	46	Chemistry of gases
1867	March	Thurs./Sat.	6	Coal gas

The research conducted in this laboratory will be described, with work performed elsewhere, in the next chapter. Here it is sufficient to indicate briefly the nature of Frankland 's lecturing task at the Royal Institution. His courses are summarised in Table 8.1:

There can be no doubt as to the popularity and success of these lectures, in large measure due to the splendid experiments and demonstrations whose preparation called for much exertion by his assistant. One piece of sound advice was conveyed in a letter from Bence Jones, dated only '30 December' but from its content almost certainly relating to his first series of Christmas Lectures at the end of 1862. Jones remarked:

> Forgive my saying: Look at the clock often & if possible end soon after the hour. Remember how rapid Faraday & Tyndall are; it is better to cut short a part than leave out a good experiment if you have it prepared.[62]

His Christmas Lectures in 1866/7 on 'The chemistry of gases' were attended by his elder daughter Maggie.[63] Her *Diary* describes (among much else) his preparation of solid carbon dioxide (1 January). On the evening of the 5th the two girls Maggie and Sophie 'played at lecturing', taking the rôles of Frankland and McLeod respectively. 'Royal Institution lectures' were to become a popular diversion and endless source of hilarity in the juvenile part of the Frankland household for many months to come. It is hardly surprising.[64]

Throughout the 1860s the rival demands of research and lecturing created a tension which Frankland was never able to resolve. When initially declining lectures at the Royal Institution he had rather unnecessarily reminded Bence Jones that research was needed 'to retrieve the national character'.[65] Despite the care needed to prepare demonstrations he had only 12 lectures as part of his annual contract, together with extra lectures at Christmas on alternate years, sharing the task with Tyndall. However he appears to have taken on a number of extra lecturing duties for which payment does not seem to have been made. He recalled:

> It was my duty to give a course of eight or ten afternoon lectures in the year, and it was also considered at that time the proper thing for each professor to give two Friday evenings, during each session.[66]

In 1862 Frankland had informed St. Bartholomew's Hospital that he would be lecturing in the spring at the Royal Institution on Tuesday and Thursday afternoons as well as the Saturdays for his contracted lectures.[67] In addition Herbert McLeod records occasional Friday lectures on combustion (1861 and 1863), the glacial epoch (1864), lactic acid research (1864), on the origin of muscular force (1866), and a Tuesday afternoon lecture in 1865 on coal, smoke and flame.[68]

Writing many years later of his responsibilities at the Royal Institution, Addiscombe Academy and St. Bartholomew's Hospital, Frankland commented:

> All these three lectureships I held simultaneously for a year or two, and at this time I had not unfrequently to give three lectures a day, one at St. Bartholomew's in the morning, a second at the Royal Institution at 3.p.m., and a third at Addiscombe at 7 p.m. I remember becoming quite at home in this triple duty, and rather wishing that it could be made quadruple, by interposing another lecture between St. Bartholomew's and the Royal Institution, as it was quite impossible to settle down to research work during this interval.[69]

Frankland was probably referring to a few Fridays in early 1861.[70] A couple of years later he was finding even a double lecturing programme too difficult, announcing that his research 'was most seriously curtailed by the combined duties of the two appointments', and relinquished all but 25 of his lectures at St. Bartholomew's Hospital. In 1864 he resigned entirely from that post.[71] No sooner

Day & Haghe Lith.rs to the Queen.

THE LABORATORIES
of the ROYAL COLLEGE of CHEMISTRY,
the first stone of which was laid by
HIS ROYAL HIGHNESS PRINCE ALBERT,
in the presence of the Council & Members.
June 16th 1846.

James Lockyer, Architect

Fig. 8.2 The Royal College of Chemistry, Oxford Street. (Author.)

had he disentangled himself from these commitments than an event took place in Germany that was to embroil him again in lecturing duties, and on a bigger scale than ever before. The Director of the Royal College of Chemistry, A. W. Hofmann, was invited back to his home country to a chair at Bonn, with responsibility for the design of new laboratories. This was soon to be followed by a call to the University of Berlin. As his absence in Prussia was not certain to be permanent a *locum* was required for the Royal College of Chemistry. Noted for his flexible attitude to additional teaching responsibilities, as well as his undoubted scientific ability, Frankland was a natural choice. Accordingly, on 17 February 1865, Sir Roderick Murchison, Director of the Royal School of Mines, was informed that the Lords of the Committee of the Council on Education wished Frankland to undertake Hofmann's duties 'pending the absence of the latter in Prussia'.[72] Therefore, when Hofmann sailed for German lands later that year, his place was taken by Edward Frankland who, contrary to popular belief, was merely a temporary replacement and had *not* been installed as Hofmann's permanent successor.

Despite the transient nature of this additional post Frankland took his responsibilities seriously. According to Herbert McLeod his lecture course (on inorganic and organic chemistry) began on 2 October and concluded on 9 January, pausing briefly for Christmas, with a punishing schedule of one lecture every day from Monday to Friday. Small wonder that he could complain that 'all Hofmann's duties are on my shoulders in addition to those of the Royal Institution'.[73] After two years of this dual responsibility the strain was beginning to show. Stenhouse wrote to Walter Crum in February 1867:

> I saw Dr. Frankland the other day who looks 10 years older than he did two years ago; he is evidently overworking himself and is greatly perplexed by Dr. Hofmann's eccentric proceedings. Hofmann's leave of absence will expire in a year and Frankland fears he will return and disposess [*sic*] him.[74]

Four days before this letter Frankland had persuaded the Royal Institution Managers to agree to reducing his lecture load to only six for that year 'to give him more leisure to pursue his researches'.[75] Writing to Roscoe in May he declared that he would give up lectures at the Royal Institution altogether next year as 'the combined work of the two chairs is too much for me', but he was not in a position to resign from either (presumably for financial reasons).[76] Hofmann continued in a state of serious indecision as to whether to return to England or not. Then matters came to head with a new development. On 25 August Faraday died, leaving vacant his Fullerian Professorship in Chemistry. Frankland, who till then had merely possessed an *ad hoc* professorship, must have been considered a natural successor. It was he who took it upon himself to present Roscoe with a precious sample of ammonium vanadate sent in 1831 by Berzelius to Faraday. The bottle was discovered during Roscoe's visit to the RI for a Friday evening Discourse on 14 February 1868; his biographer even gives Frankland Faraday's old title.[77] But promotion was far from certain. Frankland wrote to Hofmann indicating that he could readily understand his indecision, but 'my own position has become still more embarrassing since I last wrote owing to the death of my colleague Mr Faraday'. The whole professorial staff at the Royal Institution was to be recast. He requested a meeting when Hofmann was next in London.[78] This took place, but Hofmann was still unable to indicate whether he would stay in Germany or return to the Royal College of Chemistry, and this uncertainty continued into 1868. He wrote to Frankland in February of that year:

> When I last saw you in London I told you that I was nearly decided to remain in Berlin and I told you also the reasons which prevented me from giving you a definite statement of my final decision. These reasons are still in force, nevertheless I hope that I shall be able to shorten very materially the state of suspense which has lasted so long.
>
> The immediate object of this note is to ask you, whether if I hand in my resignation I should mention *your name as my natural successor*. Indeed I consider it so much as a matter of course that you are to be definitely appointed to the College that I should not have thought of doing so, unless you had expressed, when last I saw you, some misgivings, which of course I could not understand but which may be founded on some underhand causes of which I am not aware. Could you let me have a line about these questions.[79] [my italics]

Frankland's reply indicates something of his own confused frustration, and clearly tries to eliminate the impression of 'underhand' dealing in the College of Chemistry or perhaps the Royal Institution. A draft letter contains many words crossed out and reads:

> When we spoke about the College before Christmas last I did not mean to convey the impression that there was any underhand current tending to my non-appointment as your successor; on the contrary everything that has come to my knowledge on the subject exhibits a current quite in the other direction. What I meant was that my appointment could not be looked upon as certain & consequently until your decision was known and your (in the event of your not returning) your successor formally appointed, I could not possibly either resign altogether my appointment at the Royal Institution or avail myself of the very advantageous terms which were recently offered to me by the Managers on the re-construction of the staff of professors consequent

on the death of Faraday. Nevertheless I have no means of judging of the views of the Council on Education on the subject, and I should therefore feel obliged if, in the event of your resignation, you would kindly mention ~~my name~~ me as ~~your~~ a suitable successor. I need not ~~state~~ say that ~~in the event of your resignation~~, the articles you mention or any other things belonging to you in the College, & which you may desire to have, shall be immediately sent over to Berlin in ~~the~~ case you resign.[80]

Another month elapsed before this reply landed on Frankland's desk:

I have very much to apologize for having taken so much time for my final decision. It was only with great pain that I was able to make up my mind to say farewell to the country of my adoption . . . My resignation will be forwarded to Sir Roderick Murchison. Accept once more my heartfelt thanks for the difficult part which you have performed in this transaction. May every success and prosperity accompany you on your future career.[81]

Even now his suspense was not over. A note to the Department of Science and Art conveyed the information that the Professors of the Royal College of Chemistry unanimously believed that Frankland should be Hofmann's successor.[82] But behind the scenes the Lords of Committee of the Council on Education were in debate about his reliability (possibly on the grounds of his pluralistic activities but more probably in view of the uncertainties surrounding the future of the College). They were not prepared, even now, to authorise a permanent appointment but hoped that Frankland would accept the post on a provisional basis.[83] After three years of hard work at the College this must have been very hard to take. But in September 1868 Frankland indicated to Murchison that he was willing to accept the offer of a provisional appointment as successor to Hofmann at the Royal College of Chemistry and would resign from the Royal Institution Professorship on 1st November next.[84] A letter from Murchison to the Department of Science and Art expressed satisfaction at securing the services of 'so distinguished a chemist and so good a teacher'.[85]

Frankland's move from the Royal Institution to the Royal College of Chemistry raises a number of interesting questions, most obviously about his own motivation. The Royal Institution had given him so much, so why was he willing to leave it? In acknowledging the award of the Royal Society's Copley medal in 1894 Frankland expressed gratitude to the Royal Institution for the five years he spent researching there, producing over one third of all his published papers. 'By the kind arrangements of Dr Bence Jones . . . I was relieved from almost all lecture duties, and thus enabled to devote nearly all of my time to research'. He still wished to stress 'the advantage to researchers of relief from excessive occupation in teaching'.[86] Not only did the considerate Bence Jones arrange for the lecture programme to cause the minimum disruption to research, but Frankland also had no responsibilities for training laboratory students. His years there were 'a perfect elysium'.[87] On the other hand, as he well knew, the Royal College of Chemistry would place heavy demands on him for lecturing and laboratory teaching.

It has been suggested that Frankland's resignation from the Royal Institution was prompted by the need to support his growing family.[88] In other words the over-riding consideration was financial. There may be some truth in this, and it is consistent with his general attitude to income from the early days in Manchester. At the beginning of his London period Frankland had, as we have seen, undertaken both consultancy work and extra lectures at other institutions. In December 1862 he even asked Crookes, the editor of *Chemical News*, for a fee for proof-reading the reports in that journal of his Christmas Lectures. Crookes declined, but the request was quite in character.[89] Furthermore, it had taken a long time for the Royal Institution to produce the expected rise in salary. However on 3 February 1868 his salary was increased from £200 to £300 (and Tyndall's from £300 to £450). These were 'the very advantageous terms' referred to in his letter to Hofmann of 15 March. Yet it was widely known that working for the Royal Institution was less remunerative than for the Royal College of Chemistry. Attention has also been drawn to the fact that his own salary was only two-thirds that of Tyndall.[90] But Tyndall had served the Institution four years longer than Frankland, and was older by five years; there is no hint in their voluminous correspondence of any breach in a long-standing friendship through resentment on account of salary differential.

There must be a more substantial explanation for Frankland to forsake the unique world of the Royal Institution. A clue to the underlying reason may be found in a letter from Bence Jones to Frankland just after Faraday's death. He indicated that there would be no meeting of Managers until November and no change in the house probably until the end of the year. Faraday's rooms would be

Table 8.2. *Frankland's chemical appointments 1857–69*

Institution	1857	1858	1859	1860	1861	1862	1863	1864	1865	1866	1867	1868	1869
Barts. Hospital	X	X	X	X	X	X	X	X					
Addiscombe			X	X	X								
Royal Inst.					X	X	X	X	X	X	X	X	
RCC									X	X	X	X	X

Shading indicates a *tenured* appointment

divided between Frankland and Tyndall. But the big question was that of the Fullerian Professorship that Faraday had filled. Frankland must have considered himself an obvious successor, and with good reason. But it was not to be. *Bence Jones proposed to persuade the Managers to appoint as Fullerian Professor of Chemistry not Edward Frankland but William Odling.*[91]

The reason was simple: 'your present opinions about lecturing'. From this it appears that Frankland himself had engineered what was to be a fatal deterrent to further progress at the Royal Institution. By his constant attempts to avoid encroachment on to research time, by his request to the Managers to be relieved of half his lecturing responsibilities in 1867, by his complaints over the most modest requirements for public exposition, and by his avowed intention to avoid lecturing altogether in the following year Frankland had finally wearied even his most loyal supporter and ally in research. He had burned his boats behind him. A conversation in the previous summer had left an indelible impression on his friend:

> Of course if you would like it I should prefer your having the Fullerian Professorship also but that I fear from my last summer talk to you is impossible. I wish I had £1000 a year to give you both. However that cannot be.[92]

Quite apart from the financial disadvantage Frankland was now to be denied the privilege of being successor to one even more distinguished in the public mind than Hofmann himself: Michael Faraday. Not only Tyndall but now Odling was to be placed above him. It was the last straw. No wonder that he wrote two days later to Hofmann with urgency and embarrassment. The supreme irony of the situation is that by expressing so vehemently his distaste for lecturing he had landed himself a job in which that (and the hated laboratory teaching) would be pre-eminent. Yet it is to his lasting credit that he showed no bitterness to his rivals and continued to count them as friends.

So ended nearly a remarkable decade of chemical pluralism. It is summarised in Table 8.2.

It is easy to imagine that arrangements of this kind were unique to Frankland and that he was therefore either specially greedy or exceptionally able. While the latter is undoubtedly true the fact remains that in Victorian England many scientists held a plurality of appointments. An early example was that of Playfair himself, and thereafter the practice became contagious. Huxley (another of

Playfair's junior associates) is a good example, receiving £600 from the Royal School of Mines, £200 from the Normal School of Science and £700 as salmon fishery inspector.[93] Similarly John Tyndall, who served the Royal Institution from 1853 to 1887, held the additional post of Lecturer in Physics at the Royal School of Mines until 1868. But so far as chemistry was concerned there is no doubt that Frankland was the best known holder of several jobs simultaneously. In 1868 the *Chemical News* gave a brief citation from the *British Medical Journal*, as noteworthy for its imprecision in describing Odling as successor to Frankland (rather than Faraday) as for its insight into the cause of this remarkable phenomenon:

> There is some stir in the chemical world *à propos* of existing vacancies and probable promotions. Unfortunately the number of remunerative offices which can be held by scientific chemists in this country is small, and the emoluments, for the most part, are absurdly insufficient. That accomplished chemical pluralist, Dr. Frankland, having resigned one of the least lucrative of his positions, Dr. Odling will succeed him – whether wholly or in part seems not quite clear. Dr. Odling will, we believe, associate with himself a co-lecturer at St. Bartholomew's Hospital; and this will probably create a vacancy in the chemical chair of another metropolitan medical school . . .

The paragraph indicates that such 'chemical pluralism' was not a recent discovery (or invention) of historians but a well-recognised feature of Victorian Britain.[94] It may also be added that there were earlier precedents in France, with a system of multiple salaries known as *cumul*, borrowed from the clergy, and a conspicuous feature of academic life in the first half of the century. It is well exemplified by the life of J. L. Gay-Lussac.[95]

2 Man of affairs

When considering a man of such scientific talent as Frankland it is easy to overlook the wider context of that individual's development. Those very achievements for which he is most famed – in chemical research, practice and teaching – depend on many more variables than those conventionally ascribed merely to the science in question. It is an emasculated account that portrays discoveries as inevitable or as contingent only on the data which nature presents in the laboratory. With any complex personality, in science or anything else, understanding must be built upon a *gestalt* view that takes into account all available features of background as well as foreground, even though they seem at first unlikely to be relevant. In Frankland's case that must include the wide web of social relationships in which he lived. We begin by considering his immediate family and conclude with the *élite* community of science that he longed to claim as his own.

2.1 *Family*

In the eleven years between his leaving Manchester and joining the Royal College of Chemistry Frankland was able to integrate himself into the scientific community as never before and gradually emerged as a leader of considerable ability. It is fairly clear from extant diaries of his elder daughter Maggie and younger son Percy that he did this from a secure base of family life.

The Frankland family moved from Manchester to a house on Haverstock Hill in North London, between Hampstead and Chalk Farm, a couple of miles north west of Euston station. The 1861 Census for their address, 42 Park Road, indicates the two parents and four children: Margaret (8), Frederick (6), Sophia (5) and Percy (1). One other child, Lionel, was born in 1863 but lived for only just over a year. The oldest three were born in Lancashire but the youngest in Middlesex. There was also a nursery governess born in Cassel, and an English cook.[96] It was a comfortable, if unpretentious, establishment, typical of a young but upwardly mobile couple in Victorian London. These days were recaptured by his daughters in a 'Conclusion' to their father's *Sketches*:

> His children also remember with pleasure the delightful evenings when he read aloud to them 'Sandford and Merton', 'The Arabian Nights', the works of Dickens, and in later years those of Trollope.
>
> Gardening was also a hobby of his, especially in earlier years in Park Road, Haverstock Hill, where he lived for about twelve years after leaving Manchester in 1857. Although the garden here was quite small, he made it productive and grew an abundance of cherries and even peaches. With all his occupations he found time also to keep the greenhouse well stocked with flowers, and his last act every night, before going to bed in the winter, was to stoke the furnace.[97]

Music became a family pastime, especially after Frankland returned from the Paris Exhibition in 1867 with news that he had ordered a piano from Stuttgart. Noises emanating from his study next day proved to be from an accordion which he had also acquired in Paris (he bought another for Duppa). Thereafter musical afternoons or evenings were common, with the two daughters often playing the piano for their father. He enjoyed singing operatic arias or duets with Maggie. To some of these merry events scientific friends would come, including Hirst, Debus and Tyndall. By the end of 1869 these were not always an unqualified success. Having heard Frankland sing the *Vogelfänger* Tyndall rudely suggested that his going to the opera to better his singing was like sending a black man to wash himself white; on another occasion their performance found Debus 'listening with a grim and stupid look'.[98]

The family was surprisingly close-knit; diary entries and letters contain more than the usual conventional Victorian expressions of affection. The two sisters remained in touch with each other for the whole of their lives, and clearly had a great fondness for their brothers, as in the tragic scenes that Maggie recalled of Lionel's terminal illness and (later) of Fred's exile to New Zealand. Frankland seems to have been a devoted father. At this period his love was warmly returned, particularly by his eldest child Maggie. However he had exacting standards for their behaviour. Failure at school or college was frowned upon severely, earnest exhortations were given for improvement in musical performance, they were encouraged and expected to attend lectures at the Royal Institution and other places of scientific instruction. Discipline was firm. Percy recalls how, one February day, the servant accidentally knocked to the ground part of a gas light (the mantle?) and no one bothered to pick it up. Inevitably it was broken by somebody treading on it, and on his return in the evening Frankland decreed the family should spend the rest of the day in darkness; he did not repair the damage until two days later.[99]

Family holidays continued, with Windermere still exerting its call upon them. The Helm grandparents moved to Leyland by 1863 and were often visited by Frankland himself, and later by the children. But London was much nearer the south coast than the north west counties and by 1860 Frankland had bought a small cutter, the *Sophia*, which was berthed at Cowes in the Isle of Wight,- favourite holiday ground for London-based Victorians. Here, too, his mother and step-father settled for a short while, presumably providing a land-base for the family on holiday.[100] The

(a) (b)

Fig. 8.3 The children of Edward and Sophie Frankland: (a) Maggie and Fred;
(b) Sophie and Percy. (J. Bucknall Archives.)

Sophia was only big enough for two people to sleep in so the family
was necessarily excluded from the delights of extended yachting in
the Solent; in any case the children were too young at first.
However there were occasions when all five joined him for whole
days on the water, as in 1861 when he had no less than two months
of yachting holiday in the area. On one occasion he, Sophie and the
two older children were cut off by a rapidly rising tide at Scratchells
Bay in the Isle of Wight, being rescued only by Frankland and his
mate wading through deep water and carrying the others on their
backs to safety.[101]

There were many other occasions for pleasant family holidays,
mostly chronicled by the children in their diaries. The summer of
1864, for example, saw them off for another two months to Norway
whose delights Frankland had discovered in July of the previous
year. He admired it for the honesty of its inhabitants, the general
respect for the English ('God grant that English travellers may not

soon undeceive them!'), the ease of travel (driving one's own carriage at 7–9 m.p.h. at $2\frac{1}{2}$d per mile) and above all the magnificence of its scenery.[102] They sailed from Hull to Molde and proceeded to various resorts including Dombås and Lillehammer. Sometimes they would all explore the countryside on foot, though Frankland would often go salmon-fishing on his own, with Fred or with his friend the Consul Andersen. Contrary to rumour that 'he never caught a salmon in his life'[103] he appears to have been moderately successful. At other times he would go shooting, for keper or hares. For a fortnight in the middle of their stay Frankland, with a couple of male companions, went off to explore the glaciers at Jostedal, the first he had ever seen.[104] The family concluded their holiday at Christiania (Oslo) and, despite much rain, Maggie confided to her diary 'I like no place so well. Nor do I think there is any place so nice'.[105] Frankland's enthusiasm for Norway made a deep impression on his colleagues, as may be seen in the following doggerel, printed in a collection of irreverent verse in 1865:

> Frankland sat down on my right, and cheerfully looked through his glasses
> Telling of fishing exploits he had had in the summer in Norway.
> Vivid description gave he of a fight with a twenty-pound salmon,
> Of hyperbolical trout in streamlets of swift-running water;
> All the manoeuvres he told with a pathos that made it impressive –
> Yea, with a force and a fervour becoming the 'Father of Ethyl'.[106]

In 1866 Sophie Frankland and the children took the train to North Wales for seven weeks, Sophie and the girls travelling by first class and the newly arrived baby (Lionel) and nurse by second. Frankland met them at Llandudno, and they were also joined by his old friend Nicholson from Manchester, together with his wife and child. Staying at Betws-y-Coed they soon discovered the challenge of the nearby mountain Moel-Siabod, to say nothing of lesser heights. One day (17 September) Fred and his father walked up and down Snowdon without a break. By now Maggie and her sister Sophie were old enough to go off on excursions together, often with sketch-pads and pencils. Their father again spent much time reading or fishing, though he also enjoyed time spent with Nicholson and George Busk, a London surgeon and treasurer of the Royal Institution, whose family was on holiday nearby. He had to take out two days for a London trip. From this holiday the girls returned with extreme reluctance, discovering that 'London air nearly suffocated us'.[107]

The following spring marked the death of little Lionel Edward, the baby born in January of the previous year, and a summer holiday in Scotland helped in mitigating the family's grief. It began in August with the whole party travelling overnight by rail, this time third class. At Glasgow they were met by the Helms who were to join them for the holiday. From there they set off by boat to Oban, via Ardrishaig and the Crinan Canal. Again there was much exploring to do and again they were joined by the Nicholsons. In mid-August Frankland, John Nicholson and Fred took the steamer to Skye from which they returned a fortnight later. A companion for part of the time was Herbert Spencer ('author of the much despised books that Freddy bothered us so dreadfully about'[108]). Their return journey was *via* Lancaster where they stopped for Frankland to show his children the scenes of his own boyhood. Sophie and Edward then departed for a week together at Kirkby Lonsdale while the children were committed to the care of their grandparents at Leyland. Percy and his father stayed an extra week with the Helms, and it was during this week that the crucial correspondence between Frankland and Bence Jones began.[109]

Next summer (1868) was the occasion for a family holiday in the Lake District, based at Howtown on Ullswater. Again the environmentally-conscious Maggie regretted her return to London: 'At London we found all well. Except the air which is always horrid'.[110]

In 1869 they began a long series of continental vacations, with six weeks in Switzerland. Details of these and much else are recorded in the diaries of Maggie and Percy Frankland. Occasionally Frankland would go off on holiday alone, as for an idyllic few days in 1862 at the Kyles of Bute in Scotland.[111] But he was always punctilious about writing to Sophie, and would gently remonstrate with her if her own letters were too short or uninformative. As in Manchester days he found great therapy in a change of scene; as he wrote from the Isle of Wight:

> It is such a relief from the hurry and bustle and tension of London. Already I feel the strength and vigour returning which I always derive from even a few days of such utterly careless life.[112]

2.2 *Informal scientific networks*

A notable feature of all the Frankland holidays is the appearance of many figures from the fields of science and learning. This happens too often to be mere co-incidence and it is likely that Frankland viewed vacations as opportunities to socialise with his peers. Whether this was a conscious ploy to cement stronger links with the scientific community at large is at least a question worth asking. It is sufficient here to record a few of the links so established.

John Tyndall, whom he first met at Queenwood in 1847, was an old walking-companion and their camaraderie on holiday turned out in later years to have important scientific consequences. A tour of the Lake District in April 1859 was a remarkable testimony to the stamina and physical fitness of both men, as those familiar with the Cumbrian mountains will readily agree from their itineraries on successive days, all on foot (modern spelling, mountains in *italics*):

1. Ambleside – Grasmere – *Helvellyn* – Keswick
2. Keswick – *Skiddaw* – Bassenthwaite – Keswick
3. Lodore – Borrowdale – Stonethwaite – *Langdale Pikes* – Esk Hause – Wasdale Head
4. Wasdale Head – *Lingmell* – *Sca Fell Pike* – Mickeldore – *Sca Fell* – Wasdale Head
5. Wasdale Head – *Pillar* – Black Sail Pass – *Kirkfell* – *Great Gable* – Wasdale Head
6. Wasdale Head – Burnmoor Tarn – Hardknott – *Wetherlam* – *Old Man* – Coniston

The companions then proceeded by train from Windermere to Hereford where they stayed as guests of their friend Richard Dawes, the Dean.[113] For all his clerical antipathies Frankland had a deep respect for this pioneer of popular science education and mentor from Queenwood days. But it was in the Alps that Frankland and Tyndall were to experience their greatest mountaineering exploits together. We have already noticed their brief encounter in 1856, doubtfully sanctioned by Sophie (p. 188). Much more significant was their ascent of Mont Blanc in 1859 (p.412). There were many later encounters in the mountains of continental Europe.

Tyndall was by no means the only scientific companion on holiday. The attractions of Frankland's yacht seem to have brought many of the leading chemists to the Solent for its second season. They included Heinrich Debus, William Odling, Alexander Williamson and Benjamin Brodie.[114] When it came to continental holidays a different *clientèle* would be met. Through the records of these holidays flit figures like Huxley, Lubbock, Hirst, Tyndall and

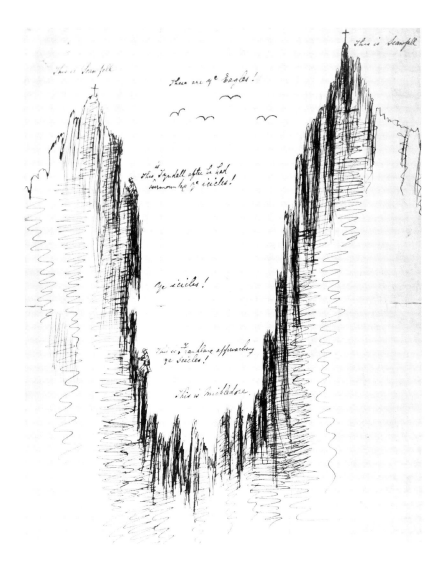

Fig. 8.4 Frankland's sketch of an adventure in Lake District mountaineering, showing himself and Tyndall crossing the ridge of Mickledore from Sca Fell Pike to Sca Fell: from a letter to his wife, April 1859. (J. Bucknall Archives.)

many others who, though not chemists, were to play an important rôle in the social history of English science in the next decade or two.

However it would be wrong to limit Frankland's social contacts with other scientists to holidays alone. In later life Frankland was much given to hospitality and now was at last able to disregard Playfair's advice to him at Putney: if he did marry he 'should take care to give no dinner-parties, but only tea-parties, which cost next to nothing'.[115] So we find a dinner-party arranged for 19 January 1867 which was a matter of some excitement for the 15-year old Maggie since the children were allowed in for dessert; it sounds like a new experience. The guests were Alexander Williamson and his wife, Hirst, Tyndall, Debus and Duppa. A fortnight later the Franklands welcomed Duppa (again), Matthiesson, Wanklyn, Müller, Perkin and Chapman. Over a year was to pass before the next dinner-party, this time upgraded by the presence of a hired waiter; in addition to Roscoe and the Russells the guests included (together with their wives) Frankland's new colleague at the Royal School of Mines, Frederick Guthrie, and W. B. Carpenter, Registrar of London University. As time went on the invitation base was broadened so (for instance) in November 1868 the guests included not only the Carpenters (again) and William Siemens and his wife, but also Bernhard Samuelson, the ironmaster and promoter of technical education, and the author Herbert Spencer. These were a good preparation for the lavish dinner parties Frankland later hosted at the Athenaeum and elsewhere.[116] In fact the Athenaeum was even in the 1860s proving to be another rendezvous for scientific friends. Frankland proposed Siemens for membership in 1871, for example.[117] Finally there was the Chemical Society Dining Club which existed during the early years of Frankland's Presidency and which, therefore, he was expected to chair. Their first dinner (21 November 1872) was at a hotel near Leicester Square, but thereafter they often met at The Café Royal, Regent Street. In 1873 Frankland was present at dinners in January, February, March, and October, missing April, May, June and November.[118]

Apart from hosting formal dinner-parties Frankland welcomed individual colleagues to his home. In the 1860s his research assistant Francis Duppa frequently 'dropped in' to see him, to go skating, for an informal meal or even to entertain the children with his painting.[119] Nor was the traffic in one direction. Frankland was often to be found visiting colleagues in their homes, as was the custom at that time. An early documented example is from 1863 when, in company with Tyndall, he called for the evening on Michael Faraday at his home in Hampton Court.[120] By far the most influential of these informal scientific networks was the X–Club, considered further in Chapter 11. He does not appear to have been a member of the convivial group of chemists known as the 'B–Hive' or 'B-Club' [121], but that did not eliminate him from their attempts at versification:

> Behold the Sage from northern county come,
> To find among ourselves a genial home,
> Scanning the science with extensive view,
> Star of the school of Saint Bartholomew!
> Thrice welcome Frankland, every one will say,
> Hail to the rising chemist of the day!
> Each year but adds new lustre to thy name,
> Each year but brings fresh trophies to thy fame.
> Father of Ethyl! that portentous birth
> Which to its very centre shook the earth;
> A shock electric thrill'd the chemic mind
> When the news spread that Frankland was confined;
> Philosophers with quickest motion sped
> To learn the news and see him brought to bed;
> Bold chemists from far-distant climes drew nigh
> To have a peep and see the baby cry.
> Some casuists indeed felt not quite sure,
> And hinted that the birth was premature,
> Said, Frankland was affected by the moon,
> And, like a seven-months' child, his offspring came too soon! [122]

Does this suggest a gentle mockery of the ambitious Frankland by contemporaries who were mostly in the middle or lower ranks of chemical achievement, and a certain scepticism about his most famous discovery?

2.3 Formal scientific institutions

One might have thought that the large and established nation-wide network afforded by the British Association for the Advancement of Science would have been attractive to Frankland. However in 1861 he wrote:

> I make a point of not going to the British Association meetings when they occur in the vacation, as the proceedings there are merely a hash of what one has been listening to at the meetings of the London Scientific Societies, and consequently they merely break in upon one's vacation to no purpose whatever. [123]

This admission underlines the great stress he placed upon keeping holidays apart from work. He did, however, break his resolution on one or two occasions, as when in 1868 he interrupted his Lake District holiday to attend the meeting in Norwich to deliver his paper on combustion under pressure.[124] His party included the Busks, Lubbocks, Spottiswoodes, Tyndall, Huxley and Hirst, all names associated with the X-Club. Others joined them and socialising appears to have been more important than attending lectures.[125] He was away for eleven days.

Then there was the Royal Society. Frankland was elected a Fellow on 2 June 1853 and served on the Council four times, including the periods 1857–9 and 1865–7. His Royal Medal of 1853 had been in recognition of his work on organometallic compounds. He served punctiliously as a referee and was a frequent attendee at scientific meetings. However it was much later that the Royal Society played a large part in his life and he exerted considerable influence over its affairs. In the 1860s the Chemical Society gave him more opportunities, and also made more demands upon him.

The Chemical Society of London was just six years old when Frankland joined it on 20 December 1847.[126] He was then at Queenwood College. He became a Life Member while at Manchester.[127] Once back in London he found it a valuable forum for his own papers and also for discussions on wide-ranging issues of which the famous debates on atomism in the late 1860s are the most memorable. He played a conspicuous part in those discussions, adopting a position of moderate conventionalism with respect to the existence of atoms.[128]

Frankland served on the Council of the Chemical Society in 1851–2 and again on his return to London for 1858–9. As Foreign Secretary (from 1861 to 1868) he found himself in a uniquely favourable position for corresponding with overseas chemists. Letters survive from several distinguished recipients of the honour of Foreign Membership: Berthelot, Cannizzaro, Fehling, Kekulé, Malaguti, Marignac and Pasteur.[129] He was also able to communicate papers of foreign authors, as in the case of Kolbe.[130] His was one of the key positions in the whole international chemical community at that time and he became well-known far outside the areas of chemistry for which he was personally renowned. In 1868 he resigned, at the height of his uncertainty over the Royal College of Chemistry. The manner of his going discloses an element of stubbornness in his character that was to display itself more dramatically later, almost, though not quite, invincible. According to the Council minutes of the Chemical Society he wrote 'expressing his desire to relinquish the post of Foreign Secretary, and, it being understood that Dr. Frankland could not be persuaded to retain the office,' it was resolved to accept his decision. Clearly there had been attempts to persuade him to change his mind, but in vain.[131]

In 1860–1 he had served the Chemical Society as Vice President and it is not surprising that efforts should have been made in the late 1860s to elect him as President.[132] It may be a matter of surprise that so ambitious a man should decline such an honour, but the fact is that he fought fiercely to avoid elevation. Consequently in 1868–70 he accepted a return to vice-presidential status and Williamson was elected to the chair of the Society. Frankland's supporters did not accept defeat too easily, however, and efforts were made by Abel and others to make him change his mind for the 1871 Presidential election. The Chemical Society's Secretary, A. V. Harcourt expressed Council's unanimous invitation and wrote that he had 'never known an expression of opinion on such a matter more decided',[133] but Frankland replied that it was 'utterly impossible'.[134] The chemist J. H. Gladstone urged him to accept so as to quash rumours that he was refusing out of pique at having been passed over years ago (which Gladstone did not believe).[135] Odling made a very similar point, suggesting that it would be most regrettable if 'the remembrance of our former difference were to be as it were stereotyped by a perpetual absence of your name from our list of Presidents'; he also added that the office might be held for one year only, and stressed the view that not all mental resolutions are best expressed in public. Would he please, 'as a personal favour', think again about it?[136] Expressions of support came in from many others.[137] Frankland's response is revealing, and the draft of a letter to Abel has even more passages crossed out and amended than his letter to Bence Jones. He is sufficiently ruffled even to get the date wrong. In essence he denies that he has the gifts needed by a President, particularly 'eloquence in scientific matters in general' and the ability to direct debates. He wrote, and then crossed out, a remark that 'when my present excessive work diminishes I am anxious to continue those experimental inquiries which have been interrupted for the present'. He did allow to stand a self-deprecating comment that it was better to leave 'the mere workers in science to pursue their investigations without interruption', adding that this 'was also Faraday's opinion & he acted upon it through life, refusing all presidentships'.[138]

On 30 January Abel acknowledged this refusal with sadness, but two days later was sending an astonished letter of delight that his friend had relented.[139] What had changed his mind in the few hours between writing the definitive letter to Harcourt on the 28th and a further communication that would have got to Abel at most four days later (with a week-end in between)? Of all the 70 or so letters that have survived from 1871 none could have reached Frankland by post in those critical four days save those from Abel and Odling. The former accepted the inevitable; the latter might have tipped the scales by the warmth and intensity of its pleading. Otherwise the only clue is a single sentence in the diary of his daughter Maggie for Sunday, 29 January:

> In the afternoon Dr. Bence Jones came to speak to Papa about the presidency of the Chemical Society.

Just that. However it was Jones' great misfortune to be too early, for Frankland did not return from Manchester till the Monday evening (the 30th).[140] So, on a scrap of unheaded paper, he wrote a letter, dated simply 'Jan. 31' and addressed 'Dear Friend'. It requested permission to ask Harcourt to withhold Frankland's letter from the Council and suggested that he could have another fortnight to make up his mind. The combined impact of this letter and that of Odling may have tipped the balance in Frankland's mind. Possibly the note from Bence Jones was taken round to Frankland early on the Tuesday morning giving him a chance to write and post a letter to Abel arriving next day, 1 February. That is as far as the existing evidence will take us, but it must be a strong possibility. In any event it is not hard to disagree with Armstrong who wrote:

> A poor debater, he had been pushed aside in his early career. In 1871 we had practically to force him to become President of the Chemical Society. I was a ring-leader. Williamson had been elected to the office not once only but for a second period – with little work to his score. Conscious of the value of his own contributions, Frankland undoubtedly resented the indifference the Society had shown to his work, in not according him public recognition.[141]

His elevation to the Presidency in fact ushered in a new phase of increasing influence, not only on the Chemical Society, but also on the community of English science as a whole.

Notes

1 W. Templeton, 'Two centuries of chemistry at St Bartholomew's ', *Chem. Brit.*, 1982, **18**, 263–7.

2 On John Stenhouse (1809–80) see *Proc. Roy. Soc.*, 1880–1, **31**, xix–xxi. After studying with Graham, Thomson and Liebig, he had been an unsuccessful applicant for the Owens College chair. He was chiefly notable for his discovery of the absorptive powers of charcoal and the invention of charcoal air filters; he sent two charcoal respirators to Frankland: J. Stenhouse to Frankland, 5 February 1855 [RFA, OU mf 01.03.0265].

3 Presumably on hearsay evidence: Frankland to F. A. Abel, 3 February 1852 [RFA, OU mf 01.06.0155]; on Abel (1827–1902) see *DNB*, J. Spiller, *J. Chem. Soc.*, 1905, **87**, 565–70 and K. R. Webb, *J. Roy. Inst. Chem.*, 1958, **82**, 147–50.

4 Cited in Templeton (note 1).

5 H. Bence Jones to Frankland, 11 July 1857 [RFA, OU mf 01.08.0561].

6 J. Paget to H. Bence Jones, n.d. [RFA, OU mf 01.08.0562]; he was to become Sir James Paget, later Vice Chancellor of London University.

7 W. F. White (Clerk to the Governors) to H. Bence Jones, 20 and 21 July 1857 [RFA, OU mf 01.08.0559 and 01.07.0064].

8 H. Bence Jones to Frankland, 22 July 1857 [RFA, OU mf 01.08.0560].

9 J. Nasmyth to Frankland, 24 November 1857 [RFA, OU mf 01.01.0054].

10 MS 'introductory address' headed 'St. Bartholomew's Hospital', n.d. [RFA, OU mf 01.04.0118]; the writing is partly Frankland's and partly that of some amanuensis, the document reading as if it were a report prepared for the press.

11 On Attfield (1835–1911) see *Proc. Roy. Soc.*, 1911–12, A**86**, xliv–xlvi, and *Proc. Inst. Chem.*, 1911, pt. ii, 27.

12 J. Attfield, letter to *Surrey Mirror*, 1 September 1899, reproduced in *Sketches*, 2nd ed., p.451. He incorrectly states that he joined Frankland in 1851; 1857 must be intended.

13 *Sketches*, 2nd ed., p.137.

14 E. W. Binney to Frankland, 18 June 1858 [RFA, OU mf 01.07.0150].

15 Copy-letter, Frankland to E. W. Binney, 28 June 1858 [RFA, OU mf 01.07.0152].

16 Copy-letter, Frankland to A. Brice, 19 May 1859 [RFA, OU mf 01.01.0295].

17 Copy-letter, Frankland to Gluten Food & Starch Co., 31 December 1860 [RFA, OU mf 01.01.0290].

18 E. Frankland and A. W. Hofmann, *J. Soc. Arts*, 1859, **7**, 661–4.

19 H. McLeod, *Diary*, reprinted as *Chemistry and theology in mid-Victorian London*, ed. F. A. J. L. James, Mansell, London, 1987: entry for 11 May, 1861.

20 A. S. Eve and C. H. Creasey, *Life and work of John Tyndall*, Macmillan, London, 1945, pp.312–13.

21 Copy-letter, Frankland to C. W. Curtis, 14 and 23 October 1862; Curtis & Harvey to Frankland, 8 July 1863; Frankland to Curtis & Harvey, 14 July 1863 [RFA, OU mf 01.08.0628, 0621, 0633 and 0617].

22 See B. Gardner, *The East India Company*, Hart-Davis, London, 1971; or, for

the years to 1820, J. Keay, *The Honourable Company: a history of the East India Company*, HarperCollins, London, 1991.

23 Report of annual prize distribution, 1848, *Illustrated London News*, 22 July 1848; the Principal's speech 'dwelt on this subject at some length, urging upon those gentlemen present who were concerned with India the great advantages to be derived from promoting instruction in civil engineering among the officers of both services, civil and military'.

24 Report of annual prize distribution, 1847, *The Times*, 13 July 1847, p.7.

25 The civil service college at Haileybury was closed in 1858, but reopened as a public school four years later.

26 H. M. Vibart, *Addiscombe: its heroes and men of note*, Constable, Westminster, 1894.

27 *Ibid.*, p.154.

28 None appears on any of the extant maps of the buildings, as in Vibart (note 26) or in the Brochure for Sale [Croydon Public Library].

29 M. Archer, 'India and natural history: the role of the East India Company 1785–1858', *History Today*, 1959, 9, 736–43.

30 M. Berman, *Social change and scientific organization: the Royal Institution 1799–1844*, Heinemann, London, 1978, pp.81–6.

31 On D. T. Ansted (1814–1880) see *DNB* and *Proc. Roy. Soc.*, 1880–1, 31, i–ii.

32 *The Times*, 17 July 1845, p.6; Putney *Prospectus* for 1848 [British Library MSS, 8364.de.24].

33 On Solly (1819–1886) see *DNB* (which gives his dates at Addiscombe as 1845–1849).

34 Vibart (note 26), p.220.

35 Lincoln to Playfair, 15 March 1845 [Imperial College Archives, Playfair papers, no. 188].

36 *Sketches*, 2nd ed., p.136.

37 Copy-letter, Frankland to Cahours, 1 January 1861 [RFA, OU mf 01.04.0011].

38 E. Solly, *Syllabus of a complete course of lectures on chemistry including its application to the arts, agriculture, and mining, prepared for the use of the Gentlemen Cadets at the Honourable East India Company's Military Seminary, Addiscombe*, Longman, London, 1849.

39 *Proc. Roy. Soc.* (note 11).

40 Frankland to Stokes, 14 January 1862 [Cambridge University Library, Stokes papers, Add. MSS 7656/F365].

41 R. Dawes to Frankland, 30 January 1861 [RFA, OU mf 01.07.0116]; he repeated these sentiments on 22 September , having previously (14 August) expressed amazement at the failure of the East India Council to transfer the Addiscombe professors to Sandhurst [*ibid.*, 01.07.0119 and 0122].

42 Vibart (note 26), p.200; Sdetky had been drawing master at the college.

43 At the time of its closure it had 151 cadets, all between 16 and 21 years of age (Data from 1861 Census, cited by D. C. H. Hobbs, 'Nineteenth century Addiscombe', Dissertation, Portsmouth Polytechnic, 1985). Those with unfinished courses were sent to Woolwich to complete them. Altogether in its 52 year life span the college had despatched about 3600 cadets into armies

in Bengal, Bombay and Madras. Amongst its most famous alumni were the Field-Marshals Lord Napier and Earl Roberts.

44 It reverted for a time to two-way traffic, as may be inferred from instructions on a Royal Institution *Syllabus* for Juvenile Lectures, December 1866: 'It is Requested, That Coachmen may be ordered to set down with their Horses' heads towards Piccadilly, and to take up towards Grafton-street'.

45 J. Nasmyth to Frankland, 23 June 1853 [RFA, OU mf 01.03.0195].

46 H. Bence Jones, *The life and letters of Faraday*, 2 vols., Longmans, Green & Co., London, 1870; *The Royal Institution: its founder and its first professors*, Longmans, Green & Co., London, 1871.

47 These and all other appointments to the RI are recorded in the Minutes of the Managers' Meetings reproduced as *Archives of the Royal Institution of Great Britain*, ed. F. Greenaway, Scholar Press, London, 1976.

48 H. Bence Jones to Frankland, 1 January 1862 [RFA, OU mf 01.03.0011].

49 H. Bence Jones to Frankland, 1 June 1862 [RFA, OU mf 01.03.0124].

50 Frankland to H. Bence Jones, 10 June 1862 [RFA, OU mf 01.03.0122].

51 H. Bence Jones to Frankland, n.d. but clearly 1862 [RFA, OU mf 01.03.0051].

52 Frankland to H. Bence Jones, 5 July 1862 [RFA, OU mf 01.03.0046].

53 H. Bence Jones to Frankland, 4 May 1863 [RFA, OU mf 01.03.0336].

54 Frankland to St. Bartholomew's Hospital, 9 October 1862 [RFA, OU mf 01.01.0032].

55 Maxwell Simpson to Frankland, 10 November 1862 [RFA, OU mf 01.03.0133].

56 W. Odling to Frankland, 13 October 1862 [RFA, OU mf 01.03.0135].

57 H. Fehling to Frankland, 20 June 1864 [RFA, OU mf 01.03.0025].

58 Frankland to H. Bence Jones, 27 March 1863 [RFA, OU mf 01.03.0048].

59 Frankland to M. Faraday, 18 November 1863 [Royal Institution Archives, Faraday papers, FL9].

60 H. Bence Jones to Frankland, 4 November 1866 [RFA, OU mf 01.04.1649].

61 Frankland to M. Faraday, 1 December 1864 [Royal Institution Archives, Faraday papers, FL9].

62 H. Bence Jones to Frankland, 30 December 18?? [RFA, OU mf 01.04.0587].

63 Maggie Frankland's copy of *Syllabus* (note 44) [JBA, OU mf 02.02.1745].

64 Maggie Frankland, *Diary*, entries for 1 and 5 January 1867 [MBA].

65 Frankland to H. Bence Jones (note 50).

66 *Sketches*, 2nd ed., p.139.

67 Frankland to St. Bartholomew's Hospital (note 54).

68 H. McLeod, *Diary* (note 19), entries for 8 March 1861, 13 February 1863, 29 January and 3 June 1864, 3 January 1865, 8 June 1866.

69 *Sketches*, 2nd ed., p.136.

70 Since Addiscombe College closed late in 1861, the first year of his temporary appointment at the Royal Institution, this must have been the only year of triple lecturing commitment. We know his lectures at St. Bartholomew's Hospital were on Monday, Wednesday and Friday mornings, so the only RI Lectures on weekdays can have been the informal Friday ones attended by McLeod.

71 Templeton (note 1), p.263.

72 N. M. McLeod to Sir R. I. Murchison, 17 February 1865 [British Geological Survey Archives, Keyworth, Royal School of Mines Letter Book, GSM 1/8, pp.325–6].

73 Copy-letter Frankland to W. Brande, 30 October 1865 [RFA, OU mf 01.01.0262].

74 J. Stenhouse to W. Crum, 8 February 1867 [Strathclyde University, Andersonian Library Archives, M43/36]; I am grateful to Dr D. G. Duff for this reference.

75 Minutes of Mamagers' Meeting, 4 February 1867, *Archives of the Royal Institution* (note 47).

76 Frankland to H. E. Roscoe, 10 May 1867 [Royal Society of Chemistry Archives].

77 T. E. Thorpe, *The Right Honourable Sir Henry E. Roscoe*, Longman, London, 1916, p.128.

78 Copy-letter Frankland to A. W. Hofmann, 20 September 1867 [RFA, OU mf 01.02.1468].

79 A. W. Hofmann to Frankland, 29 February 1868 [RFA, OU mf 01.07.0330].

80 Draft/copy-letter Frankland to A. W. Hofmann, 15 March 1868 [RFA, OU mf 01.07.0333].

81 A. W. Hofmann to Frankland, 24 April 1868 [RFA, OU mf 01.01.0533]. The news was conveyed to Frankland's assistant Herbert McLeod almost at once (McLeod, *Diary* (note 19), entry for 2 May 1868).

82 N. McLeod to the Secretary, Department of Science and Art, 16 May 1868 [British Geological Survey Archives, Keyworth, Royal School of Mines, Letter Book, GSM 1/9, pp.45–6].

83 N. McLeod to Sir R. I. Murchison, 20 July 1868 [British Geological Survey Archives, Keyworth, Royal School of Mines Letter Book, GSM 1/9, p.57].

84 Frankland to Sir R. I. Murchison, 9 September 1868 [British Geological Survey Archives, Keyworth, Royal School of Mines Letter Book, GSM 1/9, p.60]; his resignation letter to the RI (dated 9 October) is included in the Managers' Minutes for 2 November.

85 Copy-letter Sir R. I. Murchison to Department of Science and Art, 12 September 1868 [British Geological Survey Archives, Keyworth, Royal School of Mines Letter Book, GSM 1/9, p.61].

86 Frankland, typescript of speech, 30 November 1894 [RFA, OU mf 01.04.0038].

87 *Sketches*, 2nd ed., p.139.

88 S. Forgan in W. H. Brock, N. D. McMillan and R. C. Mollan (eds.), *John Tyndall: essays on a natural philosopher*, Royal Dublin Society, Dublin, 1981, p.59.

89 W. Crookes to Frankland, 2 December 1862, reprinted in E. E. Fournier d'Albe, *The life of Sir William Crookes*, Unwin, London, 1923, pp.80–1.

90 Forgan (note 88).

91 H. Bence Jones to Frankland, 18 September 1867 [RFA, OU mf 01.03.0577].

92 *Ibid.*

93 L. Huxley, *Life and letters of Thomas Henry Huxley*, Macmillan, London, 1908, vol. ii, p.289; I thank Dr Gill Parsons for first drawing this to my attention.

94 *Chemical News*, 1868, 18, 23.

95 M. P. Crosland, *Gay-Lussac, scientist and bourgeois,* Cambridge University Press, Cambridge, 1978, pp.228–32.

96 1861 Census [PRO, RG9/91]. The road is now known as Parkhill Road.

97 *Sketches,* 2nd ed., p.445.

98 Maggie Frankland, *Diary* for 1867–9 (note 64).

99 Percy Frankland, *Diary* for 24–26 February 1868 [RFA, OU mf 01.03.0970]; Maggie does not mention the incident.

100 'William Helm of Cowes, Isle of Wight, Gentleman' gave an affidavit (5 September 1860) as executor to the will of his brother Thomas Helm (d. 1854) of Claughton, Garstang, Lancashire [Lancashire Record Office, Wills A1855]. For the relationship see *Lancastrian Chemist,* p.45.

101 Frankland to J. H. Nicholson, 13 October 1861, reprinted in *Sketches,* 2nd ed., pp.286–90.

102 Frankland to J. H. Nicholson, 16 August 1863, reprinted in *Sketches,* 2nd ed., pp.337–48 (338).

103 The rumour was 'according to his daughter', Sophie: H. Debus to L. Tyndall, 14 August 1899 [Royal Institution Archive, Louisa Tyndall papers].

104 *Sketches,* 2nd ed., p.351.

105 Maggie Frankland, *Diary,* entry for 9 September 1864 (note 64).

106 'The feast of the Blues', by the author of *A chemical review,* printed for private circulation, London, 1865, in file 'B-Club' at the Royal Society of Chemistry. See also note 121.

107 Maggie Frankland, *Diary,* entry for 20 September 1866 (note 64).

108 *Ibid.,* 13 August 1867.

109 Herbert McLeod recorded on 25 September (*Diary,* note 18): 'Frankland is away and no one knows when he will return'; in fact this was to be only five days later, by which time he had presumably made the necessary mental adjustments to his disappointment.

110 *Ibid.,* 15 September 1868.

111 *Sketches,* 2nd ed., pp.321–2

112 *Sketches,* 2nd ed., p.283.

113 *Sketches,* 2nd ed., pp.375–91.

114 Frankland to Nicholson (note 101).

115 *Sketches,* 2nd ed., p.276.

116 Maggie Frankland, *Diary* for 1863–1869 (note 64).

117 W. Pole, *The life of Sir William Siemens,* Murray, London, 1888, p.271.

118 Chemical Society Dining Club, Minutes of Proceedings 14 November 1872 – 14 January 1875 [Imperial College Archives, Armstrong papers no. C–183].

119 Maggie Frankland, *Diary* (note 64).

120 M. Faraday to Mrs. Barlow, 25 September 1863 [Royal Institution Archives, Faraday papers].

121 So called after Section B of the British Association, comprising 20 members including W. Odling, H. Müller, F. A. Abel and W. J. Russell (its secretary in the later 1860s). See A. Scott, Presidential Address, *J. Chem. Soc.,* 1916, **109**, 338–68 (342–51); and a few papers in 'B-Club' file [Royal Society of Chemistry Archives]. There were several budding poets in its membership,

notably J. C. Brough and Frederick Field; Brough was editor of the short-lived periodical *The Laboratory*, and Field an industrial chemist (*Chemical News*, 1872, **26**, 142 and 1885, **51**, 190). See also note 106.

122 'A.B.' [probably J. C. Brough], *A chemical review*, Taylor & Francis, London, 1863, p.13.

123 Frankland to Nicholson (note 101).

124 Frankland, *Rep. Brit. Assoc.*, 1868, **38**, Rep. Sect. 37.

125 Frankland to Sophie Frankland, 22 August 1868, reprinted in *Sketches*, 2nd ed., pp.400–2.

126 *J. Chem. Soc.*, 1848, **1**, 41.

127 Chemical Society, receipt for £10 Life Subscription, 13 December 1856 [RFA, OU mf 01.03.0540].

128 See W. H. Brock (ed.), *The atomic debates: Brodie and the rejection of the atomic theory*, Leicester University Press, Leicester, 1967.

129 Letters to Frankland from Berthelot, 14 October 1862; Cannizzaro, 15 April 1863; Fehling, 3 and 20 June, 1864; Kekulé, 18 January and 23 December 1863; Malaguti, 11 March and 27 November 1864; Marignac, 29 March 1863; Pasteur, 21 April 1863 [RFA, OU mf 01.03.0052, 0078, 0023 and 0025, 0081 and 0080, 0069 and 0057, 0055, 0017].

130 H. Kolbe to Frankland, 24 February 1864 [RFA, OU mf 01.02.1511].

131 Chemical Society Council Minute Book, entry for 6 February 1868 (vol. iii, p.182) [Royal Society of Chemistry Archives].

132 Efforts by Herbert McLeod to induce him to be President came to nothing: 'he is very strong against it' (McLeod, *Diary* (note 18), entry for 19 February 1869). A letter from Frankland's laboratory assistant J. J. Day complained of his refusal to accept the Presidency, 'a thing which no man ought to despise'; it was 'on account of his other duties being so great': J. J. Day to H. E. Armstrong, 27 March 1869 [Imperial College Archive, Armstrong papers no. C-246].

133 A. V. Harcourt to Frankland, 26 January 1871 [RFA, OU mf 01.03.0099].

134 Frankland to A. V. Harcourt, 28 January 1871 [RFA, OU mf 01.03.0101].

135 J. H. Gladstone to Frankland, 1 February 1871 [RFA, OU mf 01.04.0357].

136 W. Odling to Frankland, 30 January 1871[RFA, OU mf 01.04.0360].

137 *E.g.*, W. F. Bassett to H. McLeod, 'Friday 2'; E. Divers to H. E. Armstrong, 30 January 1871 [RFA, OU mf 01.04.0363, 0365].

138 Draft/copy-letter Frankland to F. A. Abel, 28 January 1870 [actually 1871] [RFA, OU mf 01.03.0097].

139 F. A. Abel to Frankland, 30 January and 1 February 1871[RFA, OU mf 01.04.0355, 0366].

140 Maggie Frankland, *Diary*, entries for 29 and 30 January 1871 (note 64); Percy Frankland's *Diary* has no entries for these dates.

141 H. E. Armstrong, 'First Frankland Memorial Oration of the Lancastrian Frankland Society', *Chem. & Ind.*, 1934, **53**, 459–66 (464).

CHAPTER 9

Advances in organic chemistry

1 Organic chemistry before 1860

During Frankland's seven years at Manchester his thoughts do not seem to have been specially directed towards the chemistry of carbon compounds. Of course organic chemistry formed a minor part of his lecturing schedule and also entered into some of his industrial consultancies. The relatively small amount of 'pure' research that he was able to undertake centred on the theme of organometallic chemistry, which even in those early days was a specialist study that left largely alone the growing number of organic compounds that did not happen to contain metals.

In the 1850s, except perhaps at the Royal College of Chemistry, organic investigations were not popular in England. The accidental discovery by Perkin in 1856 of the first synthetic dyestuff, aniline purple, took some little time to be seen as a forerunner of a vast multitude of new dyes that could transform organic chemistry into a lucrative pursuit involving wealth-creation on an unimagined scale. Also, as we have seen, there was little consensus among chemists as to the atomic weights of carbon and oxygen, and thus no possibility of agreeing even on such basic matters as empirical and molecular formulae. All this was to change after the Karlsruhe Congress of 1860, but even then it was not until about 1863 that Frankland finally settled for 'modern' atomic weights.

Then of course there were those two conflicting views about the nature of even simple carbon compounds, the radical and type theories. Frankland and Kolbe were of the older 'radical' persuasion but (as has again been noted) even they did not fully agree. Yet all was not darkness, and chemists began to perceive dimly a way through Wöhler's 'jungle' of organic chemistry using the concepts of valency (Frankland), a tetravalent carbon atom (Kekulé and Frankland) and catenation of carbon (Couper and Kekulé). The one over-arching concept that was needed before any major progress could occur had still to be clearly articulated: a theory of structure. Only then could the molecular architecture of organic compounds be seriously considered. Further developments as the birth of stereochemistry and (later still) the electronic theories of organic chemistry were in the distant future. But, as we shall see, enormous strides could be taken given a simple concept of structure.

It may be worth recalling what was known by the mid-1850s. Aromatic chemistry had hardly begun, though benzene and naphthalene were known, and Hofmann was undertaking his elegant researches into aromatic nitro-compounds, amines and their derivatives. But Kekulé's hexagon structure for benzene still lay ahead in the mid-1860s. Alicyclic chemistry did not exist. Quite large numbers of natural products had been isolated and empirical formulae were proposed for most of them. Fairly simple examples include the α-amino-acids glycine and alanine, lactic acid whose empirical and molecular formulae had been established by Liebig and Wurtz respectively, and the related trio glycerol, fatty acids and cholesterol. Many simple aliphatic compounds familiar today were quite unknown, as formaldehyde, methyl ethyl ketone, malonic ester, acetoacetic ester, nitroethane and so on. From the famous synthesis of urea by Wöhler in 1828 other more general synthetic routes had been developed, such as the electrolytic syntheses by Kolbe, the transformation of nitriles into acids by Dumas, Malaguti and Leblanc and the coupling synthesis of alkanes by Wurtz. The explosion of syntheses at the hands of Berthelot (and indeed Frankland himself) was still to come in the next decade. Taken overall organic chemistry towards the end of the 1850s presented a daunting challenge and cried out for reform. It is not surprising that a distinguished twentieth century historian of science, faced with the history of organic chemistry before 1860, complained that it was 'worse than alchemy!'

As he turned his face towards a career in London it is doubtful if Frankland had any notion of the epic transformation of organic chemistry that was about to begin. It is far more certain that he had little idea of the crucial rôle that he would play in the

Fig. 9.1 A.W. Hofmann (1818–92): Frankland's predecessor at the Royal
College of Chemistry. (Author.)

unfolding drama, a rôle that today is barely understood, even by
those whose specialism is that very subject. The reasons for his
posthumous eclipse will be considered later. Now it is convenient
to recall that, much nearer to his own day, a highly influential
series of review volumes, running into six editions, commenced

its survey with this far-reaching claim:

> In the form in which it exists to-day, organic chemistry may be said to take its root in the work of Frankland at the middle of last century.[1]

In fact the author was citing Frankland's paper of 1852 in the *Philosophical Transactions*. On the basis of that seminal work Frankland was able to build a considerable edifice of theoretical organic chemistry by the middle of the 1860s, partly in the laboratories of St. Bartholomew's Hospital but much more at the Royal Institution. These were years of feverish activity, never to return, with a relatively light lecture timetable, a minimum of consultant distractions and the congenial company of a young chemist whose early death deprived Frankland of a valued friend and chemistry of a brilliant experimental worker: Baldwin Francis Duppa (1828–73).[2] He appears frequently in letters to and from members of the Frankland family, and in the diaries of both Maggie Frankland and Herbert McLeod.

Frank Duppa (as he was generally known) was a cultured man, landowner and Justice of the Peace. His passion for chemistry was such that he went to work for Hofmann in the Royal College of Chemistry, studying the sulphonation of salicylic acid and titanium tetrabromide. He then joined Perkin in his private laboratory near Maidstone, and together they effected the conversion of succinic acid into *rac*-tartaric acid.[3] He came to St. Bartholomew's Hospital in 1860 to learn gas analysis and soon became involved in the wider research programme of Edward Frankland. They became fast friends and Duppa was frequently a welcome visitor at Haverstock Hill, joining with members of the family in such diverse amusements as ice-skating, music-making or painting. In collaboration with Frankland he began work on a proposed *Dictionary of isomers*[4], with a tabular arrangement based on series C_nH_{2n+2}, C_nH_{2n} and C_nH_{2n-2} and their oxygen derivatives. A typical page[5] began:

C	H	O	$C_nH_{2n+2}O$
1	4	1	Methylic A
2	6	1	Ethylic A
3	8	1	Propylic A
4	10	1	Butylic vel Tetrylic A – Isotetrylic A
5	12	1	Amylic A
6	14	1	Caproylic A – βCaproylic A
7	16	1	Oenanthylic vel Heptylic A
8	18	1	Caprylic vel Octylic A

etc.

The symbol A stood for 'alcohol' (and B for aldehyde, C for ketone, D for acid etc.), while 'vel' [Latin *or*] gave alternative names. Unfortunately this bold enterprise came to nothing, probably because organic chemists wanted something more than mere lists, as was later provided with such distinction by Beilstein.

Duppa's laboratory collaboration with Frankland came to an end when the latter forsook the Arcadian groves of pure research at the Royal Institution and re-entered the mundane world of training students at the Royal College of Chemistry. It might have resumed when a move to South Kensington took place but by that time Duppa was seriously ill and his work was over. He died of tuberculosis in 1873 at the early age of 45, deeply mourned by Frankland who always spoke of him in terms of the highest respect. Duppa was surely one of the unsung heroes of organic chemistry, yet he is denied a place in both the *Dictionary of National Biography* and the *Dictionary of Scientific Biography*. Some aspects of his fruitful collaboration with Frankland will now be considered.

2 Further developments in organometallic chemistry

The shape of Frankland's strategy from about 1855 happens to be extremely clear. Having discovered the remarkable zinc alkyls he determined to exploit their reactivity to the utmost degree. Even in Manchester he managed to find time for a modest research effort. He applied the latest Mancunian technology of high-pressure vessels to production of diethylzinc on a large scale. Using iron digesters as supplied by Nasmyth he found that the sealed tubes of reagents were kept from bursting by the external pressure of steam in the vessel.[6] It was another example of his technical versatility. With the product he examined the effects of diethylzinc on nitric oxide from which he obtained a zinc derivative of a new substance which he called 'dinitroethylic acid',[7] represented in modern terms as:

$$Zn(C_2H_5)_2 + 2NO \rightarrow C_2H_5\text{-}Zn\text{-}O\text{-}N\text{-}N{=}O$$
$$\underset{\displaystyle C_2H_5}{\overset{\textstyle |}{}}$$

Another series of experiments concerned the action of diethylzinc on other nitrogen compounds, including ammonia, aniline, diethylamine, urea and acetamide; in each case one hydrogen atom attached to nitrogen was replaced by zinc, and ethane was evolved.[8] In his own distinctive notation of 1877:

$$2NPhH_2 \; + \; ZnEt_2 \quad = \quad \left.\begin{array}{c} NPhH \\ NPhH \end{array}\right\} Zn \; + \quad 2EtH$$

Aniline   Zincethyl   Zincphenylamine   Ethylic hydride

Little else of significance for organometallic chemistry emerged from the laboratories at Owens College. As soon as Frankland had settled in London, however, the changed conditions led to a resumption of such work. He was spurred on by a discovery made by his former pupil and assistant, James Wanklyn: the existence of alkyls derived from the alkali metals. [9] These had been prepared, though not isolated, by the action of sodium or potassium on zinc alkyls. Their non-production from metal and alkyl halides, unlike the case of zinc, was attributed to reaction between (say) ethylsodium and ethyl iodide. This explanation was confirmed in a short note by Frankland,[10] perhaps chiefly important as signalling his return to the field. At all events he was shortly afterwards invited to give the Bakerian Lecture for the Royal Society, which he did under the title 'Researches on organometallic bodies'.[11] It was chiefly remarkable for its disclosure of further discoveries related to tin and mercury compounds. In the former case results can be summarised thus:

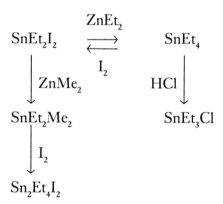

The successful methylation by dimethylzinc of the compound $SnEt_2I_2$ led Frankland to attempt an analogous reaction whereby 'mercuric methiodide' might be converted into methylethylmercury:

$$HgMeI + ZnEt_2 = HgMeEt + EtZnI.$$

However this time success eluded him. Instead a reaction took place which he represented thus [modern notation]:

$$2HgMeI + 2ZnEt_2 = 2HgEt_2 + ZnMe_2 + ZnI_2,$$

which, as he said, 'shows a mobility in the organic groups contained in these compounds which could scarcely have been anticipated'. Such a recognition of mobile alkyl groups was truly unusual at that time.

Frankland's research on organomercury compounds led to a curious incident that could have put his whole career in jeopardy.[12] Towards the end of 1865 the Parisian journal *Cosmos* carried an article reporting deplorable events at St. Bartholomew's Hospital:

> Among the organic compounds synthesised in recent times by M. Frankland is methylmercury. Doubtless requiring a large amount of this body, M. Frankland set his assistant M. C...U. (a German aged 30 years) to prepare it, and kept him at this detestable labour '*nearly three entire months!*' In consequence the poor young man became poisoned. He was admitted to one of the hospital wards on 3 February in a deplorable state; paralysis of the limbs, deafness, inflammation of the gums, scarcely able to stand, very imperfect vision, indistinct speech ... followed by intermittent delirium ... he died on 14 February![13]

The writer was a London correspondent of *Cosmos*, T. L. Phipson.[14] He referred to a second case of poisoning, a 23-year old technician who was exposed to methylmercury over 15 days. As a result 'he has become a virtual idiot, recognising no one'. The first victim was later identified as a Dr Carl Ulrich, the other as T. Sloper. The tragedy had been reported in *Chemical News*, but the *Cosmos* article raised special interest by its lurid reporting, its heading as 'a warning to young chemists' and above all its insinuation of gross negligence on the part of Frankland. It was widely cited on the Continent and at home[15], and has entered the folklore of Victorian chemistry.

Phipson, however, had made an elementary but important mistake. This was pointed out by Frankland[16] and by several friends who rallied to his defence, including Odling,[17] Hofmann[18] and Kolbe[19]. The latter wrote:

> It is completely incomprehensible, and scarcely credible, that Dr. Phipson has intentionally distorted the facts and has woven false claims into every report to the effect that both victims were Frankland's assistants or that Frankland had the methylmercury made. Frankland is one of England's top chemists and is so well known in London that everyone who is connected to chemistry and to incidents in the world of chemistry, as Phipson clearly is, knows that Frankland is a professor

of chemistry at the Royal Institution not at St. Barts. It is unthinkable that Dr. Phipson who is otherwise well educated did not know that Dr. Odling and not Frankland is at the chemistry lab at St. Barts. Secondly, both young chemists were Odling's assistants and not Frankland's, and they prepared the poisonous substance on Odling's instructions and not Frankland's. Frankland had nothing at all to do with the incident.

In fact Frankland had severed his links with St. Bartholomew's Hospital in 1864. For some reason, as Farrar and Williams remark, 'a publicity-seeking chemist of no great ability named T. L. Phipson waged a virulent press campaign against Frankland'.[20] They add that, though Frankland was completely exonerated for this particular offence,

> the most inexcusable part of the whole affair is that Frankland contributed a detailed article on organomercurials to an important chemical handbook, Watts' *Dictionary of chemistry* (1882), in which no hint is given of any possible hazard in working with these substances; indeed the taste of mercury dimethyl is described as 'faint but mawkish'!

Following his Bakerian Lecture Frankland next published on organoboron compounds. A comment in his *Experimental Researches* suggests that he now had a well-defined research programme: 'the synthesis of organic compounds by the exchange of oxygen and other negative elements for the alcohol-radicals, through the intervention of the zinc-compounds of the latter.' This programme really goes back to his 1852 paper; it can also be seen as an extension of his previous work where most of the reactions involved replacing *halogen* by alkyl. What he had in mind may be expressed as:

$$>C{=}O \xrightarrow{\text{ZnR}_2} >CR_2 \quad \text{and} \quad -OR \xrightarrow{\text{ZnR}_2} -R$$

He continues:

> As early as 1859, I endeavoured to effect substitutions of this kind in ethylic silicate, and I obtained and analysed silicopropionic acid (SiEtOHo), which obviously contains the silicon analogue of oxatyl (SiOHo); but, learning from Messrs. Friedel and Crafts that they were desirous of pursuing this branch of the subject, I left it in their hands, with, as the sequel has shown, the most happy results. I then attacked the corresponding ethylic borate.[21]

This account, with territorial claims settled with generosity and equanimity, reflects not only the manners of the participants but also the size of the field where there was, indeed, room for all and no need to duplicate work. The 'happy results' of Friedel and Crafts were the preparation in 1863 of tetraethylsilicon by the action of diethylzinc on silicon tetrachloride. They were excitedly communicated in a private letter to Frankland[22] and published soon afterwards.[23] Interestingly, a comparison between this compound and the tetraethyltin of Frankland was used by Mendeléeff in 1871 as he was predicting the properties of eka-silicon [germanium].

However Frankland's engaging picture of chemical courtesy needs some qualification. The formation of 'silicopropionic acid' is problematic and does not appear to have been published then or later. Only polymeric materials can have been obtained. More seriously, there are reasons to doubt that he was as deferential to the interests of Friedel and Crafts as he makes out. In seeking a further grant from the Royal Society in 1863 Frankland referred to previous work involving the action of diethylzinc upon trimethylboron and upon ethyl oxalate. He went on to say:

> I propose to complete this research and to extend it to the ethers of carbonic and silicic acids. *In fact I have already devoted much time to the reactions with Silicic Ether.* [my italics][24]

In other words he had not abandoned his quest for organosilicon compounds between 1859 and 1863. This letter was written *less than a month* before he heard of the success of Friedel and Crafts. The probability is that his experiments on 'ethyl silicate' were, for once, getting him nowhere and he was overtaken by rivals who used a more amenable starting material, silicon tetrachloride. It was Ladenburg, some years later, who explored successfully the reaction of orthosilicates with zinc alkyls and studied silicon analogues of organic acids.[25]

So, for whatever reason, Frankland abandoned organosilicon research and moved into the little known world of boron chemistry. Working for the first time with Duppa he produced a couple of papers on alkyl derivatives of that element.[26] They made for the first time 'boric ethide' and 'boric methide', highly reactive materials that are spontaneously inflammable in air. A summary of their most important findings is as follows:

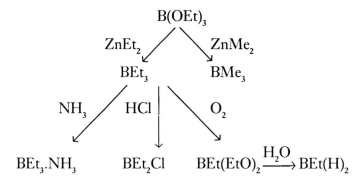

Frankland drew special attention to the action of dimethylzinc on ethyl orthoborate as indicating that (*contra* Kekulé) the organometallic reagent did not remove oxygen as such, for in that case triethylboron must have resulted. It was rather a replacement of an alkoxy group.
Frankland writes:

> These investigations [on boron compounds] were prosecuted in the very inconvenient chemical laboratory of St. Bartholomew's Hospital, and concluded in the laboratory of the Royal College of Chemistry at South Kensington. My more intimate acquaintance with these organo-boron compounds, however, was somewhat disappointing. I had expected that bodies possessing such an enormous chemical energy towards oxygen would have given rise to numerous compounds with other elements and radicals; but, although I spared no pains in trying to obtain such compounds, I was invariably thwarted in my attempts.[27]

On the assumption that organoboron compounds may be properly called organometallic, these researches concluded the assault made by Frankland on this branch of chemistry which he had made specially his own.[28] But they did not signal any wish to abandon the use of zinc alkyls as a reagent; quite the contrary. For now Frankland was more determined than ever to exploit its immense potential in organic syntheses, and never more effectively than in the study of aliphatic acids.

3 The lactic acids and questions of structure

It might be supposed that by about 1860 the nature of simple organic acids was widely understood. However it is a fact that in Kekulé's *Lehrbuch der organischen Chemie*, published in the following year, acetic acid was represented by no less than nineteen different formulae. The list did not include the formulae favoured by Frankland as,

$$C_4H_3O_2 \atop H \Big\} O_2 \quad \text{or, a little later,} \quad \Big\{ {CH_3 \atop COHo}$$

and recognises a degree of confusion scarcely credible today. As is often the case in this history, solutions to basic problems did not emanate in the first place by studies of the simplest compounds. Now Frankland was to make significant progress by a study, not of acetic acid and its homologues, but in the more complicated field occupied most prominently by lactic acid. In his search for further ways of using diethylzinc to introduce ethyl groups he started with the ethyl ester of oxalic acid, a substance long known to chemists. To his initial delight a reaction took place with much vigour, producing ethane and ethylene. However this was due to a secondary reaction. If the whole vessel were first immersed in a cold bath and then gently warmed an oily product was obtained containing zinc. When this was treated with water 'torrents' of ethane were evolved and an oily product obtained which he called 'ethylic diethoxalate' from which could be obtained the parent acid. With iodine another product was formed. The sequence was (in modern terms):

$$
\begin{array}{c}
 \quad \text{OZnEt} \qquad \text{OH} \qquad\qquad \text{OH} \\
\quad \text{ZnEt}_2 \quad | \quad \text{H}_2\text{O} \quad | \quad (1)\ \text{Ba(OH)}_2 \quad | \\
\text{COOEt} \longrightarrow \text{Et-C-Et} \longrightarrow \text{Et-C-Et} \longrightarrow \text{Et-C-Et} \\
| \qquad\qquad\quad | \qquad\qquad\quad |\quad (2)\ \text{H}_2\text{SO}_4 \quad | \\
\text{COOEt} \qquad\quad \text{COOEt} \qquad \text{COOEt} \qquad\qquad \text{COOH} \\
\quad \Big\downarrow \text{I}_2 \qquad\qquad\qquad\qquad\qquad \text{'diethoxalic} \\
\qquad\qquad\qquad\qquad\qquad\qquad\qquad \text{acid'}
\end{array}
$$

EtOOC-CEt$_2$-O-Zn-O-CEt$_2$-COOEt

It was soon found that the oxalate would react just as well if treated with a mixture of zinc and the alkyl halide. The combinations successfully tried, with their products, were described in a series of papers from February 1864 to April in the following year:[29]

Apart from the production of 29 new compounds (including all the above) this research was remarkable for several advances in theoretical organic chemistry. The first of these was recognition of the importance of *synthetic arguments leading to matters of constitution.*

Table 9.1. *Alkylations of dialkyl oxalates*

Reagents	Products
zinc+ethyl iodide+diethyl oxalate	CEt$_2$OH.COOEt
zinc+methyl iodide+dimethyl oxalate	CMe$_2$OH.COOMe
zinc+ethyl iodide+dimethyl oxalate	CEt$_2$OH.COOMe
zinc+ethyl and methyl iodides (1:1)+diethyl oxalate	CEtMeOH.COOEt
zinc+amyl iodide+diethyl oxalate	CAm$_2$OH.COOEt *and* CHAmOH.COOEt
zinc+ethyl iodide+diamyl oxalate	CEt$_2$OH.COOAm
zinc+amyl iodide+diamyl oxalate	CAm$_2$OH.COOAm

The acids derived from these esters (all of which were new) are clearly related to lactic acid, now known to be MeCHOH.COOH. Before this time the only other known member of the lactic series was glycollic acid, and their constitution was a matter of considerable controversy. Wurtz had supposed lactic acid to be dibasic, since two chlorine atoms were introduced by phosphorus chloride. Kolbe, however, considered it a peroxidic derivative of propionic acid, and the debate was as long as it was convoluted.[30]

Frankland's achievement was to use arguments based on synthesis, an important step forward but not entirely without precedent. Ladenburg pointed out many years later that 'the synthetical method constitutes a necessary complement' to the older analytical procedures and gave several examples in the decade up to this work of Frankland and Duppa. They included alanine (Strecker, 1850), mustard oil (Zincke, 1855), glycine (Perkin and Duppa, 1858), racemic acid (the same, 1860), malic acid (Kekulé, 1860), taurine (Kolbe, 1862) and guanidine (Hofmann, 1861).[31] Frankland was able to show that the nature of his products was necessarily as depicted because the known effect of zinc alkyls was to replace oxygen by alkyl. Given the oxalic acid starting point it followed that all could be regarded as alkylated hydroxy-acetic acids, so, by analogy, lactic acid was similarly constituted. Nor did he rely exclusively on syntheses: 'analytical evidence, although possessing far less weight than synthetical, may still be of service as corroborative testimony'. Thus the degradation of the dimethyloxalic ester by phosphorus chloride [into ethyl methacrylate], and the subsequent cleavage of the double bond by excess potash at 180°, yielded the salts of propionic and formic acids, the latter deriving 'from one of the semi-molecules of methyl introduced into the oxalic acid'.

The sequence was:

$$
\underset{\underset{\text{COOEt}}{|}}{\overset{\overset{\text{OH}}{|}}{\text{Me-C-Me}}} \overset{\text{PCl}_3}{\longrightarrow} \underset{\underset{\text{COOEt}}{|}}{\text{Me-C}=\text{CH}_2} \longrightarrow \underset{\underset{\text{COOH}}{|}}{\text{Me-C}=\text{CH}_2} \overset{\text{KOH}}{\longrightarrow} \underset{\underset{\text{COOH}+\text{H}_2}{|}}{\text{Me-CH}_2} + \text{H.COOH}
$$

Related to this relatively new use of synthetic arguments was a second feature of Frankland's work: recognition of important *new cases of isomerism*. Thus Frankland's 'ethylic diethoxalate' had the same molecular formula as the ethyl ester of leucic acid which Strecker had obtained by the action of nitrous acid on the amino–acid leucine:

$$\text{Me}_2\text{CH.CH}_2.\text{CH(NH}_2).\text{COOH} + \text{HNO}_2 = \text{Me}_2\text{CH.CH}_2.\text{CH(OH).COOH} + \text{H}_2\text{O} + \text{N}_2$$

   leucine   leucic acid

So confident was Frankland about the structures that he could affirm isomerism on the basis of syntheses even though the physical properties of the isomers might seem, as here, almost indistinguishable.

A third aspect of this research on the lactic acid series is that it discloses a dawning recognition of what rapidly became known as *structure*. To represent the work by modern formulae is not to imply that Frankland thought of them with that degree of structural precision. The epistemological problems underlying the whole concept of structure will be encountered later (p. 270; it is sufficient to note here that to represent his thoughts as clearly as possible he had recourse to a new system of graphic formulae introduced in 1864 by Crum Brown which, as he said, indicated the 'chemical position' of the atoms depicted, *i.e.* their relationship to one another in terms of chemical connections. Taking now the simplest cases of isomerism in this series he showed how synthesis could demonstrate the construction of the isomers and their mutual relationship. Knowing independently the structures of the cyan–hydrins giving rise to lactic and paralactic acids one could immediately see which was which (and therefore by elimination which was the third). His argument may be summarised as follows:

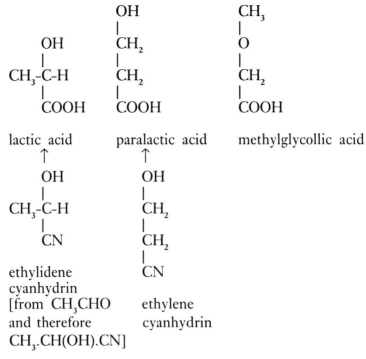

The constitution of lactic acid was confirmed by its formation from alanine with nitrous acid (the former a product from aldehyde ammonia), and by the reduction of Debus' pyruvic acid recently effected by Wislicenus[32]:

Fourthly, Frankland defined the *structural unit of acidity as 'oxatyl'* or, as we should say, carboxyl. Written as -COOH or as -COHo, it was a compound radical persisting unchanged in acetic and acrylic acid series and conferring upon them acidic character. Its hydroxyl group could be replaced:

- by metal oxides to form salts
- by alkoxy groups to form esters
- by hydrogen to form aldehydes

    – by alkyl to form ketones
    – by oxygen (semi-molecule) to form acid anhydrides, and
    – by amino to form amides.

He concluded that 'an organic acid containing n semi-molecules of oxatyl is n-basic.' Thus, within just over a year, Frankland and Duppa had given for the first time a genuinely constitutional description for an important series of organic compounds and had unwittingly helped to found the modern structural theory of organic chemistry.

4 Unsaturated acids and problems of reactivity

Having produced so many new compounds of the lactic acid family it was entirely natural for Frankland and Duppa to wish to investigate their reactions. One of the most spectacular of these was an exothermic reaction occurring when (say) the ester 'ethylic diethyloxalate' was treated with phosphorus trichloride. Analysis of the oily product revealed that a molecule of water had been removed, the product being similar to the esters of the known acrylic and crotonic acids. The acid was designated ethylcrotonic acid and formation of its ester was represented as:

$$3\begin{cases} CEt_2Ho \\ COEtO + PCl_3 \end{cases} = 3\begin{cases} CEt''Et \\ COEtO + PHOHo_2 + 3HCl, \end{cases}$$

a more modern formulation being:

$$\underset{\underset{COO.C_2H_5}{|}}{\overset{\overset{OH}{|}}{CH_3\text{-}CH_2\text{-}C\text{-}CH_2\text{-}CH_3}} \quad \overset{PCl_3}{\rightarrow} \quad \underset{\underset{COO.C_2H_5}{|}}{CH_3\text{-}CH_2\text{-}C\text{=}CH\text{-}CH_3}$$

Several new acids were formed of the type R.C(COOH)=CHR', together with esters and some metal salts[33] as in Table 9.2.

    The arguments for these structures, though simple to modern chemists, were quite revolutionary in their own day. They were, as previously, arguments based upon synthesis. But they were much more than that. Frankland and Duppa explain the relations between acrylic and lactic series by first formulating three equations:

$$\begin{cases} CEt_2Ho \\ COEtO \end{cases} -OH_2 = \begin{cases} CEt''Et \\ COEtO \end{cases}$$

$$\begin{cases} CEtMeHo \\ COEtO \end{cases} -OH_2 = \begin{cases} CEt''Me \\ COEtO \end{cases}$$

$$\begin{cases} CMe_2Ho \\ COEtO \end{cases} -OH_2 = \begin{cases} CMe''Me \\ COEtO \end{cases}$$

They then continued:

> It will be seen from these equations that, in the passage from the one series to the other, the hexad carbon type is preserved, since the abstraction of one atom of hydrogen from one of the monad radicals of the original acid, converts the latter into a dyad radical, thus saturating the bond vacated by the semimolecule of hydroxyl.

In this passage may be seen, not merely a clear argument from synthesis, but also a clear application of the laws of valency to which Frankland had made such a signal contribution thirteen years previously. The phrase 'saturating the bond' was striking in its daring novelty, dubious to those maintaining an agnostic position with regard to the very existence of atoms, abhorrent to those who deplored the whole notion of a chemical structure being accessible to experiment, but vividly expressive for any whose inhibitions did not close their minds to developments that were to make organic chemistry something much more akin to an exact science than it had ever been before.

Frankland must have experimented thoroughly with these new unsaturated acids and esters. Yet he is surprisingly silent about their capacity to undergo addition reactions or even polymerisation. He describes the penetrating smell of the esters and the acidic character of the parent acids. But one reaction seems to have exerted special fascination, for he and Duppa report it at length. This was the decomposition of the acids on being heated with potash; hydrogen is given off and two carbon-containing salts are formed. This was already well-known for acrylic acid:

$$CH_2{=}CH.COOH + 2KOH \rightarrow HCOOK + CH_3COOK + H_2$$

Frankland now took a further step forward by actually distinguishing two possible ways in which this could happen; in fact he was posing a *mechanistic* question, and once again anticipating modern understanding in a remarkable way. One could assume the simple

Table 9.2. *unsaturated acids synthesised by Frankland and Duppa*

R	R'	acid
Me	H	methacrylic
Me	Me	methylcrotonic
Et	Me	ethylcrotonic

displacement of 'methylene' by two hydrogen atoms, the expelled methylene being immediately oxidised to formic acid. Alternatively,

> the molecule of acrylic acid is entirely broken up by the separation of the positive and negative semimolecules, the oxatyl by combination with hydrogen yielding potassic formate, whilst the positive semimolecule is oxidized and forms potassic acetate.

He represents these possibilities in two different ways:

$$\begin{cases} \text{C:Me}'' : \text{H} \\ \text{COHo} \end{cases} \quad \text{and} \quad \begin{cases} \text{CMe}''\text{H} \\ \text{COHo} \end{cases}$$

In other words, he suggests cleavage of either a double or single bond:

$$CH_2{=}CH\text{-}COOH \text{ and } CH_2{=}CH{-}COOH$$

With methacrylic acid the products would be the same in either case, namely potassium propionate and formate:

$$\begin{matrix} CH_3\text{-}C{=}CH_2 \\ | \\ COOH \end{matrix} + 2KOH \rightarrow \begin{matrix} CH_3\text{-}CH_2 + HCOOK + H_2 \\ | \\ COOK \end{matrix}$$

But with methyl crotonic acid the products could be different:

$$CH_3\text{-}CH{=}CMe\text{-}COOH \rightarrow CH_3COOH + CH_3CH_2COOH$$

$$CH_3\text{-}CH{=}CMe{-}COOH \rightarrow CH_3CH_2CH_2COOH + HCOOH$$

In fact the former products are found and so cleavage of the *double* bond is proved. Thus an important feature of any reaction is established, the actual mode of decomposition. Indeed Frankland

and Duppa proceeded on the assumption that, in molecules like these, the double bond is more easily broken than a single bond, which, in the light of contemporary knowledge, was a remarkable thing to assume. As Frankland remarked in a note to Odling 'the cutting of the double bond consistently explains the reaction throughout'.[34] Ideas like that, though commonplace today, were real advances for a subject where valid generalisations on structural or mechanistic matters were comparatively rare.

5 Acetoacetic ester and its reactions

Organic chemists today need no reminder of the importance of ethyl acetoacetate (or acetoacetic ester). Many generations of students have been introduced to it as a synthetic tool of great versatility. Its usual mode of formation by the action of sodium ethoxide on ethyl acetate (the Claisen condensation) is remarkable alike for its elegance and for its mystery. Easily conducted by a competent undergraduate, the reaction nevertheless eluded satisfactory explanation until the advent of the electronic theory of organic chemistry. Furthermore, acetoacetic ester at an early stage raised interesting problems of constitution, being one of the first recognised cases of keto–enol tautomerism. This critically important compound lay at the heart of the next researches of Frankland and Duppa, though it was not recognised as such at the beginning. They entered the field in the following manner.

Having apparently exhausted the possibilities of treating esters with zinc and alkyl halides Frankland and Duppa turned in 1865 to the use of the more reactive metal sodium, possibly suggested by Wanklyn's recent discovery of alkylsodium compounds. Their first results[35] were encouraging, but their accounts are difficult for modern chemists to comprehend, partly because their term *ether* covers both ethers and esters of current usage, but also on account of some singularly barbaric formulae as

$$O_2(CO)''(C_4H_7)Et \quad \text{and} \quad O_2(CO)''(C_5H_9)Et.$$

Fortunately, however, these were soon abandoned for the conventional type formulae of the time, and later for the Crum Brown structural formulae.

The technique was generally to heat the ethyl acetate with

sodium until hydrogen ceased to be evolved, and then to allow the resulting mass to react with ethyl or methyl iodide. Heat was generated and after the reaction the resultant oil was separated and further investigated. No attempt was made to isolate the initial product from the action of the sodium on the ester.

The first discovery seems to have been that the oily products could be readily decomposed by cold dilute barium hydroxide solution to yield ketones. However attention was soon focused on the substances obtained from distillation of these ethereal oils. The products were simple esters, and it is almost certain that they were produced by carbon-carbon fission induced by basic material remaining in the final reaction product (in other words the so-called 'acid hydrolysis' of acetoacetic ester and its congeners). However for Frankland and Duppa it was – at least at first – a case of simple replacement of one or two hydrogen atoms of ethyl acetate by alkyl. In their first paper to the Royal Society they put it in the following way:

> It was proved, nearly twenty years ago, that methyl is a constituent of acetic acid; and in the year 1857 the derivation of this acid and a large number of other organic compounds from the carbonic acid or tetrad carbon type was proposed. According to this view, which is gradually receiving the assent of chemists, the constitutional formula of acetic ether is
>
> COMeEto,
>
> or, with the formula of the contained semimolecule of methyl fully developed,
>
> $$\left\{ \begin{matrix} CH_3 \\ COEto \end{matrix} \right. \quad \text{or} \quad CO(CH_3)Eto$$
>
> Thus the radical methyl, in acetic ether, contained three single atoms of hydrogen, combined with a tetrad of carbon. If one of these atoms of hydrogen be displaced by methyl, an ether having the composition of propionic ether will obviously be produced -
>
> $$\left\{ \begin{matrix} CMeH_2 \\ COEto \end{matrix} \right.$$
>
> If a second atom of hydrogen be displaced by another semimolecule of methyl, butyric ether or its isomer will, in like manner, be formed -
>
> $$\left\{ \begin{matrix} CMe_2H \\ COEto \end{matrix} \right.$$

They continued with predictions of the following products, given the appropriate alkyl halides:

$$\begin{cases} CEtH_2 \\ COEto \end{cases} \qquad \begin{cases} CEt_2H \\ COEto \end{cases} \qquad \begin{cases} CAm_2H \\ COEto \end{cases}$$

At this early stage of their researches they reported the successful isolation of the two ethylated products from ethyl esters of acetic, butyric and isocaproic acids.

In their next paper[36] Frankland and Duppa report the formation, not of acetoacetic ester itself, but of its diethyl derivative, which they called ethylic diethaceto-acetate. Ostensibly their work was stimulated by a discovery by Wurtz[37] that ethyl lactate could be converted to its ethyl ether by treatment with sodium followed by ethyl iodide. Such alkylations were clearly of great synthetic interest but they had posed several problems given the uncertain nature of lactic acid. Frankland and Duppa therefore decided to start their investigations with a simpler substance, 'the ether of a well-defined monobasic acid' and (of course) ethyl acetate was selected. They admitted cheerfully, and with masterly understatement, that the results were 'not strictly analogous' to those obtained by Wurtz, but they were without question 'highly remarkable'. In fact the research followed naturally from their previous discoveries with ethyl acetate and that is sufficient reason by itself.

As before, sodium was added to ethyl acetate and the product heated to 100° for several hours. After addition of ethyl iodide and cooling, excess water was added and the product distilled, yielding some unchanged starting material, some ether and a complex mixture from which were derived the following substances, identified as:

$$\begin{cases} COMe \\ CEt_2 \\ COEto \end{cases} \qquad \begin{cases} COMe \\ CEtH \\ COEto \end{cases} \qquad \begin{cases} CEtH_2 \\ COEto \end{cases} \qquad \begin{cases} CEt_2H \\ COEto \end{cases}$$

or, in modern terms, $CH_3.CO.CEt_2.COOEt$, $CH_3.CO.CHEt.COOEt$, $CEtH_2.COOEt$, and $CHEt_2.COOEt$ respectively.

One of the most remarkable features of the research was its sheer scale. The ethyl acetate was prepared by the action of 'sulphovinic acid' ($EtHSO_4$) on dried sodium acetate, using 6000 g. of the latter, 3600 g. of ethanol and 9000 g. of sulphuric acid. The sodium was added in pieces $1 \times 1 \times \frac{1}{4}$ in. An immense amount of care must have been put into the elemental analysis and vapour density measurements of all the products, very good agreement being obtained with the theoretical values. Their constitution was, however, derived largely from the facts of their decomposition.

Ketones were obtained by alkaline hydrolysis of the 'diethodiacetates', a process represented in the following manner:

$$\left\{ \begin{array}{l} \text{COMe} \\ \text{CEtH} \\ \text{COEto} \end{array} \right. + 2\text{KHo} = \text{COKo}_2 + \text{EtHo} + \left\{ \begin{array}{l} \text{COMe} \\ \text{CEtH}_2 \\ \text{'ethylated} \\ \text{acetone'} \end{array} \right.$$

and

$$\left\{ \begin{array}{l} \text{COMe} \\ \text{CEt}_2 \\ \text{COEto} \end{array} \right. + 2\text{KHo} = \text{COKo}_2 + \text{EtHo} + \left\{ \begin{array}{l} \text{COMe} \\ \text{CEt}_2\text{H} \\ \text{'diethylated} \\ \text{acetone'} \end{array} \right.$$

This was in accord with their general understanding of decarboxylation as in the unsaturated acids. However the larger problem of production of the 'diethodiacetates' required a more sophisticated strategy than they were able to supply; it was 'explained', they said, in the two following equations:

$$2 \left\{ \begin{array}{l} \text{CH}_3 \\ \text{COEto} + \text{Na}_2 \end{array} \right. = \left\{ \begin{array}{l} \text{COMe} \\ \text{CNaH} \\ \text{COEto} \end{array} \right. + \text{NaEto} + \text{H}_2;$$

$$\left\{ \begin{array}{l} \text{COMe} \\ \text{CNaH} \\ \text{COEto} + \text{EtI} \end{array} \right. = \left\{ \begin{array}{l} \text{COMe} \\ \text{CEtH} \\ \text{COEto} \end{array} \right. + \text{NaI}.$$

While it is easy to see how the last reaction is a simple alkylation as in many organometallic processes, it is not clear how Frankland could have imagined the first one, with formation of a four carbon chain. His postulated sodium compound was, of course, the now-familiar derivative of acetoacetic ester itself. The best explanation he and his colleague could give suggests as an intermediate 'a dyad body . . . equivalent to ethylene':

$$2\begin{cases} CH_3 \\ COEto + Na_2 \end{cases} = \quad 2''\begin{cases} CO \\ CNaH \end{cases} + 2EtHo + H_2$$

followed by a kind of 'addition' reaction:

$$\begin{cases} CH_3 \\ COEto + \end{cases} \quad {}''\begin{cases} CO \\ CNaH \end{cases} = \quad \begin{cases} CH_3 \\ CO \\ CNaH \\ COEto \end{cases} \quad or \quad \begin{cases} COMe \\ CNaH \\ COEto \end{cases}$$

Shortly after this Kolbe proposed an alternative route, where the active intermediate was the disodium derivative of ethyl acetate which then reacted with another molecule of unchanged ester to yield the monosodium salt of acetoacetic ester.[38] Frankland and Duppa subsequently conceded the superiority of Kolbe's mechanism.[39]

However when it came to explaining the formation of two–carbon chains Frankland and Duppa merely suggested straight alkylations of the methyl group without recourse to anything like acetoacetic ester:

$$\begin{cases} CH_3 \\ COEto + Na_2 \end{cases} = \quad \begin{cases} CNa_2H \\ COEto + H_2; \end{cases}$$

$$\begin{cases} CNa_2H \\ COEto + 2EtI \end{cases} = \quad \begin{cases} CEt_2H \\ COEto + 2NaI. \end{cases}$$

The weakness of this view (as Frankland and Duppa recognised) is that no sodium derivative of ethyl acetate had been isolated. Meanwhile, J. A. Geuther of Jena had been working on similar lines, though his work was published in obscure German journals.[40] Frankland and Duppa became aware of it only as their first paper was going to press, thereby incurring the wrath of Geuther for what he considered a 'childish' and 'ungentlemanly' intrusion into his own territory.[41] Amongst much else he insisted that 'carboketonic ethers' [β-ketonic esters] were intermediates, actually converting the ethyl derivative of acetoacetic ester to ethyl butyrate, using sodium ethoxide:

$$\begin{cases} COMe \\ CEtH \\ COEto \end{cases} \quad \rightarrow \quad \begin{cases} H \\ CEtH \\ COEto \end{cases}$$

The structures of many of the four–carbon chain products were specially problematic. To be sure, vapour density and elemental analysis results agreed nicely with the molecular formulae given. But to speak of 'constitutional formulae' in the era of pre-structural chemistry is to invite misconceptions, charges of '*a priorism*' and worse. The fact is that Frankland and Duppa were wrestling with genuinely structural problems in the days before structural criteria had been established. Their arguments are therefore particularly instructive.

First, they argued on the basis of the general reactivity, and the fact that these newcomers seemed to be 'compounded of the residues of ketones and carbonic acid'. So 'we have ventured to consider these bodies as the ethereal salts [esters] of peculiar acids' and wrote the parent acids and their esters in general terms as follows:

$$\begin{cases} COMe \\ C(C_nH_{2n+1})_2 \\ COHo \end{cases} \quad \text{and} \quad \begin{cases} COMe \\ C(C_nH_{2n+1})_2 \\ COEto \end{cases}$$

Secondly, they considered such formulae in the light of decarboxylation by potash to yield potassium carbonate; this was a known reaction of alkyl carbonates (as 'carbonic ether, $COEto_2$). After simple algebraic manipulation[42] the above formula for a dialkylated ester could then be rearranged in similar form as

$$CO(OC_xH_{2x-1})Eto$$

Hydrolysis of the monomethyl derivative could be expressed:

$$CO(OC_4H_7)Eto + 2KHo = COKo_2 + (C_4H_7)Ho + EtHo$$

Apart from ethanol the organic product was identified as methylated acetone [methylethyl ketone], but what was the propriety of representing this as an (unsaturated) alcohol? Certainly Kane[43] had attributed alcoholic properties to some ketones, having successfully introduced chlorine by treatment of PCl_5, but Frankland was well aware of the 'very palpable difference between the ketonic and alcoholic families' and wisely did not pursue this formulation further.

Thirdly, Frankland and Duppa found that the degradations of the carboketonic esters into either ketones or simpler esters of monobasic acids could most simply be explained on the basis of the constitution assumed:

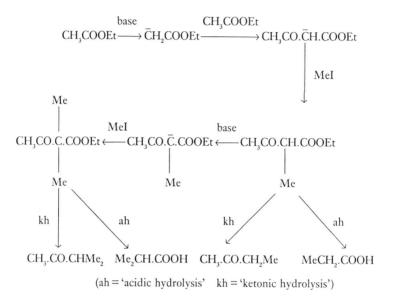

$$\begin{cases} CH(C_nH_{2n+1})_2 \longleftarrow \\ COHo \end{cases} \begin{cases} COMe \\ C(C_nH_{2n+1})_2 \\ COHo \end{cases} \longrightarrow \begin{cases} COMe \\ C(C_nH_{2n+1})_2H \end{cases}$$

'acid hydrolysis'   'ketonic hydrolysis'

In modern terms the sequence of reactions may be represented thus:

$$\underset{}{CH_3COOEt} \xrightarrow{\text{base}} \bar{C}H_2COOEt \xrightarrow{CH_3COOEt} CH_3CO.\bar{C}H.COOEt$$

MeI

$$\underset{\underset{Me}{|}}{\overset{\overset{Me}{|}}{CH_3CO.C.COOEt}} \xleftarrow{\text{MeI}} \underset{\underset{Me}{|}}{CH_3CO.\bar{C}.COOEt} \xleftarrow{\text{base}} \underset{\underset{Me}{|}}{CH_3CO.CH.COOEt}$$

kh   ah   kh   ah

CH₃.CO.CHMe₂   Me₂CH.COOH   CH₃.CO.CH₂Me   MeCH₂.COOH

(ah = 'acidic hydrolysis'   kh = 'ketonic hydrolysis')

Two shorter papers dealt with, respectively, the use of isopropyl iodide as alkylating agent[44], and the question of evolution of hydrogen when ethyl acetate is treated with sodium.[45]

The new compounds that emerged from this research included those shown in Table 9.3.

It will be noticed that they did not isolate the parent compound, acetoacetic ester itself, but assumed it to be present. This was, however, obtained by Geuther in Jena, through the action of carbon dioxide on the sodium salt. Subsequent work on the acetoacetic ester synthesis was carried out by Johannes Wislicenus at Würzburg and Ludwig Claisen at Kiel.

In 1877 Wislicenus (whose daughter Marie married a nephew of Sophie Frankland) published a paper on acetoacetic ester.[46] Although his 'inquiry was not limited to unravelling Frankland and Duppa's experiments' it has been wisely said that 'the whole subject was submitted to a critical re-examination with results which have

Table 9.3. *Products from acetoacetic ester syntheses by Frankland and Duppa*

	R	R'	Frankland's name
β-ketonic esters,	H	CH_3	ethylic methaceto-acetate
$CH_3.CO.CRR'.COOEt$	CH_3	CH_3	ethylic dimethaceto-acetate
	H	C_2H_5	ethylic ethaceto-acetate
	C_2H_5	C_2H_5	ethylic diethaceto-acetate
	H	$(CH_3)_2CH$	ethylic isopropaceto-acetate
esters of monobasic acids,	CH_3	CH_3	ethylic dimethacetate
$CHRR'.COOEt$	H	C_2H_5	ethylic ethacetate
	C_2H_5	C_2H_5	ethylic diethacetate
	H	$(CH_3)_2CH$	ethylic isopropacetate
	H	C_5H_9	ethylic amylacetate
monobasic acids,	CH_3	CH_3	dimethacetic acid
$CHRR'.COOH$	H	C_2H_5	ethacetic acid
	C_2H_5	C_2H_5	diethacetic acid
	H	$(CH_3)_2CH$	isopropacetic acid
	H	C_5H_9	amylacetic acid
ketones,	H	CH_3	methylated acetone
$CH_3CO.CHRR'$	CH_3	CH_3	dimethylated acetone
	H	C_2H_5	ethylated acetone
	C_2H_5	C_2H_5	diethylated acetone
	H	$(CH_3)_2CH$	isopropylated acetone

proved of the highest importance to synthetical organic chemistry'.[47] He refined the method of preparation, showed that only one hydrogen atom at a time could be replaced by sodium, and established the best conditions for acidic and ketonic hydrolyses.

Then in 1887 Claisen produced what Cohen has called 'the first serious contribution to a theory of the acetoacetic ester synthesis',[48] placing it in the wider context of a large number of base–catalysed condensations that he had studied and suggesting that the essential condensing agent is sodium ethoxide obtained from adventitious traces of free ethanol in the starting materials.[49]

Constitutional questions regarding acetoacetic ester and its derivatives were unresolved for many years after Frankland and Duppa had finished their work. Geuther's formulae for ethyl acetate and the sodium compound of acetoacetic ester were:

$$CH_3,CO_2 \begin{Bmatrix} OH \\ OH \end{Bmatrix} C_2H_4 \quad \text{and} \quad \begin{matrix} CH_3,CO_2 \\ CH_3,CO_2 \end{matrix} \begin{Bmatrix} NaO \\ HO \end{Bmatrix}, C_2H_4$$

Although at this stage Geuther had not fully grasped the essentials of structural theory, the above constitutional formula clearly has elements of unsaturation and of hydroxyl. It came to be regarded as the archetype of the enol formula, whereas that of Frankland and Duppa was of the keto form. Subsequent controversy cannot detain us now, but the idea of the two forms in tautomeric equilibrium (so vindicating the original authors in Jena and London) came to fruition with the isolation by Knorr in 1911 of each:[50]

$$CH_3CO.CH_2.COOEt \rightleftharpoons CH_3.C(OH) = CH.COOEt$$

It is not always remembered that we owe to Frankland and Duppa not only realisation of the synthetic potential of these compounds but also the first moves towards the keto–formula so widely used today.

6 The theory of structure

In following the work of Frankland and Duppa in aliphatic chemistry it is possible to see not only the development of formidable synthetic techniques but also a kind of intellectual pilgrimage which (from the standpoint of later structure theory) may be justly styled 'from doubt to faith'. By the time they had finished their intense period of activity on acetoacetic ester derivatives (about 1867) that theory was not only looking credible; it had become indispensable.

The so-called structure theory of organic chemistry is in its simplest form a declaration that the properties of a compound may be uniquely defined by its molecular structure, i.e. the arrangement of atoms in a molecule with respect to one another. The theory did not emerge overnight, nor can any one man be said to have been its founder, though the most plausible claimants include Butlerov and Kekulé.[51] In fact, in the early 1860s, there was a veritable rash of structural ideas derived from the later forms of the type theory, with contributions from Erlenmeyer, Crum Brown, Beilstein and several others. What had once been the haunt of positivist scepticism now appeared to become a focus for brave new speculations about the way atoms were actually arranged within a molecule. The old idea that types were purely taxonomic in function, serving chemists much as the genera and species had served the botanist, had become replaced by a notion that type formulae might conceal explanations

Fig. 9.2 August Kekulé (1829–96): Frankland's chemical opponent and arch–rival (Author.)

as well as analogies; the familiar curly brackets might have some suggestion of internal constitution. Yet most of the neo-typists, if pressed, would have shown considerable reluctance to admit that their 'constitutional formulae' were more than convenient fictions, with considerable predictive value, it is true, but not in any sense representing what was 'really' there. Partly this was a legacy from the structural agnosticism inherent in earlier type theory, and partly it was a well-founded strategy to avoid any implications that the positions of atoms in even 2-dimensional space could actually be known; the thought of 3-dimensional representation was anathema. The philosophical position adopted by the majority of these people was that of pure conventionalism.

However there was another approach to the whole matter, and that was represented uniquely and with immense vigour by Frankland's old friend Hermann Kolbe. His position has been fully and brilliantly explored by Alan Rocke[52] so it is necessary to make only the briefest of comments here. For reasons of chemical tradition, professional rivalry, nationalistic sentiment, personal idiosyncrasy, and philosophical stance he developed a strong antipathy to the writings of the *Strukturchemiker* (as he called them) and especially their leader August Kekulé. He rejected flatly their position of cautious conventionalism. For Kolbe radicals in molecules were real. A chemical structure actually existed. The irony of this naive realism is that Kolbe believed that these structures were forever beyond the reach of chemists to discover. Only by unbridled speculation – as with the *Strukturchemiker* – could imaginary 'structures' be produced and there was no guarantee that such devices had any relation to reality.

Kolbe, like Frankland, was in the tradition of electrochemical radical theory. Had they not years ago obtained methyl from electrolysis of acetic acid and established in that substance the presence of methyl and carbonic acid? Dualism, it seemed to Kolbe, gave a far surer foundation to organic chemistry than the flights of fancy deriving from the French chemists and their followers. He was, in Rocke's phrase, a 'crypto-structuralist'. It may be noticed in passing that the connection once made between Kolbe's cautious attitude to molecular structure and his alleged agnosticism in religion[53] now seems thoroughly misplaced. Kolbe, son of a Lutheran pastor and apparently sharing his faith, is in sharp contrast to his rivals who were 'younger upper-middle class urban liberals and agnostics, such as Kekulé'.[54]

In view of the incipient structuralism in his work on organic acids Frankland may be regarded as another important figure in the evolution of structure theory. In a letter to the French chemist Cahours he defines his understanding of the part played by radicals in organic molecules. The passages placed now in italics reveal a strong commitment to inorganic analogies and a firm belief in the importance of the nature of the radical for the properties of compounds in which it may be found. Both of these are in perfect harmony with ideas of Berzelius. By resting his case on the doctrine of valency, and its assumption that free atomicities must somehow be neutralised (as when radicals dimerise), Frankland was implying the real existence of radicals in a combined state and coming near to an electrochemical theory of structure (though Berzelius would not have liked dimeric molecules of the kind H_2 and R_2). Part of his letter reads [my italics]:

We seem chiefly to differ in our definition of a radical which you define to be a body capable of combining directly with certain elements, regenerating the compounds from which it was isolated. On the other hand *I consider a radical to be simply a body which is isolated from its compounds with Cl, I &c by reactions analogous to those by which its inorganic analogues are separated from the same elements*; if the radical thus isolated recombines with these elements without conditions of heat or light which break up its constitution, then, like you, I also require that it should regenerate the original compounds. But I do not consider as an essential property of a radical the power to combine with O, Cl &c. without the aid of heat or light, otherwise even hydrogen itself would be excluded from the radical title.

At a heat of 500°C. hydrogen does not combine with O. At this temperature ethyl cannot exist. All compound radicals giving true compounds with chlorine do so in the dark. But chlorine has no action upon ethyl in the dark.

I quite agree with your high estimate of Dumas' types. I have always considered his view of a chemical type as one of the most profoundly philosophical ideas ever enunciated, but he unfortunately impaired its truth by asserting that *the properties of compounds depend only upon the position and not upon the character of their constituent elements.* Carbonic acid (CO_4) and oxide (CO_2) are my types for nearly all organic compounds. *I do not believe in any isolated radicals.* Every compound of carbon, with few exceptions, is produced by the substitution of one or more atoms of O in CO_4 or CO_2 by other and analogous carbon molecules and finally the so called alcohol radicals are procured by the substitution of their last atom of oxygen by the true radical itself. Thus methyl is

$$C\begin{cases} H \\ H \\ H \\ CH_3 \end{cases}$$

Or in other words the group CH_3 being unstable requires a fourth molecular space to be occupied which can be done by O, H, Cl or by CH_3; the latter case is a type of all monatomic radicals in their so called isolated condition. On this view H is also of course a double molecule

$\left.\begin{smallmatrix}H \\ H\end{smallmatrix}\right\}$. Similarly a diatomic radical in combination is nothing more than CO_4 in which 2 atoms of oxygen are replaced by positive molecules and I regard such a radical, in its isolated condition, as a double molecule in which each single molecule stands to the other in the position of a diatomic body thus occupying the two vacant places in each molecule. But I shall weary you with these rough ideas which require yet a good deal of work to be done before they can be either established or refuted. I had hoped that the organic compounds of boron would have furnished some analogies in this direction but unfortunately that element has at length decidedly declared itself as teratomic instead of tetratomic. I am however just now engaged with organo-silicon compounds from which I hope to get some insight into these matters.[55]

In fact, as we have seen, he had less success with organosilicon compounds than with their boron counterparts. It was the research on organic acids that drew him into a structuralism that professed to know the detailed inter-atomic relations within quite complicated molecules. Shortly after this he was lecturing on natural organic bases at the Royal College of Chemistry; his notes tersely confess 'little known as to internal structure of these alkaloids',[56] but the very admission testifies to his faith in the meaningfulness of the phrase 'internal structure' and the possibility of knowing it in simpler cases.

Frankland's location within the minority stream of electrochemical dualism might suggest that he identified with the structural ideas of Kolbe rather than of Kekulé. There is some evidence to support this. Writing to Kolbe in 1871 he said that he had long regarded the ideas on chemical constitution associated with Gerhardt, Kekulé, Erlenmeyer and others as 'absurd', as worthy of Kolbe's 'severest criticism', and greatly inferior to his own system using 'the very neutral word bond'.[57] A few months earlier Kolbe had assured Frankland of his delight that 'we agree over structure formulae' and differ from the *Strukturchemiker* of the Kekulé school who 'deny the actual existence in compounds of compound radicals', and for whom methyl, ethyl etc. were only ideal entities of great convenience.[58] A belief in the reality of radicals had informed their early work, was totally absent in Kekulé, and continued to insulate both of them from the Kekulé school. In matters of detail Frankland agreed with Kolbe's criticisms of structures suggested by Baeyer who 'either cannot write clearly or does not think it worth his while to write clearly'.[59] Long after this, in 1884, he confided to Ost that Kolbe was, 'if the title can be assigned to any one man, the father of *Strukturchemie*'.[60]

On the other hand Frankland was more in debt to the French school than he may have cared to admit at first. His lecture notes refer to textbooks by Wurtz and Kekulé as source material.[61] Concerning the all-important question of the structure of benzene Frankland appears to be in debt to Kekulé. However caution is necessary. In his *Lecture Notes* Frankland used Kekulé's hexagon formula for benzene, on which basis it has been stated that he 'soon came out for the theory'.[62] However the first edition of this book gave a linear structure (as below),[63] the hexagon appearing only in the second edition of 1872[64]:

1866 formula

1872 formula

Even then Frankland remarked that such a structure would predict *four* dihydroxybenzenes whereas *three* were actually known, a fact that 'has been overlooked by Kekulé'.[65] Privately Frankland admitted a preference for Ladenburg's prism formula (which does predict three isomers), but said he had been over-persuaded by his assistant Japp, 'a fanatical disciple of Kekulé'.[66] As we have seen earlier (p. 134) there was a long-standing gulf of misunderstanding and suspicion between the two chemists.

When, years later, Kolbe sought to enlist his support against *Strukturtheorie* Frankland was horrified, and emphatically stated his view that

> Chemistry owes its progress from empiricism to exact science entirely [so far as theoretical conceptions are concerned] to Strukturchemie. Without Strukturchemie there is no science of chemistry.[67]

Frankland's own views were, to say the least, at times inconsistent. On the very basic question 'who founded structural chemistry?' he gave conflicting answers. Contrast his declaration to Ost (above) with this announcement to von Meyer in 1892: 'I have always considered Kekulé to be the founder of *Strukturchemie*'.[68]

So what does it all mean? Surely that Frankland, more than anyone else, managed to combine the analytical realism of radical theory with the imaginative hypothesis construction of the neo-typists. He seems to have adopted a modified conventionalism which, while admitting in theory that atoms might not exist[69], in practice constructed research programmes on the basis that they did; as he once remarked at a Chemical Society debate on atomism he was 'no blind believer in atoms' though it was right to use the hypothesis as a 'kind of ladder to assist the chemist'.[70] For Frankland 'Structure theory' seems primarily to have meant the abundant use of graphic, or constitutional, formulae as appeared first in his papers on acetoacetic ester and shortly afterwards in his *Lecture Notes*. In Frankland's view they enable 'constitution to be expressed in the simplest and clearest manner' and are merely 'symbolic expressions of atomicity'.[71] It was this application of structure theory that gave to Frankland yet another major rôle in Victorian chemistry, to be explored in the following chapter.

Yet Frankland's massive contributions to the rise and development of the theory of structure should not obscure the achievements on which they were built, his work in organic synthesis. For, as Sir George Porter once observed, Frankland was one of the great founders of synthetic organic chemistry.[72]

Notes

1 A. W. Stewart, *Recent advances in organic chemistry*, 3rd ed. Longmans Green, London, 1918, p.1 (and in 6th ed., 1931).
2 On Duppa see *J. Chem. Soc.*, 1874, **27**, 1199–1200, and *Proc. Roy. Soc.*, 1874, **21**, vi–ix.
3 W. H. Perkin and B. F. Duppa, *J. Chem. Soc.*, 1860, **13**, 102–6.
4 B. F. Duppa to Frankland, 2 January 1872 [RFA, OU mf 01.04.0671].
5 B. F. Duppa, MS accompanying letter (note 4) [RFA, OU mf 01.04.0663].
6 Frankland, *Phil. Trans.*, 1855, **145**, 259–75.

7 Frankland, *Phil. Trans.*, 1856, **147**, 59–78; *c.f.* I. S. Butler and M. L. Newbury, *J. Coord. Chem.*, 1976, **6**, 195.

8 Frankland, *Proc. Roy. Soc.*, 1856–7, **8**, 502–6.

9 J. A. Wanklyn, *Proc. Roy. Soc.*, 1858, **9**, 341–5.

10 Frankland, *Proc. Roy. Soc.*, 1858, **9**, 345–7.

11 Frankland, *Phil. Trans.*, 1859, **149**, 401–15.

12 The incident has been discussed by (a) W. V. Farrar and A. R. Williams in C. A. McAuliffe (ed.), *The chemistry of mercury*, Macmillan, London, 1977, pp.32–3, and (b) from the point of view of modern safety regulations by F. Dewhurst, 'Edward Frankland and COSHH', *Chem. Brit.*, 1989, **25**, 702–6 (where Duppa is wrongly described as Frankland's 'laboratory assistant').

13 T. L. Phipson, 'Avis aux jeunes chimistes', *Cosmos*, November 1865, 548–9.

14 Thomas Lamb Phipson (1833–1908) had obtained a doctorate at Brussels in 1855, and after editing *Cosmos* in Paris, directed an analytical laboratory at Putney (at 4, The Cedars, almost on the site of Frankland's earlier exploits at the now demolished College of Engineering). He became a Fellow of the Chemical Society in 1862 and was a prolific author of short papers as well as an accomplished amateur violinist: C. J. Bouverie, *The scientific and literary works of Dr. T. L. Phipson, F.C.S.*, with short biographical notice, Wertheimer, Lea & Co., London, 1884.

15 *E.g.*, *Chem. News*, 1866, **13**, 7, 23, 35, 47, 59, 84.

16 Frankland, *Cosmos*, November 1865, 564–5 (the charges were repudiated as 'pure invention', a stronger draft version referring to successful preparation of the offending compounds by Duppa and himself at the Royal Institution [RFA, OU mf 01.03.0371]); draft letter 27 November 1865 to *Medical Times & Gazette* [RFA, OU mf 01.03.0376].

17 Copy-letters W. Odling to *Cosmos*, 20 November 1865 and *c.*2 December 1865 [RFA, OU mf 01.03.0374 and 01.04.1013].

18 A. W. Hofmann, *ibid.*, pp.7–8; he also defended Odling on the grounds that no one was then aware of the poisonous nature of organomercurials.

19 H. Kolbe, *Hessische Morgenzeitung*, 30 December 1865 [also in RFA, OU mf 01.03.0385].

20 Farrar and Williams (note 12a). The reason for this vendetta against Frankland is obscure, but may have been connected with his recommended rejection of a paper on the mineral trikazite submitted by Phipson to the Royal Society: Referee's report, 10 January 1863 [Royal Society Archives, RR5/185]. He also acquired the dubious distinction of a resolution by the Chemical Society Council declining to publish another paper (on sulphocyanides): Minute Book of Chemical Society, minutes for 18 June 1874 [Royal Society of Chemistry Archives]; Frankland was not present.

21 Frankland, *Experimental Researches*, p.299.

22 C. Friedel to Frankland, 18 March 1863 [RFA, OU mf 01.03.0061].

23 C. Friedel and J. M. Crafts, *Annalen der Chem.*, 1863, **127**, 28; 1865, **136**, 203.

24 Copy-letter Frankland to Royal Society, 14 February 1863 [RFA, OU mf 01.04.0028].

25 A. Ladenburg, *Annalen der Chem.*, 1874, **173**, 143–66.

26 E. Frankland and B. F. Duppa, *Proc. Roy. Soc.*, 1860, **10**, 568–70; Frankland, *Phil. Trans.*, 1862, **152**, 167–83.

27 *Sketches*, 2nd ed., p.210.

28 A summary of Frankland's chief research papers in organometallic chemistry is given below:

date	reference	new organometallic compounds
17 June 1849	*J. Chem. Soc.*, 1849, **2**, 265	$ZnMe_2$, $ZnEt_2$
18 Feb. 1850	*J. Chem. Soc.*, 1850, **3**, 30	[impure $ZnAm_2$]
10 May 1852	*Phil. Trans.*, 1852, **142**, 417	$SnEt_2I_2$, $SnEt_2Cl_2$, $SnEt_2O$, $SnEt_2S$
9 Feb. 1855	*Phil. Trans.*, 1855, **165**, 259	
19 June 1856	*Phil. Trans.*, 1856, **167**, 59	$EtZnO.NEt.NO$
18 June 1857	*Proc. Roy. Soc.*, 1857, **8**, 502	
17 June 1858	*Proc. Roy. Soc.*, 1858, **9**, 345	
17 Feb. 1859	*Phil. Trans.*, 1859, **149**, 401	$SnEt_4$, $SnEt_2Me_2$, $ZnMe_2 . Et_2O$, $ZnEt_2 . Et_2O$
15 May 1862*	*Phil. Trans.*, 1862, **152**, 167	BEt_3, BEt_2Cl, $BEt(EtO)_2$, $BEt_3.NH_3$, BMe_3, $BMe_3.NH_3$
19 Nov. 1863*	*J. Chem. Soc.*, 1863, **16**, 415	$HgMe_2$, $HgAm_2$, $HgAmI$, $HgAmCl$
3 Dec. 1863*	*J. Chem. Soc.*, 1863, **17**, 29	$ZnAm_2$
15 June 1876	*Proc. Roy. Soc.*, 1876, **25**, 165	$BEt_2 EtO$, $BEtEtO.OH$, $BEt(OH)_2$

* = jointly with B. F. Duppa

29 E. Frankland and B. F. Duppa, *Proc. Roy. Soc.*, 1864, **13**, 140–2; 1865, **14**, 17–19, 79–83, 83–6 and 191–98.

30 A good summary is in C. Schorlemmer, *The rise and development of organic chemistry*, 2nd ed., ed. A. Smithells, Macmillan, London, 1894, pp.122–130; and in A. J. Rocke, *The Quiet Revolution: Hermann Kolbe and the science of organic chemistry*, University of California Press, Berkeley, 1993, pp.218–24.

31 A. Ladenburg, *Lectures on the history of the development of chemistry since the time of Lavoisier*, (trans. L. Dobbin), Alembic Club, Edinburgh, 1905, p.289; interestingly, he does not mention Frankland and Duppa; for a brief general discussion see C. A. Russell, 'The changing rôle of synthesis in organic chemistry', *Ambix*, 1987, **34**, 169–80.

32 J. Wislicenus, *Annalen der Chem.*, 1863, **127**, 145.

33 E. Frankland and B. F. Duppa, *J. Chem. Soc.*, 1865, **13**, 133–56.

34 Frankland to W. Odling, 18 February 1868 [Royal Society of Chemistry Archives]; the letter is freely illustrated with graphic formulae.

35 E. Frankland and B. F. Duppa, *Proc. Roy. Soc.*, 1865, **14**, 198–204.

36 E. Frankland and B. F. Duppa, *Phil. Trans.*, 1865, **156**, 37–72.

37 A. Wurtz, *Ann. Chim. Phys.*, 1860, **109**, 161–90.

38 H. Kolbe, *Zeitsch. Chem.*, 1867, **10**, 636–40; if Kolbe had suggested a *mono*sodium derivative of ethyl acetate (or more accurately its carbanion) he would have been much nearer to our understanding today.

39 E. Frankland and B. F. Duppa, *ibid.*, 1868, **11**, 60–4.

40 J. A. Geuther, *Göttingen Nachrichten*, 1863, 281–96; a battered copy of this paper survives in the Frankland archive [RFA, OU mf 01.03.0450].

41 J. A. Geuther, *Jenaische Zeitschrift für Medicin*, 1866, **2**, 387–420; 1868, **4**, 241–63 and 570–7.

42 The algebraic transformation is thus: the second formula can be rewritten as
$$CO[CO + CH_3 + C + C_{2n}H_{4n+2}]Eto,$$
$$or, \ CO[O + C_{3+2n} + H_{3+4n+2}]Eto;$$
$$let \ x = 3 + 2n, \ so \ 2n = x - 3.$$
The formula then becomes
$$CO[O + C_x + H_{3+2x-6+2}]Eto, \ or \ CO[OC_xH_{2x-1}]Eto.$$

43 R. Kane, *Ann. Phys. und Chem.*, 1838, **44**, 473–94.

44 E. Frankland and B. F. Duppa, *J. Chem. Soc.*, 1867, **20**, 102–116.

45 E. Frankland and B. F. Duppa, *Proc. Roy. Soc.*, 1870, **18**, 228–30.

46 J. Wislicenus, *Annalen der Chem.*, 1877, **186**, 161–228.

47 J. B. Cohen, *Organic chemistry for advanced students*, Arnold, London, Part I, 'Reactions', 4th ed., 1924, p.262.

48 *Ibid.*, p.263.

49 L. Claisen, *Berichte*, 1887, **20**, 646–50.

50 L. Knorr, *Berichte*, 1911, **44**, 1138–57.

51 On their claims see C. A. Russell, *The history of valency*, Leicester University Press, 1971, pp.146–52; Kekulé is so described in a printed Appeal for his statue, *Aufruf zur Errichtung eines Denkmals für August Kekulé in Bonn*, 1897 [RFA, OU mf 01.01.0831].

52 Rocke (note 30).

53 A connection made by, *inter alia*, the author (*The history of valency*, note 51, p.146) on the basis of an unsubstantiated allegation by W. V. and K. R. Farrar, *Proc Chem. Soc.*, 1959, 289, and a dubious reference by Frankland himself (*Sketches*, 2nd ed., p.50).

54 A. J. Rocke, 'Kolbe versus the transcendental chemists: the emergence of classical organic chemistry', *Ambix*, 1987, **34**, 156–68 (164).

55 Frankland to A. A. T. Cahours, 1 January 1861 [RFA, OU mf 01.04.0011].

56 MS Lecture notes, Lecture 28 [RFA, OU mf 01.05.0590].

57 Frankland to H. Kolbe, 3 December 1871 [Deutsches Museum Archive, Munich, Urkundsammlung Nr. 3566].

58 H. Kolbe to Frankland, 21 April 1871 [RFA, OU mf 01.03.0600].

59 Frankland to Kolbe (note 57).

60 Frankland, obituary note for Kolbe accompanying letter to H. Ost, 20 December 1884 [Deutsches Museum Archive, Munich, Urkundsammlung Nr. 3576].

61 MS Lecture Notes, Lecture 28 (note 56).

62 A. J. Rocke, 'Hypothesis and experiment in the early development of Kekulé's benzene theory', *Ann. Sci.*, 1985, **42**, 355–81 (374).

63 Frankland, *Lecture notes for chemical students, embracing mineral and organic chemistry*, van Voorst, London, 1866, p.240. R. Anschütz (who prints Frankland's structure upside down) thinks Frankland may not then have heard of Kekulé's 1865 paper (*August Kekulé*, Verlag Chemie, Berlin, 1929, vol. i, p. 582).

64 Frankland, *Lecture notes for chemical students*, van Voorst, London, 2nd ed., 1872, vol. ii, *Organic chemistry*, pp. 44 *etc.*; Anschütz (note 63) ignores this apparent change of view by Frankland.

65 *Ibid.*, p.75.

66 Frankland to H. Kolbe, 6 March 1881 [Deutsches Museum Archive, Munich, Urkundsammlung Nr. 3570].

67 Frankland to H. Kolbe, 23 September 1883 [Deutsches Museum Archive, Munich, Urkundsammlung Nr.3573].

68 Copy letter Frankland to E. von Meyer, 28 January 1892 [RFA, OU mf 01.01.0852].

69 He privately confessed to Kolbe 'probably no such things as atoms exist', Frankland to Kolbe, 3 December 1871 [Deutsches Museum Archive, Munich, Urkundsammlung Nr. 3566].

70 *J. Chem. Soc.*, 1869, **22**, 435; see W. H. Brock (ed.), *The atomic debates: Brodie and the rejection of the atomic theory*, Leicester University Press, Leicester, 1967.

71 Frankland to Kolbe (note 57).

72 G. Porter, 'The chemical bond since Frankland', *Proc. Roy. Inst.*, 1965, **40**, 384–96 (387).

CHAPTER 10

The communication of chemistry

1 The birth of Frankland's notation

There were many ironies in the life of Edward Frankland, but
perhaps none more remarkable than the following. Although he
was reluctant to speak in public and (on his own admission) a poor
lecturer, he was to become one of the great chemical communicators
of the nineteenth century. A basically shy individual who never
courted publicity, he achieved this distinction largely through his
papers and books. The former reflected his enthusiasm for research
and the latter may not have been unconnected with financial
rewards for mass communication. But behind each lay a passion for
the kind of education which had been denied to him throughout his
youth but in which he saw real advantages for the individual and
the nation.

Central to his strategy was the need for an all-embracing view of
chemistry coupled with a dazzlingly simple mode of representing
that view on paper. A major aspect of his later chemical world-view
was the theory of structure, but the problem remained of language
and notation. Given that the individual atoms were held together
by valency bonds (another doctrine attributable to Frankland) it
became clear in the 1860s that all manner of possibilities existed for
imaginatively representing molecules for heuristic purposes. Thus
Hofmann illustrated a lecture at the Royal Institution with models
made of croquet balls and rods.[1] They became known as Hofmann's
'glyptic formulae'. More simply such arrangements may, in
principle, be expressed on paper. The first appearance of such

'structural formulae' appears to have been in the unlikely context of a thesis for a medical doctorate,[2] submitted in 1861. The candidate (who was successful) was Alexander Crum Brown (1838–1922), member of a famous Scottish family of literati and divines and former student of Kolbe.[3] These formulae first appeared in print three years later.[4] The following examples show clearly the existence of two isomeric propyl alcohols:

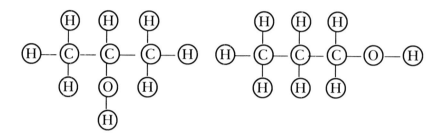

Within a few months of seeing these formulae Frankland, with Duppa at his side, was using them in a discussion of the products obtained from acetoacetic ester syntheses. Possibly 'the details of his intellectual odyssey are not known'[5] but it is not hard to suggest two reasons. To a pioneer of valency theory and inventor of the chemical term 'bond' such an incarnation of his seminal ideas must have seemed a heaven-sent gift at precisely the time when it was most needed. He was now plunged into the intensive teaching programme of the Royal College of Chemistry and needed all the heuristic skills available. Secondly, he and Duppa were wrestling with the constitutions of compound after compound produced in their acetoacetic ester researches. Such was their complexity that the simple type formulae hitherto employed were obscuring rather than illuminating their subject. And so we find, at the end of their second paper, a discussion of ethylic isopropaceto–acetate and its derived ketone concluding with the words: 'In order to prevent any misapprehension of our views concerning the constitution of these bodies, we hereby subjoin their graphic formulae'.[6] They speak for themselves:

$$\begin{cases} \text{COMe} \\ \text{C(CMe}_2\text{H)H} \\ \text{COEto} \end{cases} = \begin{cases} \text{COMe} \\ \text{C}\beta\text{PrH} \\ \text{COEto} \end{cases} =$$

```
                                    H
                                    |
                        H      H—C—H
                        |        |
                   H—C—H      C=O
                        |        |
                   H—C————————C—H
                        |        |
                   H—C—H      C=O
                        |        |
                        H        O
                                 |
                              H—C—H
                                 |
                              H—C—H
                                 |
                                 H
```

These formulae appeared in January 1867. As late as the previous September the *Journal of the Chemical Society* had carried a paper on 'Contributions to the notation of organic and inorganic compounds'[7] in which Frankland argued strongly for formulae in the forms of the older type theory. However he was now investing them with a fresh meaning, or rather, perhaps, a newly restricted one. The familiar curly bracket was now intended simply 'for expressing combination between two or more elements which are placed perpendicularly with regard to each other and next to the bracket in a formula'. But he was moving towards a linear representation, as in 'rare cases where oxygen links together two elements or radicals in the same line'. His example was 'diacetic glycol':

$$\begin{cases} \text{CH}_2\text{-O-CMeO} \\ \text{CH}_2\text{-O-CMeO} \end{cases}$$

However a few months before this article appeared he had written to Crum Brown, applauding his new symbolism and declaring that 'the water type, after doing good service, is quite worn out'.[8] A letter by return of post from the delighted Scotsman gave details of an Edinburgh firm of printers who could make castings for printing the new formulae, in 'two sizes of circles' to contain the atomic symbols.[9] Shortly afterwards Frankland's textbook *Lecture notes for chemical students* appeared in the book shops, adorned with the new formulae.

In making such innovations he acknowledged the source of his ideas:

> I have entirely adopted the graphic notation of Crum Brown which appears to me to possess several important advantages over that first proposed by Kekulé.[10]

The system of Kekulé referred to was the 'sausage'-type formula introduced in his *Lehrbuch*[11] and exemplified by the following representation of acetic acid:

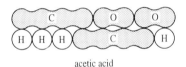

acetic acid

It will be noted that links between adjacent atoms are only in the vertical direction and that double bonds (as in the $>C=O$ group) have two points of contact between the two atoms. Such formulae were, however, little used even by their creator and Frankland was right to perceive advantages in his own system. At first Kekulé found the Crum Brown system unacceptable, particularly because it was incapable of representing a triple bond (as all bonds are confined to one plane)[12]. To solve this problem he proposed a tetrahedral structure for the carbon atom, seven years before the more famous announcement by his former student van't Hoff,[13] though five years after it had been tentatively suggested by Butlerov (in order to account for the alleged isomerism between $C_2H_5.H$ and $CH_3.CH_3$).[14] A mechanical version of Kekulé's model was exhibited in 1867 at a science convention at Frankfurt and used in his lectures at Bonn.[15] Meanwhile, another Edinburgh chemist, one of Playfair's students, James Dewar, constructed a different mechanical model on the basis of Crum Brown's formulae. He sent it to Kekulé with the result that he was invited to spend a summer in his laboratory, used it to predict isomers of benzene and displayed it at a British Association meeting at Norwich in 1868.[16] He 'concluded by advocating the judicious use of such a model in the tuition of organic chemistry'.[17]

In fact Dewar was right, for representations of 3-dimensional molecules on paper are always difficult, and as nothing whatever was known to suggest how the different atoms might be disposed in space, more problems were created than solved. So it is not surprising to find Kekulé using the linear type of formulae even in the paper that advocated his alternative model, and within two years he was permitting himself graphic formulae like the following[18]:

$$H_2C\!-\!OH$$
$$|$$
$$H\ C\!-\!CH_3$$
$$|$$
$$H_2C\!-\!OH$$

However it is interesting to note that a book by an American author used both new and old symbolism (ascribed respectively to Crum Brown and Kekulé).[19]

Other German chemists slowly adopted the Frankland/Crum Brown notation, particularly those in the dyestuff industry for whom they offered the only means of achieving anything like an adequate understanding of the new and complex molecules accumulating daily in their laboratories. Even so the first extensive use of structural formulae in the literature of colouring matters does not appear before the early 1880s.[20] There was one notable case of explicit opposition. On sending a copy of his Lecture notes to Kolbe in 1867 Frankland met with a hostile response, for Kolbe felt the graphic formulae were expendable, doubtful and positively misleading to students who might infer too much from them.[21] This was despite the disclaimer by Frankland that 'the graphic formulae are intended to represent neither the shape of the molecules nor the [supposed] relative positions of the constituent [hypothetical] atoms'.[22] Kolbe maintained his scepticism to the end, greatly to the detriment of his own reputation as a leading organic chemist.[23]

A study of the way in which 'Frankland's notation' was received in England discloses a curious dichotomy. In the learned journals, particularly that of the Chemical Society, it took over a decade before type formulae were finally displaced and old and new structures appeared side by side well into the 1870s. Among the reasons for slow acceptance must be the fog of atomic scepticism that permeated the Chemical Society in the late 1860s and left behind it a legacy of distrust for anything smacking of structural dogmatism. It is remarkable that when Frankland himself issued in 1877 a huge volume of his own researches[24] he edited most of the formulae into what he deemed to be an acceptable modern form (for which reason the book's value as historical source material is severely limited); but the form chosen was usually the bracketed structures of type theory. In the world of chemical *research*, academic or industrial, the conservative reluctance to take advantage of the new notation is clearly a problem requiring further study. But the world of chemical *education*, especially in England, presents so sharp a contrast that one might be forgiven for supposing an unbridgeable chasm within the chemical community. The ready acceptance by students of 'Frankland's notation' is a phenomenon of equal interest to sociologist and chemist.

2 Frankland and popular chemical education

2.1 A new audience for chemistry

The years just following the Napoleonic wars had witnessed a great deal of social unrest, political activity on behalf of the newly vocal working classes, and establishment of the Mechanics' Institutes where science could be available to all.[25] Unfortunately, for a variety of reasons, this early promise was not fulfilled and by the 1860s warning voices about the general neglect of science in England became a clamour and renewed efforts were made to foster the idea of popular scientific education.[26]

The Department of Science and Art had been founded in 1853 as a fulfilment of the vision of Prince Albert for a centralised institution to deal with science and its applications ('art'). In this enterprise he was abetted by Lyon Playfair and Henry Cole, a civil servant who had served on the managing committee for the Great Exhibition that year and later became Secretary to the Department.

It took under its wing both the School of Mines and the Royal College of Chemistry. For its first six years the DSA attempted to establish classes for working class adults, though without great success. Then, in 1859, its fortunes and direction were changed by the arrival of a new Inspector, J. F. D. Donnelly.[27] Shortly afterwards the system of 'payment by results' was established whereby teachers of candidates who succeeded in passing the Department's examinations were financially rewarded, a teacher receiving £4 for each pass obtained by his students. Though later acquiring unenviable notoriety in orthodox educational circles, the system put firmly on the map that most distinctive of late-Victorian institutions, the written examination with rewards for success.[28] It also had the merit of perceptibly raising standards of instruction for the increasing numbers of artisans anxious to learn science. Once a certain level of competence was reached an evening lecture could be sure to attract large numbers of mechanics, solicitors' clerks, apprentices and others eager to share in the benefits of a scientific training.

Some indication of the scale of success may be gauged from the number of candidates sitting the Department's examination. In 1860 there were 104 candidates, next year there were 1000 and twenty years later (1881) 48 498; nearly a ten-fold increase in the first year and a 50-fold increase over the next twenty years. One shrewd, if amateur, observer commented that by about 1867 the tide had begun to turn.[29] That year also saw publication of Farrar's *Essays on a liberal education* (advocating the science of common things), a British Association Report promoting scientific education in schools, and the Paris Exhibition which, in some places at least, was a clarion call for the revival of British science.

Soon the movement spawned what can only be called 'mass meetings'. In 1866/7 Roscoe addressed an audience numbering up to 4000 in a course of 13 lectures delivered in Manchester[30] while it was said that popular lecturers like Tyndall and Huxley were 'as much sought after as Spurgeon', the Baptist preacher then at the peak of his fame.[31] In all this whirl of activity one figure was conspicuously absent, Edward Frankland. As we have seen from his Manchester days the great public meeting was not his *métier* and he does not appear to have addressed any large gatherings in the 1870s, though he did claim to have delivered courses of lectures to working men in the 600-seat lecture theatre in Jermyn Street with 'many hundreds' refused admission for lack of room.[32] However Frankland's skill as a communicator took other forms, not least in the promulgation of his valency doctrine and notation. He developed his own distinctive strategy.

2.2 Frankland's strategy

There can be little doubt that Frankland's activities in the field of popular education during the late 1860s and through the 1870s were part of a concerted campaign. Determined to raise the value of science in British public esteem he saw its introduction into mass education as an unavoidable duty. As we shall see (Chapter 11) the campaign had deep roots in a Victorian ideology that did much more than inspire popular education. In the case of Edward Frankland it was also rooted in his personal experience.[33] His own lack of formal scientific training had, in a curious way, been of enormous benefit. It had thrown him upon his own resources, encouraged him to exercise to the full his native ingenuity and assured him of the virtue – and realistic possibilities – of self-help. The emphasis on 'training' in his autobiographical *Sketches* is sufficient testimony to his passion in this respect. At Queenwood, Manchester, and a range of metropolitan institutions as varied as Addiscombe and the Royal Institution he had acquired many didactic skills. His lifelong friendship with Richard Dawes profoundly influenced him in the cause of 'science for the people'. He had been forced to communicate with farmers, schoolboys, artisans of all kinds, budding doctors and soldiers and the fashionable audiences at Albemarle Street. From the late 1860s, as Hofmann's successor at the Royal College of Chemistry, he was now faced with classes of full-time students who *wanted* to learn chemistry and were prepared to dedicate their careers to its pursuit.

It was in the context of the Royal College of Chemistry classes that the value of Frankland's notation for teaching purposes first became apparent. Dewar's remark about the mechanical version proved to be prophetic. Frankland himself wrote in 1877 of his system:

> For more than ten years I have had considerable experience in the use of it, by large classes of chemical students, and have been well satisfied with the result. By means of this method, students arrive at the same point in their curriculum, in about two thirds of the time necessary under the older systems.[34]

So the first aspect of his strategy fell into place: *the writing of a textbook*. Practising chemists had already had a preview of Frankland's ideas in his paper to the Chemical Society on 'Contributions to the notation of organic and inorganic compounds',[35] but they had been spared the ball-and-stick formulae that were the next logical step. In England Frankland's *Lecture notes* of 1866 became the chief propaganda vehicle for his new notation and the notions of valency and structure that lay behind it.[36] Having introduced his theory of atomicity (valency) he continued:

> To give a concrete expression to these facts, the atom of hydrogen may be represented as having only one point of attachment or *bond* by which it can be united with any other element, zinc as having two such bonds, boron three, and so on. Thus the atoms of these elements may be graphically represented in the following manner:

$$- \text{H} \qquad - \text{Zn} - \qquad \overset{|}{\underset{\diagup\diagdown}{\text{B}}} \qquad - \overset{|}{\underset{|}{\text{C}}} - \textit{etc.}$$

He went on to argue that:

> No element, either alone or in combination, can exist with any of its bonds disconnected, and hence *the molecules of all elements with an odd number of bonds are generally diatomic and always polyatomic.*

and illustrated his point with structures like

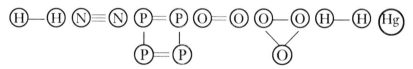

His distinctive terminology was illustrated by nitrogen or sulphur:

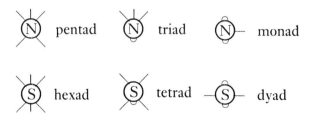

The two chlorides of iron could be represented thus:

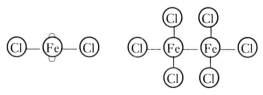

The second edition (1872) omits the enclosing circles and has an unmistakably modern look. The remainder of the book takes the student through the whole of chemistry with atomicity and structure as the unifying principles.

Thus Frankland helped not only to spread his own ideas but, as he would have admitted, made chemistry accessible to a far larger number of people than would have been possible before. That, however, was only one plank of his strategic platform. A second was at least as important.

This second part of his strategy lay in *control of examinations*. From Frankland the author we turn to Frankland the Examiner for the Department of Science and Art. When he succeeded Hofmann at the Royal College of Chemistry in 1865 Frankland inherited a number of responsibilities not strictly connected with the college. Amongst these was the rôle of examiner in chemistry to the Department of Science and Art. From 1865 to 1876 he performed this function alone; for his last four years (1877–80) he worked in harness with Roscoe. Thereafter Roscoe continued the task with W. J. Russell.

In accepting this responsibility – however reluctantly in the first place – Frankland was to find himself in a position of immense influence. By storming the heights of the examination system he was able to determine both the curriculum and the manner of teaching it for 15 critical years in the history of English technical education.[37]

The curriculum adopted by Hofmann had been very theoretical, with much stress on calculation but no concession to modern theories of structure. Frankland continued a theoretical emphasis but with graphic notation and valency well to the fore. He was sufficiently concerned with this aspect of the curriculum that he made a special point in 1868 of enquiring how many students were still using the old notation.[38]

His syllabus for Elementary Inorganic Chemistry in 1873 is remarkable for its brevity and comprehensiveness:

> Definition of chemistry. Simple and compound matter. Different modes of chemical action. Combining weights. Volume weights. Principles of chemical nomenclature. Symbolic notation. Chemical formulae. Chemical equations. Atomicity of elements. Simple and compound radicals. Classification of all elements into metals and non-metals, into positive and negative elements. Classification according to atomicity.[39]

The question paper for that year[40] was as follows:

1. If I allow a current of steam to blow through some iron nails heated to redness in a crucible, what happens? If I substitute copper nails for the iron ones in the above experiment, how will this affect the result?

2. I put the end of a piece of platinum wire into the flame of a candle; what takes place? What will be the difference in the result if I replace the wire of platinum by one of magnesium?

3. Describe minutely what occurs when hydrochloric acid and ammonia gases are mixed together. Give a sketch of the apparatus you would employ to show this experiment.

4. Describe the apparatus, exclusive of the battery, required to electrolyze water, acidulated with sulphuric acid, and make out a sketch of it. What are the names and relative volumes of the gases given off, and how would you demonstrate their characteristic properties?

5. How would you prepare chlorine, and how would you show that a jet of hydrogen burns in chlorine? Give a sketch of all the apparatus you would employ.

6. Mention the colour of the following liquids:
    1. Solution of potassic nitrite acidified with sulphuric acid.
    2. Solution of potassic permanganate.
    3. Dilute sulphuric acid tinted with magenta.If solution no. 1 be added to solution no. 3, what alteration of tint is observed, and what produces the change? If solution no. 1 be added to solution no. 2, what change of colour occurs, and why?

7. What is the weight of a litre of a gaseous compound, the molecule of which contains $COCl_2$?

8. Draw the graphic formulae of the following compounds, and mention the atomicity of each element contained therein:-
    $HCl, H_2O_2, SO_2, BOHo, \begin{cases} OCl, NOHo \\ OHo \end{cases}$.

9. How would you separate:-
    1. Sand from sugar;
    2. Chalk from salt;
    3. Charcoal from nitre;
    4. Glass from boric acid;
    5. Sulphur from water?

10. What do you mean by the term compound radical? Illustrate your answer by writing out the names and formulae of a few compound radicals.

11. Classify the following elements into metals and non-metals, and into positive and negative elements:-Al, Ca, C, Cl, Cu, F, H, I, Fe, Pb, Mn, Hg, N, O, P, K, Si, Ag, Na, S, and Zn.

12. Express in words the meaning of the following equation:

$$2NO_2 + 2(OH) = 2NO_2(HO).$$

This paper is strikingly reminiscent of many set for the next hundred years, up to the advent of Nuffield and other innovative schemes. Yet in 1873 this very paper had a failure rate of 933 out of 3404 candidates (about 27%). One need not speculate why. A successful paper would have had to recall a large number of observations (questions 2 to 6 especially) and some actual laboratory experience (particularly in 9). Frankland's Report concluded:

> I have to state that it would, in my opinion, be desirable to call the attention of such Science teachers as give instruction in chemistry, to the circumstance, that the unsatisfactory results of the examinations in the Elementary Stage of Inorganic Chemistry are obviously due chiefly to the want of sufficient experimental illustrations in the classes. It would be well to bring to the notice of such teachers that the performance of the following experiments, at least, ought to be witnessed by every pupil who comes up in the Elementary Stage, and that for the future it will be assumed that such has been the case.[41]

There followed a list of 109 simple experiments and an impassioned appeal for teachers to require pupils to make notes during lectures and to write out an abstract at home. As he said,

> a large proportion of the Elementary papers create in my mind a strong impression that the candidates are, for the first time in their lives, attempting to express their ideas in writing.[42]

In this emphasis on demonstration-experiments Frankland was ahead of most of his contemporaries and breaking new ground in educational practice. As with his graphic notation this innovation is completely taken for granted today.

Given Frankland's predilection for theory-laden chemistry it is perhaps surprising that abstract theory is called for only in questions 7, 8, 11 and 12, and the graphic notation only in 8. He clearly thought that elementary students should first have a general grasp and 'feel' for chemistry before embarking on detailed theoretical explanation. Even so such novices were quickly to have a taste of such heady delights. At the Second or Advanced Stage they were much more prominent. Taking again the year 1873[43] we find in the Inorganic paper the following question (which appeared in almost the same form in 1870):

> Define empirical, rational and constitutional formulae, and illustrate your definitions by examples.

The Elementary Organic paper included the question:

> Draw the graphic formulae of the following compounds:-
> acetic acid, oxatyl, ethylic alcohol, ethylic ether, acetic aldehyde, and lactic acid,

and the Advanced Organic paper the following:

7. Develop completely the formula $CBuH_3$, and reduce to a single

   line the formula $\begin{cases} CH_3 \\ (CH_2)_6. \\ CH_3 \end{cases}$

8. Give the graphic formulae of the following bodies:-
   glycylic ether, monoacetic glycol, pyruvic acid, and boric methide.

9. Define the following families of organic compounds:-
   radicals, alcohols, ethers, aldehydes, acids, anhydrides, ketones, and ethereal salts.

10. Give the names and formulae of some of the most important natural alkaloids.

Finally it is worth noting that at the Third or Honours degree level at least one question was asked about recent research. In 1873 candidates were asked about Roscoe's work on tungsten (mostly in the previous year) and about Deacon's process for regenerating chlorine from waste hydrochloric acid in the alkali industry; it had been introduced only about five years previously. This procedure has been categorised as a 'rather crude method' to test current awareness[44] but did reflect a genuine concern to connect education with research.

In all the activity associated with these examinations it is worth reflecting on Frankland's precise rôle. That he was not involved in teaching the students is obvious; all the DSA examiners were busy people with other things to do. He certainly set the papers but he seems not to have done any marking, using his assistants for that purpose. Thus in 1868 Herbert McLeod was given the task, and promised that if he did it next year he would receive £50 for 1000 papers.[45] Nor was monitoring always a problem for Frankland: 'he did not want to look them over as he had found nothing to alter in the previous ones'.[46] Next year Valentin was drafted in to help McLeod with about 3000 scripts between them.[47] And in 1870 there were four 'examiners': McLeod, Valentin, Maxwell Simpson and Alexander Pedler[48]. They met with Frankland for a briefing session at South Kensington, where 'we arranged about the value of the answers'.[49]

All these examinations were taken in the examination hall as written papers. As Frankland reminisced long afterwards 'the practical or laboratory teaching of science' in England may be traced back to the 1840s and to Queenwood College.[50] Elsewhere it was slowly taken up, but the problems of scale with the DSA courses deferred its introduction for many years. Although Frankland had been appealing for laboratory-based courses in chemistry since at least 1869 (and had persuaded the Department of Science and Art to finance some laboratory facilities), practical examinations had to wait until 1878. At first they were optional extra tests at advanced and honours levels. They remained the only practical examinations in British science education outside the universities until 1892.

Practical tests or not, the importance of laboratory experience was paramount. For this to be possible, let alone safe, there was need for another development. This was *teacher training in laboratory skills*, the third part of Frankland's strategy that came into play in the late 1860s.

It began in 1869 when Frankland threw open the doors of the Oxford Street laboratory of the Royal College of Chemistry and invited teachers to spend a week there from 9 to 5.30 each day. The teachers were not only from DSA courses but from schools and other places where practical chemistry was valued. It gave to Frankland an unrivalled opportunity not only to display his own superlative experimental skills but also to inspire others to acquire the same. In 1870 the plan was extended and 120 teachers arrived in early July and the course extended over three weeks as no more than 40 could be accommodated in the laboratory at one time.[51] His lecture course was published in the form of notes taken by one of his students from 1872, G. Chaloner (who taught at Birkbeck College).[52] Here, in six blocks, were the 109 experiments that Frankland considered essential for any teacher:

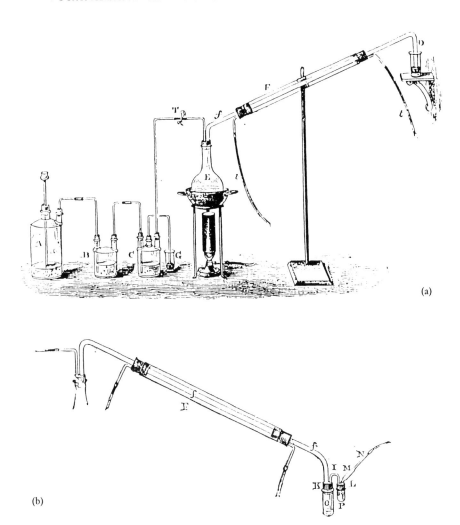

Fig. 10.1 Frankland's apparatus for (a) preparing, and (b) isolating, diethylzinc. The whole operation is conducted in atmosphere of CO_2, generated in A and dried by sulphuric acid in B and C. (E. Frankland, *How to teach chemistry*, pp. 20–1.)

1. Chemical forces; examination of water
2. Hydrogen; oxygen; diethylzinc; composition of water
3. Properties of water; ozone; hydrogen peroxide; hydrochloric acid and chlorine
4. Chlorine oxy-compounds; boron; carbon and its oxides; nitrogen
5. Oxy-compounds of nitrogen; ammonia
6. Atomicity; sulphur and its compounds

An example of his technique is in the paragraph below, the numbers being those of the relevant experiments:[53]

> Oxygen re-lights a glowing wick (27). Use cedar splints, or green tapers coloured with copper; the oxide of copper causes the wick to glow longer. Phosphorus burns brilliantly (28), and when the heat volatilises it a flash of light fills the whole vessel, owing to the points of contact between phosphorus and oxygen being indefinitely increased. For this experiment a dry flask of oxygen is best, and the phosphorus must be dried. Moisture causes spurting, and will probably break the flask. Iron or steel must be burnt in oxygen (29).

The apparatus was, where possible, ingeniously devised from the simplest components but some of it was considerably more complex, as for the preparation of diethylzinc[54], though it did 'not involve the use of any apparatus beyond what most science teachers can command' (Fig. 10.1).

At the end of the book are advertisements for scientific equipment. Among them was a notice from M. Jackson, of 65 Barbican, who was prepared to provide:

	£	s.	d.
The Complete Set of Apparatus and Chemicals, for performing the 109 Experiments, recommended by DR. FRANKLAND in the 'Science Directory', packed in Box, price	10	10	0

Nor was Mr Jackson to have the monopoly. The laboratory furnishers Mottershead & Co. of Manchester offered to supply (*price lists free on application*):

> All the Apparatus and Chemicals required to perform the 109 Experiments recommended by DR. FRANKLAND. – *Science Directory*, p. 107 (29th Edition).

And from Newcastle-upon-Tyne the following information was provided by a firm known throughout the North East for its promotion of modern chemistry:

> MAWSON AND SWAN are also prepared to supply a set of Apparatus and the Chemicals for the Illustration of Professor FRANKLAND'S 'How to teach Chemistry.'
> *Prices and particulars on application.*

Fig. 10.2 The reaction between chlorine and steam. Frankland's alleged apparatus for demonstrating this displacement reaction at high temperature: $2Cl_2 + 2H_2O = 4HCl + O_2$. (E. Frankland, *How to teach chemistry*, p. 8.)

Today very few of the experiments described would be permitted by safety regulations. The preparation of diethylzinc, for example, is a hazardous proceeding at the best of times, not least on account of the inflammability of the product and its decomposition by water. No fume cupboard was suggested for preparations of hydrogen sulphide, nitrogen dioxide, chlorine and other toxic gases; indeed the only concession to the effects of an 'irritating gas' such as chlorine was a liquid trap, as caustic soda solution (with all the dangers of inadvertent 'suck-back').[55] Some of the illustrations were inaccurate; the worst example was the apparatus for the reaction between chlorine and steam (Fig. 10.2). Any attempt to reproduce in the laboratory the apparatus as depicted would have had disastrous results. Not only was one flask (at X) omitted but the flask (Y) for generating chlorine was deprived of its safety funnel.

However the generally lax attitude to safety in Victorian laboratories, though unacceptable today, did enable a large number of exciting and informative experiments to be seen by multitudes. With certain Health and Safety Regulations verging on the paranoid, modern students have missed out on a great deal more than slight danger to their persons. The great achievement of

Frankland, and his later imitators, was to bring practical chemistry within reach of all.

2.3 The reception of Frankland's new chemistry

Assessment of the impact of Frankland's ideas on British chemistry generally may be based on a variety of sources. Evidence that they were of considerable influence comes from an unexpected quarter, the often ephemeral literature of the movement for popular science. This was far removed from the usual forums in which issues of grave import were debated with high seriousness by the chemical establishment.

One such source is a relatively new organ of popular culture, a periodical known as the *English Mechanic*. Founded in 1865 and published weekly at one penny, it claimed to be the first regular publication to cater for 'the millions who lay claim to the title at the head of our journal'[56] and was a resounding success. By 1870 it had some justification for claiming to be 'a monument of co-operative industry', with at least 1000 correspondents and hardly any question of scientific interest not capable of being asked and answered in its columns. Modestly it concluded:

> As an organ for the distribution of knowledge appertaining to the everyday wants and inquiries of scores of thousands of people the *English Mechanic* stands without a rival.[57]

It soon incorporated the *Mirror of Science and Art* and continued well into the twentieth century until it was finally absorbed itself. To open the pages at random is to glimpse a Victorian world of science[58] that has largely eluded historians of mainstream science. It is a wonderful mixture of bizarre recipes (worming the cat, preserving green peas, curing tetanus by tobacco), marvellous contraptions (burglar alarms, washing machines, gas-fired steam tricycles), medical advice (including a prescient article on smoking and heart disease) and quite often some erudite discussions of scientific theory.

As might be expected the *English Mechanic* concentrated on engineering more than any other subject. Yet, in terms of frequency, chemical matters were not far behind. This is hardly surprising for, as Donnelly remarked in 1869, 'it would be scarcely too much to say that where science was spoken of it was generally supposed to mean chemistry.'[59] As early as 1868 the magazine produced a course of lessons on chemistry, mainly inorganic;[60] requests came rapidly for lessons in organic chemistry and even in the new spectroscopy.[61] Within these pages[62] came one of the earliest uses in English literature of the word 'valency', if not the very first.[63] And, sure enough, it was here that the Crum Brown/Frankland formulae made an early appearance in the late 1860s.

Of course it was not all plain sailing. At first there was genuine difficulty in getting the mind round such new symbols and vociferous protests were lodged with the editor. One correspondent objected to 'Dr Owen [sic] Brown and Dr. Frankland's glyptic, which I must decline to have anything to do with, being rather too ridiculous to myself, and equalling those gentlemen's imaginary pronged atoms'.[64] The same writer did not include Frankland's *Lecture notes* in his list of recommended books (preferring Miller, Fownes, Odling, Wilson and Watts' *Dictionary*). An article on 'The modern chemical notation' by F. Hurter came from a more informed chemical source and is interesting for its negative attitude to the new symbols and terminology. It used the terms 'atomicity' and 'basicity', as well as type and 'sausage' formulae.[65] But perhaps the real problem was neither the conservatism of academic chemists nor the gullibility of the untrained but simply the sheer variety of styles available. As one writer complained, 'formulae and nomenclature are at the present time woefully confused'. Though he himself used them,

> graphic formulae are seldom seen in any textbook on elementary chemistry, but specimens are given in Miller's *Elements*, vol. 3, Frankland's *Lecture notes for chemical students*, and in Buckmaster's *Inorganic chemistry*. This species of formulae is utterly ignored by some chemists.[66]

In 1872 there was an explicit admission that 'right or wrong' Frankland's system of chemistry with his 'atomicities' was taught in Government science classes.[67] A stream of reviews in the *English Mechanic* and other journals indicated something of the new industry that had arisen to provide additional books for the DSA examination. Other writers followed Frankland's example, especially his own disciples. Textbooks by Miller[68] and Valentin[69] used the new notation. The former was revised by Frankland's demonstrator Herbert McLeod, while the latter was the work of his assistant at

South Kensington.[70] Another book by Miller, *Introduction to the study of inorganic chemistry,* was hailed as 'one of the best' of its kind,[71] while *Inorganic chemistry for elementary classes* by W. A. Snaith, specifically written for the DSA examinations, was welcomed with the remark that it was a condensation of a book by J. C. Buckmaster which used modern symbols.[72] An *Introduction to scientific chemistry* by F. A. Barff covered most of the questions set between 1865 and 1871.[73] A slightly later book[74] offered solutions to DSA examination questions in chemistry, Elementary and Advanced, from 1870 to 1880; it freely uses Frankland's symbols and concept of atomicity in its answers. Despite all the competition Frankland's own *Lecture notes* continued to flourish. By 1876 volume I had already sold 6000 copies.

The *English Mechanic* was not the only forum for discussion. *The Engineer* referred – apparently in some surprise – to a general disbelief in atoms among chemists, including Frankland.[75] It later complained that a lucid exposition by Crum Brown of his form of notation[76] was of 'an abstruse character'.[77] The *Chemical News*[78] (founded in 1859 by Crookes) retained a popular appeal in its early years and provided regular reviews of textbooks. While Frankland's books were well received the unfortunate Buckmaster was savaged for having produced a book full of 'bad grammar, confused statements, bad chemistry and false chemistry'.[79] Even the august journal *Nature* permitted reviews like that of Roscoe[80] objecting to a book by Geuther[81] on the grounds of excessive use of 'quantivalence' (which applies only to gases!) and of calling chlorine a 'heptad' because it may be found in molecules like $HClO_4$. However, an anonymous review of Valentin's book, reflecting on its use of the Frankland notation, observes that 'it appears that of late years this system has gained much ground'.[82]

One of the most percipient critics of the new approach was the amateur chemist and pioneer of popular technical education Thomas Twining.[83] In an influential book he argued that chemical beginners need to know about less than half the number of elements, can dispense with symbols and equivalents and should certainly have a detailed knowledge of properties before going into questions of valencies.[84] He argued the case for a chemistry foundation course, queried whether university professors were necessarily the right people to plan popular courses, and produced an alternative syllabus for which Frankland's book would be recommended only at Grade IV.[85] In a pamphlet on the Department of Science and Art examinations he was scathing about the failure (as he saw it) of academics to identify what 'would have suited the requirements and capabilities of the common Artizan', having in mind particularly organic chemists who were quite unable 'to stoop to the low cultural level and mundane requirements of the Artizan class'. He approved of Barff's textbook because of its view of symbols and equivalents as 'almost insufferable impediments'.[86] It is hard to believe that Frankland was not included in his denunciation. If so, his new approach was under attack from two quarters: the academics of the Chemical Society and elsewhere, and the protagonists of artisan interests.

After Frankland left the examination scene the papers of the Department of Science and Art (and their imitators) began to lose their characteristic emphasis. Partly this was because chemistry itself was changing. Thus inorganic chemistry teaching was revolutionised by the general acceptance of the Periodic Table of Mendeléeff, even though Frankland had anticipated some of his conclusions. Gradually the limitations became clear of a notation that was unable to distinguish different kinds of bonding, or which found itself in real difficulties with co-ordination compounds. Organic chemistry was particularly suited to Frankland's graphic formulae yet this branch of the subject attracted only just over 20% of the candidates,- it was much less popular in England than in Germany. Perhaps the demise was inevitable of theory-laden syllabi that yet contrived to pack in a vast amount of factual material. It was severely undermined by the advent of the heuristic method of 'chemistry by discovery', ushered in by H. E. Armstrong in 1884.[87] It is ironic that perhaps the most devoted of Frankland's followers should have been responsible for diminishing his master's great influence on chemical education.

It was left to the amateur poets of the B-Club, however, to express most vigorously the popular reaction to Frankland's new symbolism:[88]

> Though Frankland's notation commands admiration,
>   As something exceedingly clever,
> And Mr. Kay Shuttleworth praises its subtle worth,
>   I give it up sadly for ever;
> Its brackets and braces, and dashes and spaces,
>   And letters decreased and augmented

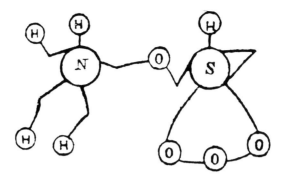

Fig. 10.3 Contemporary skit on Frankland's concept of a chemical bond. One of many such caricatures in the literature of the 'B-Club'. (Royal Society of Chemistry Archives: B-Club file.)

> Are grimly suggestive of lunes to make restive
>   A chemical printer demented.
>
> I've tried hard, but vainly, to realise plainly
>   Those bonds of atomic connexion,
> Which Crum Brown's clear vision discerns with precision
>   Projecting in every direction.
> In fine, I'm confounded with doctrines expounded
>   By writers on chemical statics,
> Whom jokers unruly may designate truly
>   As modern atomic fanatics
>
>   and so on.

Nor was their frustration confined to doggerel; the formula for ammonium sulphate was skittishly represented as in Fig. 10.3, with nitrogen as a pentad and sulphur as a hexad.

Even that august grouping of scientists, the X-Club, permitted itself a moment of light-hearted banter:

> In the beginning was the atom, and the atom was without form and void, and darkness was upon the face of the substance. And the spirit of Frankland moved on the face of the substance, and he said, Let there be an atom: and there was an atom; and he saw that it was good. And the atom and its shadow were the first edition; and Frankland said, Let there be a bond, &c., &c.[89]

Yet Frankland's instinct was perfectly correct, and learners did find the notation of great service, given a little time to get used to it. Very soon page after page of the *English Mechanic* was adorned with structures of varying degrees of credibility, all in the notation of

Edward Frankland. From about 1870 that estimable journal was offering hospitality to a host of (presumably) young writers – mostly long forgotten today – who lacked the inhibitions of their elders and betters in the Chemical Society but found in the use of such formulae a new and vivid understanding of some fundamental principles of chemistry. Today we take for granted this system of representing chemical compounds on paper, particularly in classical organic chemistry. It is well to recognise the crucial rôle played in the process by Frankland. Just as elementary schools had no precedent to deter them from practical work, so the new popular movement was not inhibited from using valency and structure. On those grounds it was permanently indebted to Edward Frankland.

3 Teaching at the Royal College of Chemistry

Despite his excursions into mass education Frankland remained first and foremost a research scientist employed to train budding chemists at the Royal College of Chemistry. Here, as we have seen, he applied the same pedagogic principles in teaching chemical theory and, in ten years or so, was confident of their effectiveness. Since 1846 the College had occupied the modest premises of 315 Oxford Street in London's West End. By the time Frankland took over the student numbers had stabilised at around 40, of whom rather less than half were seriously studying subjects other than chemistry. The intake was deliberately limited since 40 was the maximum number of places in the single teaching laboratory. So serious was the shortage of space that the only alternative accommodation, Frankland's private research laboratory, was within a few years given up to 'some half dozen of the better students'. There was no balance room, so weighing had to be done in the library where a floor vibrating through the constant passage of people put an effective stop to accurate readings. There was no room for expansion, unless it were to be upwards. In testimony to the Royal Commission on Scientific Instruction (1871) Frankland did not recommend any further outlay on the building, situated as it was on a very busy street and too near the residential area of Hanover Square that lay just behind it.[90]

Only chemistry was taught at the Oxford Street premises. For other subjects, and for many lectures in chemistry, students had to walk to the more spacious quarters of the School of Mines at Jermyn Street, a mile or so to the south. Even there, however, the resident departments complained of shortage of space and Frankland found ready allies in Huxley (biology) and Guthrie (physics). The three had long cast covetous eyes at the plot of land that the Commissioners of the 1851 Exhibition had purchased on the outskirts of London but within a very few minutes' walk of the new underground railway station at what was to be called South Kensington. They were all too willing to help in the realisation of Prince Albert's dream of a great complex in which science and the 'arts' (technology) could flourish side by side. When the autocratic Director of the School of Mines, Murchison, died in 1871 it seemed that their time had come, for he had been on the side of those other professors who feared absorption within a larger unit at South Kensington.

For some time there had been perfunctory attempts to erect a grand new laboratory building at South Kensington. In early 1869 Frankland and McLeod began to look at plans of the laboratories at Berlin, Bonn and Leipzig with a view to formulating detailed ideas for South Kensington. Two months later McLeod reported that, in the absence of government money, the half-finished building was to be left as it was for a whole year, to the great annoyance of the builder who would have scaffolding worth £2000 doing nothing. By May 1870, twelve months later, work recommenced with a promised grant of £20 000. Yet little seemed to happen. Another year passed and Frankland, always a shrewd critic of official ineptitude, expressed himself strongly when obliged to turn down an offer from his old friend Kolbe for some collaborative research. It was a delightful prospect, but

at present, however, work for me is impossible; because I have actually no bench to work at and no time to work. My private laboratory is filled with students who cannot find room in the large laboratory. Our government appears to be altogether incapable of finishing the new Laboratories at South Kensington. At present they are scarcely doing one month's work at them in a year. England has become the slowest country in the world for things of this kind. It is 5 years since these laboratories were commenced and they certainly will not be finished next year at this time.[91]

His public protests were hardly less muted. Reporting in 1872 he brought the issue into the national arena of public attitudes to science training:

Fig. 10.4 The Museum of Practical Geology, at Jermyn Street from 1851.
(Author.)

I have again both the gratification and the pain to state that the college
has been constantly overcrowded with most urgent and diligent
students, who for lack of space and the appliances of modern chemical
laboratories have received very inadequate instruction, more especially
in those higher branches of the science which are so essential to the
applications of chemistry to arts and manufactures. I can only
reconcile myself to the continuation of this imperfect, and, for the
present day, entirely inadequate instruction by the reflection that
there exists in the Kingdom no other chemical laboratory which does

not lack in an equal degree the necessary accommodation and appliances for pursuing the study of practical chemistry. For many years past it has been impossible for English students satisfactorily to pursue in this country that higher department of chemical training which includes invention and discovery, and there are not wanting instances of such students proceeding from this college to the noble State laboratories of Germany and Switzerland to continue their training beyond the elementary instruction given in their native land.[92]

The communication of chemistry in Britain was fatally flawed by public indifference and government inertia. It was to become a favourite theme in the years ahead. To Kolbe he confided:

When I urge upon Politicians here the disgraceful position of science in this country as compared with Germany, they reply contemptuously "what has science done for German commerce and manufactures? To whom are due the invention of the two greatest modern chemical manufactures – Paraffin oil and Analine [sic] colours? Are they not due to Englishmen? Why, if all this science does her any good, does Germany still dread the competition of our manufacturers, and impose enormous duties on our goods, and why, notwithstanding these high protective duties, and the greater cost of labour and coal in this country, is Germany still our best customer for chemicals and steel? Why is Germany obliged to apply to England to get her capital supplied with Water and Gas?["] These are awkward questions, & when I say "There is a want of enterprise in Germany (to which the wretched condition of the streets of Berlin & the overcrowding bear testimony) the reply of course is Yes! and this want of enterprise is caused by too much attention to, and dependence upon, abstract science". If you could furnish me with some good statistical proofs of the decided superiority of German chemical manufactures, it would have more weight with most people here than any appeals to the national duty as regards science.[93]

The sheer practical difficulties of communicating chemistry under these conditions helped to crystallise in his mind an idea of national scientific declension that was to politicise his science and shape his own life in the years ahead. Meanwhile, however, the Council of Education had received the following advice from the Council of the School of Mines that consisted of the professoriate and had acquired Murchison's management function:

As there appears to be no prospect of obtaining the necessary extension of premises, the Council are of opinion that it would be advisable to transfer the instruction in Physics, Chemistry, and Natural History to the buildings in South Kensington, where it is understood that adequate accommodation may be obtained.[94]

That did it. To quote from Frankland's letter to Kolbe on New Year's Day 1873:

> My laboratory here progresses towards completion very slowly, but it is crowded with students &, in about a fortnight, I hope to get to work in my private laboratory which will now be soon so far completed as to allow work to be done in it.[95]

By 1874 he was able to report 'a great influx of students in experimental science has occurred. Indeed the increase has in the chemistry department exhausted at one bound the capacity of the space provided'.[96] This, indeed, is what he had predicted. When asked by the Devonshire Commission in 1871 about the adequacy of the proposed new building at South Kensington he replied that it would be satisfactory, and comparable to the best German examples, 'if entirely devoted to chemistry',[97] In fact, of course, it was to be shared with Biology and Physics Five weeks after giving evidence he wrote to Lockyer:

> I was over the new building at South Kensington today trying to find out how the School of Mines was to be squeezed into it, but I cannot see how it is to be done. The lecture theatre seems to occupy about half the whole building as it runs up from basement to roof, and of course a separate lecture theatre is absolutely necessary for chemistry.[98]

However conditions were considerably improved. For example, though he had wanted three gas analysis laboratories he was still able to secure two, together with a physical instrument room off his private laboratory.[99] It is doubtful if accommodation would *ever* have been sufficient for Frankland's wants, simply because he filled all available space with students. The new laboratories are now part of the Henry Cole Wing of the Victoria & Albert Museum.

What, then, of his teaching? Frankland himself admitted that 'public speaking is not much in my line'.[100] We are fortunate to have a detailed assessment from one of his most critical, though admiring, students, H. E. Armstrong. He once remarked, 'Frankland had no art of oratory but every word told. He was the man of deeds'.[101] Or to put it another way,

> His manner of lecturing was the very antithesis of Hofmann's – altogether lacking in humour and sparkle; but his lectures were more impressive on account of the wealth of fact and the attention he gave to the experimental side.[102]

The amount of such attention can be gauged by the lists of experiments he required the long-suffering McLeod to prepare, day after day and week after week. It will also be apparent from the advice given to teachers, for Frankland was never a man to shrink from taking his own medicine.

Such a man was naturally more at home in the laboratory. But much depended on the student being helped, as Armstrong recalled again:

> He had none of Hofmann's magnetic force nor his wonderful faculty, when in the laboratory, of getting behind a student's work by means of a few searching questions; there is no doubt that one was a leader of men whilst the other was an individual worker . . . Frankland had no such tendency as Hofmann; there was no trace of the slave driver in him; and he had peculiar difficulty in getting behind a student's work and finding out what had been done, whilst Hofmann saw in a flash whether his wish had been met. Only those could work effectively with Frankland who needed little if any guidance; but at times he would step in and give invaluable advice or still more important assistance by coming to the worker's aid with his wonderful manipulative skill. He had passed through an exceptionally severe and thorough school of training and was a master of his craft on the practical side.[103]

Of course this is only one man's opinion but it agrees strongly with a general impression of this shy but talented individualist. Yet we know that Frankland was concerned to help his students in their subsequent careers. He presented the Devonshire Commission with an exhaustive list of the destinies of former RCC students, indicating 366 who had particularly distinguished themselves (in the compilation of which he admits help from older students). It is impossible to say how many references he gave, but (to take a single example) no less than eight were provided for those who sought admission to the new Institute of Chemistry in its first full year (1878).[104]

Thanks to careful recording by the doughty Herbert McLeod we know also the content of his lecture course, varying little over several years.[105] For 1867/8 it was as shown in Table 10.1.

Unlike many more academic curricula this one has a strongly applied emphasis in its treatment of metals and minerals, and in the emphasis on coal gas. This becomes even clearer on inspection of the lecture notes that have fortunately survived. They also indicate the modernity of his approach to spectrum analysis and (of course) notation. A page of his notes is reproduced as Fig. 10.5 (p. 310).[106]

Finally, in all his teaching at the Royal College of Chemistry Frankland was highly indebted to his assistants. Herbert McLeod

Table 10.1. *Frankland's lectures at the Royal College of Chemistry 1867/8*

Week beginning	Monday	Wednesday	Friday
25 November	H_2S, H_2S_2, SO_2	H_2SO_4	S, Se, Te, Br
2 December	I, F, Si	Si, Sn, Ti, P	P
9 December	P, As	Sb	metals
16 December	spectrum analysis	metals	metals
6 January	K	Na, Li, Cs, Rb, Tl, Ag*	waters
13 January	waters, Cd, Hg	Hg, Cu, Au, Al	metals
20 January	organic chem. introduct.	organic analysis	nitrogen estimation
27 January	gas analysis	vapour densities	formulae, classification
3 February	organic radicals	ethylene	acetylene
10 February	cyanides, oxatyl	cyanides, acids	marsh gases
17 February	paraffins, benzene	coal gas	coal gas
24 February	coal gas	coal gas	alcohol
2 March	alcohols	glycol, glycerol	alkyl halides
9 March	– †	aldehydes	fatty acids
16 March	oleic, lactic acids	succinic acid	acid anhydrides
23 March	ketones, salts, amines	amines	organometallic compds.

†Frankland away *extra lecture added Thursday on Ba, Sr, Ca, Mg, Zn

had been a junior assistant at the College from 1858 until 1860 when he became Assistant Chemist. Of him Armstrong remarked that Hofmann 'was ably supported by his indefatigable lecture assistant, McLeod, whose ever ready and conscientious fingers afforded deft compensation for the peculiar manipulative deficiencies of his master'.[107] Given that Frankland did not suffer from those deficiencies the alliance with his assistant was most powerful, lecture experiments being carefully rehearsed and performed with unparalleled panache. A deeply religious person of High Anglican persuasion, McLeod was Frankland's right hand man and was, many years later, to write his obituary notice in the *Journal* of the Chemical Society.[108] His *Diary* offers an invaluable insight into the social worlds of Victorian science and religion. In 1871 he left to become Professor of Physical Science (later Chemistry) at the Royal Indian Engineering College at Cooper's Hill, near Staines,[109] his place at South Kensington being taken by Alexander Pedler.

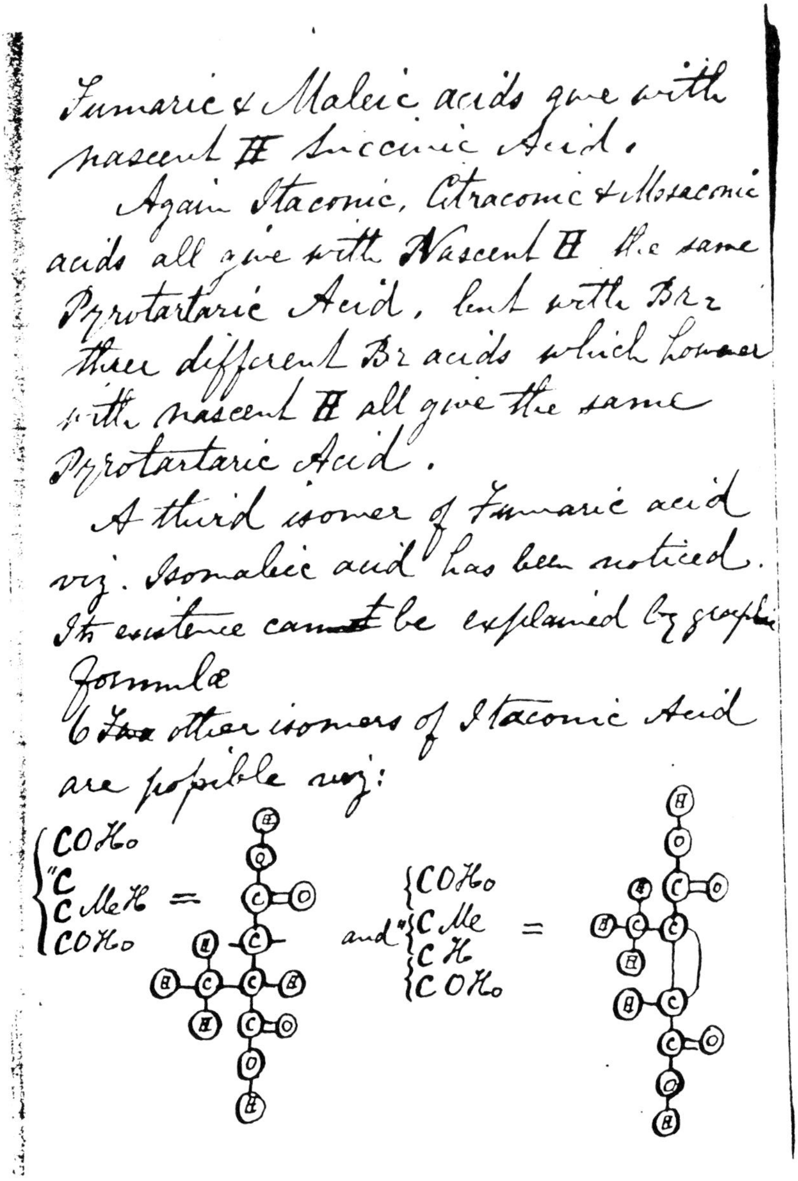
Fumaric & Maleic acids give with nascent H Succinic Acid.
Again Itaconic, Citraconic & Mesaconic acids all give with Nascent H the same Pyrotartaric Acid, but with Br_2 three different Br acids which however with nascent H all give the same Pyrotartaric Acid.
A third isomer of Fumaric acid viz. Isomaleic acid has been noticed. Its existence cannot be explained by graphic formulae
6 ~~Two~~ other isomers of Itaconic Acid are possible viz:

Fig. 10.5 Page from Frankland's lecture notes (R. Frankland Archives.)

With several other professors McLeod was declared redundant in 1900 owing to the contracting rôle of science in the curriculum at Cooper's Hill. After a huge public row they were allowed to receive pensions. McLeod then became editor of the *Royal Society Catalogue of scientific papers.* He studied the rôle of manganese dioxide in the potassium chlorate decomposition and is chiefly remembered today for the McLeod pressure gauge (for gases at very low pressures).[110]

Frankland's other assistant for his early years at the Royal College of Chemistry was W. G. Valentin. He was in charge of the laboratory and was also recalled by Armstrong:

> The laboratory instruction was in the hands of Valentin, a particularly skilful and exact analyst; instead of always tutoring us, he taught us to be self-reliant and set the best possible example by constantly working before our eyes – he had a good consulting practice – often telling us what he was doing.[111]

Opinion seems to have been unanimous as to Valentin's technical skill and his kindness to students. He devised a method for estimating sulphur in coal gas.

Each assistant had to give occasional lectures, or even short series. Each was involved in writing activities, Valentin producing his own textbooks and McLeod helping to prepare the new edition of Frankland's *Lecture Notes*. There was a strong bond of loyalty between the three chemists, the two assistants sometimes joining dinner parties at Frankland's home. When Valentin was seriously ill in 1875 Frankland took time off to visit him.[112] He died four years later.[113]

The move to South Kensington marked the end of the Royal College of Chemistry as a semi-independent part of the Royal School of Mines, for chemistry, physics and biology were now effectively one unit while the rest of the RSM remained at Jermyn Street. This was recognised by their amalgamation in 1881 as the Normal School of Science, with a special remit to train science teachers. The somewhat infelicitous name was replaced in 1890 by 'the Royal College of Science', though by that time Frankland had retired. This institution, together with the rest of the old Royal School of Mines, and the City & Guilds Institute were amalgamated in 1907 as the Imperial College of Science and Technology.[114] The great chemical tradition of that institution, with dozens of world-class alumni including two Nobel Laureates, may be traced back to the pioneering days of Hofmann and Frankland.

Notes

1 A. W. Hofmann, *Chem. News*, 1865, **12**, 166–9, 175–9, 187–90.
2 A. Crum Brown, MD thesis, University of Edinburgh, 1861, 'The theory of chemical combination'.

3 On Crum Brown see *DNB, DSB* and 'J. W.', *J. Chem. Soc.*, 1923, **123**, 3422–3431.

4 A. Crum Brown, *Trans. Roy. Soc. Edinburgh*, 1864, **23**, 707–719.

5 A. J. Rocke, *The Quiet Revolution: Hermann Kolbe and the science of organic chemistry*, University of California Press, Berkeley, 1993, p.313.

6 E. Frankland and B. F. Duppa, *J. Chem. Soc.*, 1867, **20**, 102–116 (111).

7 Frankland, *J. Chem. Soc.*, 1866, **19**, 372–395.

8 Frankland to Crum Brown, 4 June 1866, reproduced in *J. Chem. Soc.*, 1923, **123**, 3422.

9 A. Crum Brown to Frankland, 5 June 1866 [RFA, OU mf 01.04.1266].

10 Frankland, *Lecture notes for chemical students, embracing mineral and organic chemistry*, van Voorst, London, 1866, p.v.

11 F. A. Kekulé, *Lehrbuch der organischen Chemie*, Erlangen, 1861, vol. i, p.165n.

12 F. A. Kekulé, *Zeitsch. f. Chem.*, 1867, [2], **3**, 214.

13 J. H. van't Hoff, *Voorstel tot uitbreiding der . . . structuur formules in de ruimte*, Utrecht, 1874.

14 A. M. Butlerov, *Zeitsch. f. Chem.*, 1862, **5**, 300–1.

15 R. Anschütz, *August Kekulé*, Verlag Chemie, Berlin, 1929, vol. i, p.359.

16 J. Dewar, *Proc. Chem. Soc.*, 1897, **13**, 239; H. McLeod, *Diary*, reprinted as *Chemistry and theology in mid-Victorian London*, ed. F. A. J. L. James, Mansell, London, 1987: entry for 25 August 1868.

17 J. Dewar, *Rep. Brit. Assoc. Adv. Sci.*, 1868, p. 36.

18 F. A. Kekulé, *Berichte*, 1869, **2**, 365.

19 J. P. Cooke, *First principles of chemical philosophy*, London, 1870; Cooke, who taught at Harvard, had already written a number of influential textbooks in chemistry (B. V. Lewenstein, 'To improve our knowledge in nature and arts: a history of chemical education in the United States', *J. Chem. Educ.*, 1989, **66**, 37–44).

20 M. R. Fox, *Dye-makers of Great Britain 1856–1976*, ICI, Manchester, 1987, p.11; he cites G. Schultz and P. Julius, *Tabellarische Übersicht der künstlichen organischen Farbstoffe*, Berlin, 1888, as 'the first serious attempt to provide a comprehensive index of dyes' on the basis of structural formulae (p.15).

21 H. Kolbe to Frankland, 9 July 1867 [RFA, OU mf 01.04.1374].

22 Frankland, *Lecture notes* (note 10), vol. i, 1st ed., p.25. The two words in square brackets were added to the otherwise identical passage in the 2nd edition (1872) and doubtless reflect the rising tide of atomic scepticism in Britain.

23 See A. J. Rocke (note 5).

24 *Experimental Researches.*

25 For a summary see C. A. Russell, *Science and social change 1700–1900*, Macmillan, London, 1983, pp.152–173.

26 On the general response to such pressures see D. S. L. Cardwell, *The organisation of science in England*, 2nd ed., Heinemann, London, 1972.

27 On Donnelly see *DNB*, M. Reeks, *History of the Royal School of Mines*, Royal School of Mines, London, 1920, pp.126–130, and W. H. G. Armytage, 'J. F. D. Donnelly: pioneer in vocational education', *Vocational Aspect*, 1950, **2**, 6–21.

28 On Victorian examinations see J. Roach, *Public examinations in England 1850–1900*, Cambridge University Press, Cambridge, 1971.

29 T. Twining, *Technical training, being a suggestive sketch of a national system of industrial instruction, founded on a general diffusion of practical science among the people*, London, 1874, p. xiv.

30 H. E. Roscoe, letter to editor 23 November 1869, *Nature*, 1869, 1, 138; four of those lectures were on chemistry and a regular chemical class was started 'under an able Government science master'.

31 G. Fraser, *Nature*, 1871, 4, 120.

32 *Sketches*, 2nd ed., pp.140–1; he adds that when these were transferred to the new site the audience was entirely different (shop-keepers and their dependants), as 'the working man of the East End did not seem to be able to find his way to South Kensington'.

33 *Lancastrian Chemist*, especially chapters 4, 6, 7 and 8.

34 *Experimental Researches*, p. 3.

35 Frankland (note 7).

36 Frankland, *Lecture notes for chemical students, embracing mineral and organic chemistry*, van Voorst, London, 1866, and 2nd ed., vol. i, 1872, pp. 17–23.

37 F. E. Foden, 'Popular science examinations of the nineteenth century, *J. Roy. Inst. Chem.*, 1963, 87, 6–9; G. Birley, 'The chemistry examinations of the Department of Science and Art', *School Science Rev.*, 1985, 66, 546–551.

38 H. McLeod, *Diary* (note 16): entry for 17 August 1868.

39 Department of Science and Art, *Directory*, 1873.

40 Reproduced in G. N. Stoker and E. G. Hooper (eds.), Science and Art Department, South Kensington, *Chemistry*, W. Stewart & Co., London, n.d. [*c*.1880], pp.38–42.

41 *Ibid.*, p.x.

42 *Ibid.*, p.xvi.

43 *Ibid.*, Inorganic Part p.46, Organic Part pp.37, 40–2.

44 Birley (note 37).

45 H. McLeod, *Diary* (note 16), entry for 11 July 1868.

46 *Ibid.*, entry for 15 June 1868.

47 *Ibid.*, entry for 30 April 1869.

48 Later Sir Alexander Pedler (1849–1918), Vice Chancellor of Calcutta University and a Vice President of both the Chemical Society and the Institute of Chemistry (*Who was Who*). In 1871 he was a student at the Royal School of Mines, having previously been a chemical assistant at the Royal Institution, appointed at £60 p.a. in 1867 (Minutes of the Managers' Meetings reproduced as *Archives of the Royal Institution of Great Britain*, ed. F. Greenaway, Scholar Press, London, 1976, minute for 16 December 1867).

49 McLeod *Diary* (note 16), entry for 4 May 1870.

50 Frankland, *Westminster Budget*, 15 December 1893.

51 Frankland, Minutes of evidence, *Royal Commission on Scientific Instruction*, *P.P.*, 1872, xxv, p.361 (14 February 1871, Qu.1536). Many teachers also attended a 10-week evening lecture course which attracted between 200 and 250 students altogether; but they did not have the benefit of laboratory instruction (*ibid.*, p.362).

52 Frankland, *How to teach chemistry: hints to science teachers and students*, ed. G. Chaloner, London, 1875.

53 *Ibid.*, p.14.
54 *Ibid.*, pp.20–1.
55 *Ibid.*, p.9.
56 *English Mechanic*, 1865, **1**, 2.
57 *Ibid.*, 1870, **11**, 1.
58 For an entertaining anthology of articles from the *English Mechanic* see A. Sutton, *A Victorian world of science*, Hilger, Bristol, 1986.
59 J. F. D. Donnelly, *Directory of Department of Science and Art* , 1868–9, p.63.
60 E. A. Young, *English Mechanic*, 1868, **7**, 47, 92, 145, 314 *etc.*
61 *Ibid.*, 1868, **7**, 213.
62 'Urban', *ibid.*, 1869, **8**, 561: a correspondent (G. E. Davis) is advised 'to use the at present adopted term valency instead of atomicity, and dyvalent instead of dyatomic'. Subsequent issues contained enthusiastic and uninhibited discussions on questions of valency terminology.
63 C. A. Russell, *The history of valency*, Leicester University Press, Leicester, 1971, p. 88.
64 'Urban', *English Mechanic*, 1869, **9**, 222 *etc.*
65 F. Hurter, *ibid.*, 1870, **11**, 9; Hurter was later to achieve fame as chief chemist to the United Alkali Company where, many years later, he gave 'disastrous' advice on the future of electrolysis and a 'rather negative' presentation of his conclusions (W. J. Reader, *Imperial Chemical Industries: a history*, Oxford University Press, London, 1970, vol. i, p.117).
66 G. E. Davis, *English Mechanic*, 1870, **11**, 99; this was possibly the future Alkali Inspector of that name (1850–1907) and founding editor of the *Chemical Trade Journal*.
67 *English Mechanic*, 1872, **14**, 460.
68 W. A. Miller, *Elements of chemistry, theoretical and practical*, vols. i and ii, 5th ed. (rev. H. McLeod), London, 1872–4.
69 W. G. Valentin, *Introduction to organic chemistry*, London, 1873.
70 Frankland showed a great interest in the published work of his assistants. Thus while on holiday in the Scottish Trossachs he spent two hours on the mountain Ben An looking over a manuscript by Valentin on analytical chemistry (Frankland to Maggie Frankland, 17 June 1870 [JBA, OU mf 02.01.0200]).
71 *English Mechanic*, 1871, **12**, 341.
72 *Ibid.*, 1871, **12**, 462.
73 *Ibid.*, 1872, **15**, 792.
74 G. N. Stoker and E. G. Hooper, *Chemistry*, Stewart's Educational Series, London, *c*.1881.
75 *Engineer*, 1867, **24**, 483.
76 A. Crum Brown, *Phil. Mag.*, 1869, **37**, 395.
77 *Engineer*, 1869, **27**, 251.
78 See W. H. Brock, 'The *Chemical News*, 1859–1932', *Bull. Hist. Chem.*, 1992, **12**, 30–35.
79 *Chemical News*, 1866, **14**, 176,
80 H. E. Roscoe, *Nature*, 1869, **1**, 165.

81 J. A. Geuther, *Lehrbuch der Chemie gegründet auf die Wertigkeit der Elemente*, Jena, 1870.

82 *Nature*, 1873, 7, 160.

83 On Twining (1806–1895) see *DNB*; he was grandson of the tea merchant Richard Twining.

84 Twining (note 29), pp. 5, 27, 64, 69.

85 *Ibid.*, pp. 22, 28, 176.

86 T. Twining, *The elementary science examinations of the Science and Art Department*, privately printed, London, 1877 [City & Guilds Archive], cited in Foden (note 37).

87 See, *e.g.*, E. W. Jenkins, 'H. E. Armstrong, heurism and the common sense of science', *Durham Research Review*, 1976, 8, 21–26.

88 J. C. Brough, 'Modern chemistry' (1868), reproduced by A. Scott, *J. Chem. Soc.*, 1916, 109, 344–5.

89 *Minutes* of X-Club, 9 January 1868 [JBA, OU mf 02.02.1706].

90 Frankland, Minutes of Evidence (note 51), pp.355, 360–1.

91 Frankland to Kolbe, 3 December 1871 [Deutsches Museum Archive, Munich, Urkundsammlung Nr. 3566].

92 Cited in M. Reeks (note 27) pp.111–2.

93 Frankland to H. Kolbe, 1 January 1873 [Deutsches Museum Archive, Munich, Urkundsammlung Nr.3567].

94 Cited in Reeks (note 27), p.111.

95 Frankland to Kolbe (note 93).

96 Cited in Reeks (note 27), p.112.

97 Frankland, Minutes of Evidence (note 51), p.362.

98 Frankland to Lockyer, 23 March 1871 [Exeter University Archives, Lockyer correspondence].

99 Frankland to Kolbe (note 93).

100 Frankland to J. Diggens, 19 October 1891, concerning an invitation to speak at the opening of the Storey Institute in Lancaster; he nevertheless agreed to second a vote of thanks to the founder [Lancaster Public Library, MS 7142].

101 H. E. Armstrong, 'Our need to honour Huxley's will' (1933), in W. H. Brock (ed.), *H. E. Armstrong and the teaching of science 1880–1930*, Cambridge University Press, Cambridge, 1973, p.65.

102 H. E. Armstrong, *Pre-Kensington history of the Royal College of Science and the university problem*, Royal College of Science, London, 1920, p.6.

103 *Ibid.*

104 They were W. W. Abney, W. G. Bell, H. S. Billing, C. E. Bouchier, R. Cowper, W. P. Graham, G. C. Hoffmann, W. Pearce, though the list is probably incomplete [Royal Society of Chemistry Archives, Royal Institute of Chemistry records]; I am grateful to Dr. G. K. Roberts for this information.

105 McLeod *Diary* (note 16).

106 Frankland, Lecture notes [RFA, OU mf 01.05.0545].

107 Armstrong (note 102), p.3.

108 H. McLeod, *J. Chem. Soc.*, 1905, 87, 574–590.

109 J. G. P. Cameron, *A short history of the Royal Indian Engineering College at Cooper's Hill*, privately printed, London, 1960.

110 On McLeod (1841–1923) see *Proc. Roy. Soc.*, 1924, **A105**, x–xi.

111 Armstrong (note 102), pp.3–4.

112 Maggie Frankland, *Diary*, 10 April 1875 [MBA].

113 On Valentin (1829–1879) see *J. Chem. Soc.*, 1880, 37, 260, and *Chem. News*, 1879, **39**, 204.

114 See M. Argles, *South Kensington to Robbins: an account of English technical and scientific education since 1851*, Longmans, 1964.

CHAPTER 11

The X-club and beyond

Frankland's links of friendship within the scientific community extended beyond the official institutions at Burlington House or Albemarle Street. No comradeship was more lasting, or perhaps more formative on his thinking, than that which he encountered within a shadowy organisation known to its members and critics as simply the X-Club. It has attracted some considerable attention from scholars in the last few years.[1] The Club consisted at first of eight members, increased to nine by the addition of Spottiswoode within its first few weeks. Thereafter its membership remained unchanged. The nine were:

George Busk (1808–86): surgeon and anatomist; secretary of Linnean Society
Edward Frankland (1825–99): chemist; professor at the Royal Institution
Thomas Archer Hirst (1830–92): mathematician; master at University College School
Thomas Henry Huxley (1825–95): biologist; lecturer at Royal School of Mines
Joseph Hooker (1817–1911): botanist; assistant secretary at Kew Gardens
John Lubbock [later Lord Avebury] (1834–1913): biologist; banker
William Spottiswoode (1825–83): publisher and mathematician
Herbert Spencer (1820–1903): author of *Social Statics* and other books
John Tyndall (1820–93): physicist; professor at the Royal Institution

For some years growing bonds of friendship had been formed between several of these men. Thus Frankland had known Tyndall since Queenwood days, and as we shall see (Chapter 12), he encountered Hirst, Hooker and Huxley during expeditions to the Alps in the 1850s. Similarly, as Roy MacLeod has pointed out, Spencer became a close friend of the Busk family and through them met Lubbock, while Spottiswoode was another friend of Tyndall.[2] Huxley knew everybody but was becoming frustrated at the increasing difficulty of meeting his friends on a social basis. Accordingly he suggested to Hooker that some kind of regular meeting should be organised, and on 3 November 1864, the X-Club was born. It did not receive that name immediately, though various other suggestions (such as the 'Blastodermic Club') were eventually rejected in its favour. The title may have been suggested by Mrs Busk or by the mathematicians; authorities disagree. Its constitution was as brief as its name. As Huxley recalled, 'the proposal of some genius among us, that we should have no rule but the unwritten law not to have any, was carried by acclamation'.[3] They took turns to preside at the meetings and brief minutes[4] were taken.

The Club initially met at St. George's Hotel, opposite the Royal Institution in Albemarle Street. That remained the usual venue for 20 years or so, after which the Athenaeum became the preferred meeting place. Generally they met on the first Thursdays from October to May, attending the monthly Royal Society meeting two hours later. In the early years the June meeting, on a Saturday, was combined with a weekend trip to the country, with x's being accompanied by their wives (y's). At the London meetings they usually dined at 6 and often invited distinguished guests to join them. These included Charles Darwin, the mathematician W. K. Clifford, the Chancellor of the Exchequer Robert Lowe, *etc.*

It does not take long to discern the unifying feature of their friendship. They were to a man dedicated to that system of belief embodied in the phrase 'scientific naturalism': the primacy of science over all other forms of knowledge and the replacement of ecclesiastical control over society by that of a scientific 'clerisy' (Coleridge's word for a secular group of learned men wielding power much in the manner of the Victorian clergy). Thus Hirst could write that 'the bond that united us was devotion to science, pure and free, untrammelled by religious dogmas'.[5] In similar vein Frankland observed that 'all these colleagues of mine . . . were of one mind on theological topics'.[6] If this was really the major unifying belief it goes far to explaining the development of Frankland's religious attitudes that will be discussed later.

The X-Club was, of course, a strictly all-male affair. Anything else would have been unthinkable in Victorian times. The first

Fig. 11.1 George Busk. A member of the X-Club. (J. Bucknall Archives.)

scientific society to open its doors to women was the Geological, and that did not do so until 1904. Huxley's own attitude was that women were simply not fitted for science:

> The best of women are apt to be a little weak in the great practical arts of give-and-take, and putting up with a beating, and a little too strong in their belief in the efficiency of government,

but he located the hub of the problem in their circumstances rather than their inherent abilities: 'men learn about these things in the ordinary course of business; women have no chance in home life'.[7] And education was a key issue. Women were, as he put it, sunk 'in mere ignorant parsonese superstitions'.[8] But the practical effect was the same, whatever the reason. No women ever darkened the doors of the X-Club, with the conspicuous exceptions of wives on

Fig. 11.2 Herbert Spencer. A member of the X-Club. (Author.)

summer excursions. While convalescing in Bournemouth Huxley confessed to Frankland the (for him) surprising opinion that 'I am always ready to vote for the admission of ladies to the Royal Society Fellowship',[9] but it was quite safe to make such remarks – especially in private – for the Society showed no inclination for such liberality until after World War I.

Frankland maintained a similar position. After his colleague Duppa died he expressed uncertainty as to his widow's feelings as 'it is so rarely that women attain to philosophical conception of these things'.[10] Almost echoing Huxley he told his elder son that 'women should not have votes until they show themselves to be more independent of clerical influence'.[11] Yet his chauvinism was limited as is evident from the case of Mary Owens. A letter[12] from Miss Owens to Frankland was despatched from Cincinnati in 1883. The young lady had received the Normal School prospectus and proceeded to ask several searching questions, concluding her epistle:

> Are there any ladies in your laboratory working chemically?
> Eager to have my hopes and anticipations that ladies are admitted to
> your laboratory, I await your reply and am
> Most Respectfully yours
> Mary E. Owens

Frankland must have replied positively for she next writes announcing a late arrival in England and, 'as I shall be alone in London' asks for addresses where she can apply for a room.[13] She returned in due course to her home city where she taught in several 'ladies' schools' and where Frankland met her during a tour of America in the following year. Evidently she had experienced Frankland hospitality during her time in London, for she sent presents to the children and he reported to his wife that 'her parents expressed, over and over again, their gratitude for all we had done for their daughter'.[14] The X-Club would have been cautiously pleased to have one of its members associated with scientific instruction for a woman, especially one from the USA. That same member, Edward Frankland, was also the senior author of the first paper in the *Journal* of the Chemical Society to have a lady as co-author, Miss Lucy Halcrow.[15] It was about the action of air on peaty water[16] and was one of the papers discussed at the longest meeting on record of the Chemical Society.[17] Miss Halcrow was a teacher in training and had also investigated the effects of freezing polluted water (when most of the impurities remained in the unfrozen water).[18]

There are two extreme views of the X-Club. The first of these regards them as a highly influential cabal seeking to manipulate British science and public opinion as far as they could. MacLeod has dubbed them an 'Albemarle Street conspiracy'. All the members of the Club had already acquired distinction in their own fields. All (except Spencer) were Fellows of the Royal Society, three were to become Presidents and others, including Frankland, were to hold office in its executive. It was widely, and perhaps understandably, assumed that they exerted great though unseen influence in the world of science. A popular opinion, allegedly heard at the Athenaeum, may not have been too far from the truth. It emerged in a conversation supposedly overheard by Huxley:

> 'Do you know anything about the X-Club?'
> 'Why – I have heard of it'.
> 'What do they do?'
> 'Well, they govern scientific affairs, and really, on the whole, they don't do it badly'.[19]

The opposite view is what may be called the 'friendship model', ostensibly espoused by Huxley. Always opposed to adding to their number, even after death had taken its toll of some of them, he wrote to Frankland:

> As to the filling up of the vacancies in the X – I am disposed to take Tyndall's view of the matter. Our little club had no very definite object beyond preventing a few men who were united by strong personal sympathies from drifting apart, by the pressure of busy lives.
>
> Nobody could have foreseen or expected twenty odd years ago when we first met, that we were destined to play the parts we have since played – and it is, in the nature of things, impossible that any of the new members proposed (much as we may like & respect them all) can carry on the work that has so strangely fallen to us.
>
> An axe with a new head & a new handle may be the same axe in one sense but it is not the familiar friend with which one has cut one's way through wood and briar.[20]

Just over a year later he returned to the theme:

> I never could see the use of enlarging the X & continuing its existence after we all drop off.
>
> The club has never had any purpose except the purely personal object of bringing together a few friends who did not want to drift apart. It has happened that these cronies have developed into big-wigs of various kinds – & therefore the club has incidentally – I might say accidentally – had a good deal of influence in the scientific world. But if I had to propose to a man to join & he were to say 'Well, what is your object?' I should have to reply like the needy knife-grinder – 'Object, God bless you sir, we've none to show'.[21]

Against this rather appealing 'friendship model' stand an immense number of phenomena attributable to the X-Club, as in its effects on the Royal Society (Chapter 15), its permeation of many influential scientific institutions, the homogeneity of its guest-lists, private comments of some of its members and above all the discussions recorded, albeit with brevity, in its minute book. It is very hard indeed to believe that Huxley was not being deliberately disingenuous in the two letters to Frankland quoted above. He was trying at all costs to restrict the circle (for reasons that can only be guessed), and denial of any other rationale than pure friendship was one way to do it.

It will not, however, do to argue for an absence of conspiracy on the basis that there were many people outside the X-Club who held similar views, that success often eluded their grasp or that no concerted opposition can be identified. All of these things *could* be true at the same time as the X-Club were conspiring together. On the other hand a full-blooded conspiracy theory is equally hard to maintain. Conspirators are more likely to operate behind closed doors yet there was often no attempt to conduct conversations in private at the Athenaeum, for example. Their letters to one another suggest common purpose, certainly, but hardly a whiff of conspiratorial planning. The conversation cited above is not well documented and may well be apocryphal.

There appears to be a better way to understand the X-Club. That is to see it as a flexible institution held together by genuine bonds of friendship (as Huxley maintained), but as one which gave like-minded men a social power-base for individual assaults on the establishment. Ideas would certainly be generated as the conversation (and the wine) flowed freely; new policies could develop in after-dinner gossip; suggestions for filling important posts in science could emerge *en passant*; above all in a shared meal with old-established friends there would be a recharging of emotional batteries, a renewed sense of common purpose and (when it was most needed) encouragement for the faint-hearted. Then indeed 'conspiracies' might be an apt word to describe the resolutions that spontaneously emerged in such congenial company.

This is borne out by detailed inspection of the minutes, a mixture of the trivial and important. It also fits with the known ways in which X-Club ideals were disseminated: usually by *a few individuals acting separately or in concert*, very rarely indeed by the whole group together. These men would then try to capture positions of high influence for the cause or else nominate safe people to those positions, or they would pull the strings in committees and councils, write letters to the press and in a dozen other ways promote the cause of scientific naturalism in Britain.

The X-Club had several targets to attack, one of the most obvious being a profound public ignorance of science, painfully demonstrated by British failure at the Great Exhibition of 1867 (winning only ten out of the ninety prizes). As Frankland knew all too well German science was clearly ahead. When, in the early 1870s, a Royal Commission on Scientific Instruction and the Advancement of Science was set up, under the chairmanship of the Duke of Devonshire, the X-Club were ready; amongst its Commissioners were Huxley and Lubbock. Frankland was called upon to testify, and did so with much stress on the need for Britain to emulate France and Germany by establishing strong links between teaching and research.[22] After the eighth report of the Devonshire Commission (on 'Government and Science') a Ministry of Science was set up, and in 1876 the Royal Society grant-in-aid was increased from £1000 to £5000 per annum. Meanwhile a committee of the British Association, set up for a similar purpose to the Devonshire Committee, included Frankland, Hirst, Huxley and Tyndall. Can this be mere coincidence?

The possibility that Tyndall might accept a chair at Oxford, or that Lubbock might stand as MP for London University, were amongst their topics for earnest discussion. The British Association was often in their sights: Hooker was proposed as President but declined; Presidents for the separate Sections were discussed 'and the list provisionally filled'; the position of ethology at the BAAS was mooted, as was the Association's grant to Kew. Other matters included elections to the Metropolitan School Board, presidency of the Linnean Society and governance of the new Natural History Museum at South Kensington. Above all it was the Royal Society that received most attention, as in their scrutiny of the list of hopeful candidates for nomination of a President. In these activities, as Chapter 15 will show, Frankland played a full part. Then there was the growing need for science to acquire some measure of parity with law, medicine and (of course) the Church. It needed to be recognised as a profession. Of all the chemists who might have presided over the first professional association for scientists (1877) it is remarkable that the chosen one was the only chemical member of the X-Club, Edward Frankland.

THE CHEMIST AND DRUGGIST PORTRAIT GALLERY.

W. B. CARPENTER.

W. CROOKES.

C. R. DARWIN.

E. FRANKLAND.

T. H. HUXLEY.

SIR JOHN LUBBOCK.

J. TYNDALL.

Fig. 11.3 *'Chemist & Druggist* Portrait Gallery': how Frankland was seen by his fellow-chemists, with distinguished scientific naturalists for company. (*Chemist & Druggist*, 14 March 1877.)

The other target upon which the X-Club trained their artillery was organised religion. According to a biographer of Huxley 'one of the prime motivations governing the activities of X-Club members was to ward off attacks on science by conservative theologians'.[23]

There is little doubt that they, with others of like mind, pursued a protracted campaign against religion in many of its forms, seeing it as the enemy of free scientific progress and consistently portraying it as a defeated foe in a conflict which, it must be said, was largely of their own fertile invention.[24]

The X-Club discussed Huxley's anti-theological lectures with evident approval[25] and the first guest they ever invited to dinner was the heretical Bishop J. W. Colenso of Natal[26] (whose son Frank was later to marry Frankland's daughter Sophie). One crucial political issue of the 1880s was the question of Home Rule for Ireland, with its probable concessions to the Roman Catholic Church (a matter on which the Ulsterman Tyndall felt passionately). The Club resolved to oppose any such move and to support the Unionist cause. Huxley was absent, though was not thought to agree on this matter. So Frankland wrote to him:

> At the X it was considered advisable to draw up a sort of scientific declaration in favour of the maintenance of the union and Tyndall was deputed to draw it, strong but in moderate language, and we all agreed to sign. It was thought that nearly every scientist of note would sign it.[27]

At an early stage in the onslaught of the X-Club on the religious establishment a Sunday Lecture Society was established, in imitation of the church's Sunday schools, with meetings at four o' clock in the afternoon, and with special emphasis on science and its benefits to mankind. Sometimes a 'Hymn to Creation' was sung. Such things happened elsewhere, of course, but what is remarkable is that no less than five of the Society's ten vice-presidents were members of the X-Club (Frankland, Huxley, Spencer, Spottiswoode and Tyndall). Another was Darwin himself.

The X-Club was a curious mixture of formality and informality. The meeting dates, set well in advance, often had to be altered, about which there seems to have been an endless exchange of letters. Tyndall once wrote that he had overlooked the date of the next meeting, no one had reminded him, and he had engaged to invite Francis Galton to dinner at home. Either he would bring his guest to the X-Club or they would forfeit his company.[28] Charges were usually paid promptly unless the members were out of the country or unless, as once happened to Huxley, the 'little bill' had been lost.[29]

This rather engaging informality was sometimes absent. On several occasions Huxley wrote to Frankland with clear instructions.

In 1887 he indicated his possible absence, adding that he would be 'much obliged if you will kindly see that the dinner is all right and take the orders of the brethren as to the occasion of the next meeting'.[30] A few months later his request was less peremptory: 'will you be so kind as to look after the X next Thursday March 1?'[31] The niceties of apology for absence had to be observed, and apparently Frankland was not deemed a suitable agent, so Huxley wrote to him saying that he had also 'written to Hooker so as to make use of somebody as an apologist'.[32]

These letters of the 1880s confirm the view that the driving-force behind the X-Club was still Huxley. That had been so from the beginning. At the club's first meeting they discussed a new journal the *Reader*, edited by Lockyer and with potential as a forum for their views but, alas, making serious losses. Huxley joined the board and five X-Club members came up with cash.[33] Other forays followed but it was not until the 1870s that the fraternity, with Huxley at their head, reached the zenith of their power. Adrian Desmond recaptures the spirit of their renewed crusading zeal: 'The X marched behind him. The club was back in formation, their strange freemasonry having survived a civil war'.[34]

The first casualty of the X-Club agenda was the end of the weekend outings, to places like Burnham Beeches. This arose through the demise of two of the wives, Frances Hooker and Sophie Frankland, in 1874. In the years that followed contention became more common than before. Frankland as we have seen (p. 235) disagreed with Huxley over the value of attending British Association meetings, while Huxley and Hooker found themselves at odds with other members on several important issues. Huxley and Tyndall were at loggerheads with the rest on the desirability of increasing their numbers as death removed one long-standing member after another. On this issue they won. Frankland called it 'obstinacy'.[35] But the biggest division of all arose from the involvement of Frankland in 'trade'. In 1888 Hirst and Hooker had 'a rather warm controversy with Frankland' about the adversarial system of British justice in which scientists have to appear as witnesses. Frankland had apparently condemned Dewar for 'doing what he himself has done for years', acting as a scientific witness.[36] Now Frankland had been given, along with all the others, a whimsical nickname. Others included the Xalted Huxley, the Xcentric Tyndall, the Xtravagant Hirst and so on; there was also the Xpert Frankland, – not a bad name when one considered that is precisely what a scientific witness is supposed to be. At a meeting of the X-Club in 1885 Huxley inveighed against 'politics, scandal, and the three classes of witnesses – liars, d----d liars and experts'.[37] It is unlikely to have been a personal attack on Frankland but there is no doubt as to its general drift. The issue raised its head on several other occasions, and Huxley returned to it in 1889. Writing to Hooker about the Royal Society he referred to the exorcism of the old 'aristocratic flunkeyism'. However,

> the danger now is that of the entry of seven devils worse than the first, in the shape of rich engineers, chemical traders, and 'experts' (who have sold their souls for a good price), and who find it helps them to appear to the public as if they were men of science.[38]

This again is too strong to apply primarily to Frankland, but there is no doubt that he displayed at least some of those characteristics that Huxley found so objectionable. The same question may have lurked behind Hirst's acid comment that at an X-Club meeting in the same year Frankland 'talked far too much, and very loosely as usual'.[39]

Numbers dwindled from the average of seven. When Herbert Spencer heard that Spottiswoode had cancelled the meeting altogether, he decided to 'get up a dinner on my own'.[40] On a later occasion Hirst recorded that only 'Huxley and Frankland were at the Athenaeum, and expected their dinner, so I gave orders at once', the meal for three costing £1:12:9d.[41] The spirit of survival was poignantly illustrated by a remark from Huxley when, in 1887, Hooker, Tyndall and he dined alone at the X-Club:

> We three old fogies voted unanimously that we were ready to pit ourselves against any three youngsters of the present generation in walking, climbing or head work – and give them odds. I hope you are in the same competitive frame of mind! [42]

Frankland was not the most visibly active of the Club though some of his contributions are recorded. He proposed that action should be taken to hasten the publication of *Philosophical Transactions*,[43] and joined Hirst and Spottiswoode in intending to raise before the Council of the Royal Society the unsatisfactory mode of its election process.[44] Yet there is reason to believe that he, of all the others, placed the highest value on the Club. He cherished an album of portraits of its members[45] and devoted a complete chapter of his autobiography to its activities.[46] The Club met for 240 times, of which Frankland was present for 186, the highest attendance of all. It was, of course, true that he was still enjoying robust health.

The X-Club met for the last time on 10 March 1892, with Hooker and Frankland as the only members. Next month Frankland alone arrived; he departed at 8 p.m. As a last desperate measure to hold the Club together, when old age and infirmity were keeping many away, Frankland evidently offered overnight hospitality as an encouragement, for Hooker replied thus:

> I will do what I can to keep up the X-Club dinners, if the members will only let me know with some degree of probability whether or no they can attend & there are enough of them to call a meeting for. It is clear that at my age (76th year), & with my throat, I should not attempt dining out in London in winter, & in summer it is impossible to get the brotherhood together.
>
> Your proposal is a very tempting one, but then it is not all of us who can put up 5 or even 2 at once. I will however bear the thing in mind.[47]

But in vain. The X-Club was no more.

Why then was Frankland so devoted to a Club which, on occasion, rejected his commercial activity and showed little sympathy for his consultant work? Perhaps the true explanation lies deep in the psychological needs of a man who felt a keen sense of social insecurity all his life. The long shadow of his illegitimacy, and the need to protect himself and his relatives by absolute silence on the matter, were always with him. So was the compulsion to succeed as well as the Gorsts (p. 516). But here, in the X-Club, he was among his scientific peers and (in his eyes) social superiors. If they would accept him on equal terms he would feel accepted in a manner, and to a degree, that his adult experience had never known. Membership of this *élite* fraternity was a signal that he had 'arrived'. And in scientific naturalism he need fear neither the frown of the religious nor the contempt of the high-born. He was his own man and the only values that mattered were those of the science to which he was devoting his life. To him the X-Club was a symbol of lasting friendship, largely untroubled by controversies of the kind that were raging outside over pollution, water analysis and student training, and blessedly immune from family squabbles. This can only be a hypothesis but it seems to fit the available facts.

Through his active membership of the X-Club Frankland was drawn into the world of scientific naturalism to which he appears to have once been a complete stranger (in his Manchester days). This meant a wide range of new contacts to be nourished and maintained in addition to all those from industry and the chemical world. Most memorable will have been his meeting and exchanges with Charles Darwin himself. Despite their common membership of the Royal Society and friendship with Huxley and other X-Club members their paths do not seem to have intersected significantly until the 1870s. This is not really surprising. There could hardly be a greater contrast than that between Darwin and Frankland: the one virtually a recluse at his home in Kent, venturing out only when he had to and glad when socialising with near-strangers was over; the other frenetically active, travelling all over the country on water analysis business and never happier on holiday than when he was on the move. Yet their paths did cross in the 1870s, and the initiative was largely Darwin's as he needed help from Frankland in a new area of science.

The subject in question was a part of plant physiology and concerned the *Drosera*, or Sundews, several species of insectivorous plants that flower in July and attract insects by secreting a glistening, sticky fluid. This comes from glands situated on long slender stalks ('tentacles' Darwin called them) on the edges of the leaf. As an insect descends for its meal it is trapped by tentacles bending over it and destroyed by a thick viscous liquid from the centre of the plant, thus providing the latter with much of its food. It was to these plants that Darwin turned on 23 August 1872, the day after he finished proof-reading for his book *Expression of the Emotions*.[48] He was soon deflected by illness, but returned to the subject next summer and found himself in need of chemical advice. For some reason he turned to Frankland. For a year the two men corresponded and a fairly full collection of their letters has survived (mostly in two separate archives). Darwin requested meetings on several occasions and there is no reason to believe that he did not visit Frankland in his laboratory.

The great question was how the plant managed to digest its victim. Their exchanges started with a formal letter beginning ' My dear Sir, I am going to beg a great favour of you . . .', and continuing with a request to be put in touch with a competent chemical analyst. Darwin had tried experiments with little cubes of gelatine or egg-white instead of insects and watched with fascination as they, too, succumbed to dissolution by the insect-destroying liquid. It was known to be acidic, but what was the acid? He needed a chemist on hand. Frankland replied that it would be a privilege to help but it would need a lot of liquid. Darwin thought he could collect enough within a fortnight, but Frankland pointed out that by then he was off to Switzerland for the summer holiday. However he began a series of despatches from his laboratory: distilled water, 'neutral' litmus papers, soda, phosphate of lime, oleic acid and 'weak sewage' (of which Frankland hoped the *Drosera* might possibly turn out to be an indicator). Eventually he did the analyses himself, confirming his suspicion that hydrochloric acid was not involved (Darwin had thought it possible) and suggesting that one or more of the lower fatty acids might be responsible.

A further series of experiments was probably suggested when the two men met in the spring of 1874. They involved Frankland feeding his caged bullfinch at home with crocus buds, concluding the bird's selective skill must be hereditary. Darwin's response, as always, was most grateful and concluded with a post-script: 'Your simple experiment of giving the flowers to caged birds never occurred to me. See what it is to be an experimental chemist!!'. In fact the most conspicuous feature of Darwin's letters to Frankland was an almost obsequious gratitude. While in character with his moods at the time it also suggests that Frankland had yet to make any warm impression on the community of Darwin's admirers, though was obviously a man to be respected.[49]

The two men evidently became closer in the next few years. A number of letters survive from Darwin to Frankland but most of those in the reverse direction seem to have been lost. Darwin's book on *Drosera*, published as *Insectivorous plants*, was finished in 1875 and he now turned to other matters. Soil quality was one, and Frankland was drawn into a series of investigations on the relative advantages of natural and burnt soil. Another issue was the alkalinity of water left for a long time on pelargonium leaves. Darwin expressed thanks for Frankland's 'invaluable assistance', and anxiety lest a bottle of distilled water sent him had been stolen *en route* under the impression that it was alcoholic. Now he addressed him as 'Dear Dr. Frankland' and gave a sense of greater familiarity, twice requesting an opportunity to call on him.[50] The impression is fortified by the fact that now the great man was invited to Frankland's home (as opposed to calling in at the laboratory). On Sunday, 26 April 1876, something like a state visit was arranged, described thus in Maggie's *Diary*:

In the mng at 10 o'clock Mr. Darwin arrived & was received by the whole family assembled in the drawing room (which was prearranged). The birds were also in the room & some primroses and cowslips so Papa repeated the experiment before Mr. Darwin. For almost the first time the bullfinch eat [*sic*] the corolla instead of the ovary of these flowers, but it soon showed its decided predilection for the latter. The canary on the other hand pecked the flowers indiscriminately. Mr. Darwin said that one does not value a discovery according to its importance but according to the length of time one has puzzled over it. He himself had rarely been so pleased as when he found out the reason of the difference between primroses viz that the primroses had different stamens one having a pistil and another stamens. He called the bullfinch 'the classical bird' and several times remarked how ugly it was in comparison to a cock bullfinch, rather a strange thing for a naturalist to say; one would think nothing in nature could be ugly in his eyes! Papa then told him that Nannie and Luis [her cousins] were the daughters of Prof. Fick a great admirer of his and who wrote the pamphlet upon 'Einfluss der Wissenschaft auf das Recht'[51]. Mr. Darwin said he had been much impressed with the said pamphlet, and that he wished them to give his compliments to their father and tell him so. After several bows on all sides Mr. Darwin went into the boudoir with Papa to discuss some matter with Papa.[52]

Whatever 'some matter' may have been the visit can only have confirmed to Frankland the gratifying opinion that he was now accepted within the wider community of scientific naturalists, for was not Darwin the greatest of them all?[53]

With the further passage of time Frankland's circle of close acquaintances enlarged to include others of like mind to the belligerent scientific naturalists. Guest-lists for his dinner parties at home read like a roll-call of the faithful: Carpenter, Clifford, Debus, Duppa, Samuelson, Lyell, Müller, to say nothing of every member of the X-Club (with the strange exception of Lubbock, though Frankland went to stay with him). Time after time several of these individuals would meet up before or after lectures at the Royal Institution, at soirées of the Royal Society or at non-scientific functions where liberal-minded gentlemen expected to be seen. If Frankland *did* have a strategy for permeating into the heart of the liberal scientific establishment it has to be said that it was most successful. A joyful letter to his daughter about the British Association meeting at York in 1881 positively drips with names of the scientific naturalists he was meeting, especially those of his beloved X-Club.[54]

However it was not sufficient to be seen with the right people or even to share part of their agenda. The espousal of a scientific world-view (whatever that might be) was clearly desirable if only to distinguish oneself from 'mere' classicists or establishment politicians. But it was not enough. There was also the question of religion. The X-Club members made no attempt to conceal their general and corporate antipathy to matters theological. Sometimes this was made evident by the strident trumpetings of a Huxley, sometimes by the discreet silence of a Lubbock. The term 'agnostic' was freely employed, but not unambiguously adopted by all. Where agreement seems to have been solid is in the defence of scientific theory and practice against the meddlesome interference from clergy and others who felt called upon to declaim on matters for which they were extremely ill-qualified. That sort of threat inflamed the fraternity's passions to boiling point.

The position taken on these matters by Edward Frankland is most interesting. On the whole he is fairly reticent and his religious position is not always clear. It is as well to start with his most considered statement, that in his autobiographical *Sketches*. He devotes one chapter[55] to 'Religion' and recalls something of his youthful experience of formal Anglicanism followed by something very like an evangelical conversion in a Congregational Church and subsequent service among the poor with a large Independent Chapel in London. He makes clear that his youthful faith was gradually eroded by encounter with scepticism in Germany (though he quite incorrectly calls Kolbe an 'Agnostic') and Queenwood where disenchantment set in with a reading of Tom Paine's *Age of Reason*. Thereafter slippage into full-blown scepticism was gradual but definite.

All this was written half a century after the events described and in several respects is historically questionable. But that is not the point. The main importance of Frankland's very *a posteriori* account is that he gave it. He wanted his readers to see him in a certain light, as a lapsed believer, a full-blown unbeliever, a sceptic. He was lining up with the X-Club and the rest. Consequently he felt it right to take this one opportunity for expressing some trenchant opinions about the Christian faith. Speaking of the X-Club he wrote:

> As is well-known, several of these men have exercised by their writings a profound influence upon the religious thought of their age. Joining with them the name of the author of *The Origin of Species* it is not too

much to say that Darwin, Huxley and Spencer are the three great modern evangelists whose literary work will guide the thoughts and actions of men long after the teachings of the four older evangelists have become obsolete.

It is an open question, taking into account all the cruel tortures, wars, lunacies and miseries for which it is responsible how far the Christian religion has been a blessing or the reverse to mankind; but of this there can be no longer any doubt that two of the corollaries of religion, viz.

(1) That marriages are made in heaven, and
(2) That Providence regulates the number of children,

are among the most fertile sources of human misery. They are the cause of that reckless improvidence which is now sapping the foundations of civilisation.[56]

It is not pertinent at this point to smile at the facile optimism enshrined in his first paragraph, nor to wonder how his experience of two marriages and seven children relates to the second. More marvellous, perhaps, is the motivation that led this mild-mannered septuagenarian to indulge in such an isolated and vitriolic attack. Nowhere else in all his writings, private or public, has anything remotely like this been found. Not surprisingly his two daughters included it among the passages for excision in the second edition (so very few people today have ever read it). The explanation that most readily fits the facts is that it was, in a sense, a ritual declaration of solidarity with his long-remembered colleagues of the X-Club, and of dedication to the cause of scientific naturalism. It was a ruse unlikely to appeal to most scientific leaders in Britain, and many of the mathematical physicists would have viewed it with a mixture of shock and distaste. Not even his German scientific relatives would go this far. But it fitted with his cherished self-image and who can deny him the right to express it? That was how he wanted his family and close friends to remember him. He can have had no idea how much it would embarrass them.

The reality of Frankland's belief was both less extreme and more complex than this passage would suggest. As we have seen from his Manchester Inaugural Address Frankland was either much more a theist in the 1850s than he would like us to believe, or he was shamelessly playing to the gallery. The latter is much less probable. The possibility must exist, of course, that in the next few years he radically changed his position. In a note sent to Galton in answer to the statistician's queries addressed to members of the scientific community he wrote tersely 'I have belonged in succession to the Church of England, Independent and Unitarian churches'.[57] This was indeed the case and in the mid-1870s he and his family were often to be seen at various Unitarian chapels. The lack of doctrinal specificity in such places would have pleased him well.

Frankland sometimes attended the 'Church of Progress' in St George's Hall. This strange institution, presided over by a Mr Baxter Langley, was concerned with imparting useful knowledge to the masses on physical, social and moral topics. One evening meeting, attended by Fred Frankland and his father, focused on a lecture 'Dealing with drunkenness'. Fred felt that 'to call it a "Church" . . . seems to me not only a perversion of language but even a positive mischief, since it thus entirely excludes the worship of the Supreme Being'. Frankland actually agreed, but said the title was given 'solely because the law required it' and they should have waited until parliamentary consent had been given to Sunday lectures generally.[58] It thus appears as a precursor to the Sunday Lecture Society.

By the 1870s it is clear that Frankland had acquired many of the characteristics of a Huxleyite ideology. Most conspicuous of these was an almost violent anti-clericalism. Writing to Huxley in 1869 Frankland noted with glee that the former's *Lay Sermon* on 'Protoplasm' will 'frighten the parsons more than anything they have encountered for a long time'.[59] In similar vein his son Fred records his positive reactions to a sermon on the Ecumenical Council. When his father insisted that 'science, and science alone, is our protection against all this sort of superstition', Fred identifies the latter as the ascendancy of the Papal system.[60] It was not only the political power of the church that he objected to, but also what he saw as its use of superstition to blackmail men (and especially women) into believing. There is no doubt that he sincerely held that science had dispelled the mists of superstition and ignorance that had held humanity in bondage until the nineteenth century. He once visited his dying friend Duppa at his home near Maidstone. He recorded with evident approval that 'he told me when we were alone together how thankful he felt that his scientific pursuits had freed him from religious superstition & enabled him to contemplate the steady approach of death with calmness & absence of fear'.[61] Nevertheless Frankland was a hundred miles from the radical atheism of a Bradlaugh and nearly as far from the agnosticism of a Huxley. Evidence comes from a number of directions.

He did not entirely abstain from church-going, though sometimes attended for entirely non-religious reasons. On his first visit to Norway he coyly reports that on Sunday morning 'I went to church (!), chiefly to see the building, which is very ancient and curious'.[62] However, after Sophie's death he would be regularly seen at the Unitarian church over the next six months. Usually, however, Frankland stayed at home while the family went to church. He would be seen even at Anglican churches on special occasions, as at the BAAS meeting in York when he went to the Minster to hear the Bishop of Manchester reply to addresses by Huxley and Lubbock. It was, he said, 'probably the largest congregation ever assembled in this magnificent cathedral'.[63] And towards the end of his life he sometimes attended St Peter's Church in Vere Street where he appreciated the sermons of the Rev. T. Bonney.

Despite his visible rejection of public forms of Christianity Frankland retained a strong friendship with several leading clerics, most of whom probably shared his rejection of what later came to be called 'fundamentalism'. With Richard Dawes of Kings Somborne, and later Dean of Hereford, he enjoyed 'a friendship which lasted without any interruption or misunderstanding until the day of his death in the year 1867'. Dawes had no children, but Frankland relates that he 'always called me his son', reckoning him to have been 'one of the most benevolent and large-minded men that ever adorned the Church of England'.[64] When visiting Dawes at the Deanery in 1859 Frankland and Tyndall went 'like good boys' to hear him preach at the Cathedral.[65] The compliment was more than repaid when the Dean later became 'a constant hearer' of his two friends at the Royal Institution and Jermyn Street.[66]

Frankland was willing to commend members of the Anglican church provided they were seen as non-traditional and posing no threat to science by dogmatic restrictions. His view of Dean Stanley was typical enough: 'he was one of the very few remaining links binding the Church of England to the thinking portion of the community'.[67]

All of this was very reminiscent of the protestations of Tyndall or A. D. White. The former's notorious address to the BAAS at Belfast in 1874 purported to identify the enemy as 'all schemes and systems which infringe upon the domain of science', and that included all religious bodies attempting to curtail the power and authority of science. White, who achieved a greater long-term notoriety for his polemic tract masquerading as history,[68] professed to have a deep affection for the church whose reputation he so assiduously sought to undermine. He denied 'the slightest feeling of hostility towards the clergy', claiming that 'among them are many of my dearest friends'.[69] Even Huxley saw the Church of England as 'that great and powerful instrument for good and evil'.[70] Such ambivalence was equally evident in Edward Frankland. It goes some way to explaining his delight at the novels of Trollope with their merciless exposure of the power-hungry church establishment.

When it came to his colleagues Frankland formed harmonious relationships with Christian believers of several kinds. It was much more than mere toleration of detested beliefs. Thus he went out of his way to inform his assistant W. J. Russell of an appointment at a 'Christian College'. Russell was uncertain about any credal demands that it might make and sought Frankland's advice. He declared 'I could not swear that I firmly believe in the 39 articles &c', considering himself 'more christian than church man'.[71] Then there was Frankland's devoted assistant Herbert McLeod who, as often as not, would slip away to attend one of the High Church services he so passionately loved. He was one of the signatories of the Scientists' Declaration of 1865, affirming the consistency of science and scripture, and lived to write the fullest obituary notice Frankland ever had. Nowhere in his expansive *Diaries* is there a hint of *odium theologicum* between assistant and professor. On a comparable academic level to Frankland was the physical chemist John Hall Gladstone, at one time President of both Chemical and Physical Societies.[72] He was an evangelical Congregationalist whose last public act was to preside at a meeting of the Christian Evidence Society. He was often a guest at Frankland's house and ran a Bible class for young men to which Percy Frankland went with the approval of his father.

Yet it was within his own family that the ideology of the X-Club generated the greatest tensions. If he did try to impose religious doubts upon his children he appears to have been singularly unsuccessful, though it seem unlikely that his opposition extended beyond objections to religious observance. The nearest incident that discloses a conversation with any kind of theological content was in October 1872, when Maggie's *Diary* records:

> Percy went to church in the morning & in the afternoon to Dr. Gladstone's. Papa very kindly gave us a little Bible class in the evening. We read the first 2 chapters of Genesis. Papa told us the different interpretations given to these chapters. He then read a good many short pieces from Longfellow.[73]

It was when ecclesiastical practice raised its head that the sparks flew. For some years Maggie had expressed a wish for confirmation but her father had adamantly refused to grant permission. In the end she took the law into her own hands, but only at the age of 24. Even the loyalties of Maggie were strained to the uttermost, and it was on account of these religious differences that she was about to leave home. Similar antipathy to Anglican practices was displayed during a visit from the water analyst C. M. Tidy whose family had strong links with the Franklands, and who was often at their house. On one occasion at dinner, to Frankland's intense surprise, he revealed himself as a High Churchman and proceeded to extol Pusey. Maggie, an excited observer of the fireworks that followed, confessed 'great delight' but added that 'Papa was dumbfounded'. The end of the conversation was lost as the ladies rose to leave the room.[74]

In fact all his children displayed considerable interest in the Christian religion (voluntarily attending High Anglican churches, reading the Bible and religious books at home, even writing long theological tomes in the case of Fred who professed conversion to Christianity in the early 1870s). An example of their interest comes in a letter from Sophie to her sister Maggie about a packed Easter Day service at St Stephen's church in London, where they were accompanied by Percy and their grandfather William Helm:

> We had of course that beautiful Easter hymn 'Jesus Christ is risen today, Hallelujah'. Didn't you think of darling Fred in church? or how he would be thinking of us, & joining spiritually in our worship of Christ our risen Saviour?[75]

Such sentiments might seem mawkish today but they illustrate the strength of the family ties within the Frankland household and a simple piety that gave them substance. They would be unthinkable in the context of the X-Club, but by very contrast raise rather important questions about its most diligent attender.

Notes

1 (a) J. V. Jensen, 'The X-Club: Fraternity of Victorian Scientists', *Brit. J. Hist. Sci.*, 1970, **5**, 63–72;
(b) R. M. MacLeod, 'The X-Club: a social network of science in late-Victorian England', *Notes and Records Roy. Soc.*, 1970, **24**, 305–22; (c) R. Barton, 'The X-Club: Science, Religion and Social Change in Victorian England', University of Pennsylvania PhD Dissertation, 1976; (d) *Idem*, 'An influential set of chaps: the X-Club and Royal Society politics 1864–1885', *Brit. J. Hist. Sci.*, 1990, **23**, 53–81; (e) A. J. Harrison, 'Scientific naturalists and the government of the Royal Society 1850–1900', Open University PhD thesis, 1988, chapter 6 (pp.167–213).

2 MacLeod (note 1), pp.307–9.

3 T. H. Huxley, cited in L. Huxley, *Life and letters of Sir Joseph Dalton Hooker, O.M., G.C.S.I*, Murray, London, 1918, p.539.

4 Tyndall's copy of these exists in the Royal Institution, and Frankland's own set has also survived [JBA, OU mf 02.02.1706]; the minutes for 1864 are in the hand of Frankland, not Hirst as Barton suggests (note 1(d) p. 63*n*).

5 T. A. Hirst, *Diary*, 6 November 1864, p.1702, reprinted in W. H. Brock and R. M. MacLeod, *Natural knowledge in social context: the journals of Thomas Archer Hirst F.R.S. (1830–1892)*, Mansell, London, 1980.

6 *Sketches*, 2nd ed., 1901, p. 51.

7 T. H. Huxley to J. F. Donnelly, 11 November 1894, in L. Huxley, *Life and letters of Thomas Henry Huxley*, Macmillan, London, 2nd ed., 1908, vol. iii, p.342.

8 T. H. Huxley, cited in A. Desmond, *Huxley: the Devil's disciple*, Joseph, London, 1994, p.273.

9 T. H. Huxley to Frankland, 14 February 1888 [RFA, OU mf 01.04.1731]. Ladies were not in fact admitted until 1945.

10 Frankland to J. Tyndall, 14 November 1873 [Royal Institution Archives, Tyndall papers, 9/E4.18].

11 Fred Frankland, *Diary*, 20 August 1870 [Alexander Turnbull Library, Wellington, New Zealand, MS. 801].

12 M. E. Owens to Frankland, 15 June 1883 [RFA, OU mf 01.04.1485].

13 M. E. Owens to Frankland, 18 September 1883 [RFA, OU mf 01.04.1487].

14 *Sketches*, 2nd ed., pp.428–31.

15 M. R. S. Creese, 'British women of the nineteenth and early twentieth centuries who contributed to research in the chemical sciences', *Brit. J. Hist. Sci.*, 1991, **24**, 275–305 (303).

16 L. Halcrow and E. Frankland, *J. Chem. Soc.*, 1880, **37**, 506–17.

17 Frankland to J. Tyndall, 21 May 1880 [Royal Institution Archives, Tyndall papers, 9/E6.21].

18 Frankland, Report to the Secretary, Department of Science and Art , 29 April 1881; printed proofs (corrected) [RFA, OU mf 01.03.0484].

19 T. H. Huxley, *Nineteenth Century*, 1894, **35**, 11.

20 T. H. Huxley to Frankland, 8 December 1886 [RFA, OU mf 01.02.0481].

21 T. H. Huxley to Frankland, 3 February 1888 [RFA, OU mf 01.02.0414].

22 *Report of the Royal Commission on scientific instruction and the advancement of science*, *P.P.* 1872 [c.536], xxv, pp.50–1 [Qu. 832–5, 7 June 1870], and p.366 [Qu. 5792, 14 February 1871]; also, 1874 [c.1087], xxii, pp.135–8 [Qu.11053, 11108, 8 May 1872].

23 J. V. Jensen, *Thomas Henry Huxley: communicating for science*, University of Delaware Press, Newark, NJ, 1991, p.160.

24 See C. A. Russell, 'The conflict metaphor and its social origins', *Science and Christian Belief*, 1989, 1, 3–26.

25 *Minutes* of X-Club, 4 January 1866 (note 4).

26 *Ibid.*, 8 December 1864 and 6 April 1865.

27 Frankland to T. H. Huxley, 5 November 1887 [Imperial College Archives, Huxley papers 16/272].

28 J. Tyndall to Frankland, '3 March' [RFA, OU mf 01.07.0324].

29 T. H. Huxley to Frankland, 28 October 1879 [RFA, OU mf 01.02.0935].

30 T. H. Huxley to Frankland, 31 March 1887 [RFA, OU mf 01.02.0464].

31 T. H. Huxley to Frankland, 25 February 1888 [RFA, OU mf 01.02.0494].

32 T. H. Huxley to Frankland, 2 November 1887 [RFA, OU mf 01.02.0432].

33 Desmond (note 8), p.330.

34 *Ibid.*, p.371.

35 *Sketches*, 2nd ed., p.162.

36 T. A. Hirst, *Diary*, 8 November 1888 (note 5).

37 *Minutes* of X-Club, 2 December 1885 (note 4).

38 T. H. Huxley to J. D. Hooker, 26 March 1889, reproduced in L. Huxley (note 7), vol. iii, p.120.

39 *Minutes* of X-Club, 11 April 1889 (note 4).

40 H. Spencer to Frankland, n.d. [JBA, OU mf 02.02.1508].

41 T. A. Hirst to J. D. Hooker, 3 January 1890 [JBA, OU mf 02.02.1732].

42 T. H. Huxley to Frankland, 11 April 1887 [RFA, OU mf 01.02.0478].

43 *Minutes* of X-Club, 1 February 1866 (note 4).

44 *Ibid.*, 3 May 1866.

45 Portrait album [RFA, OU mf 01.04.2309].

46 *Sketches*, 2nd ed., pp.148–63.

47 J. D. Hooker to Frankland, 15 August 1892 [RFA, OU mf 01.02.0729].

48 F. Darwin, *Charles Darwin: his life told in an autobiographical chapter, and in a selected series of his published letters*, Murray, London, 2nd ed., 1902, pp.47, 321.

49 C. Darwin to Frankland, 10 November 1872, 18 October 1873 [JBA OU mf 02.02.1575, 1576]; 9 March, 12, 16 July, 21, 23, 29 September, 13 October 1873, 12, 17, 22, 28 April, 14, 20 May, 22 July, 31 August, 11 October 1874 [RFA, OU mf 01.08.0434, 0486, 0489, 0457, 0451, 0454, 0439, 0428, 0431, 0432, 0436, 0480, 01.02.1027, 01.04.0583, 0585, 1217]; Frankland to Darwin, 15, 17 July, 22, 27 September, 10, 16 October 1873, 15, 26, 30 April , 9 October 1874 [Cambridge University Library, Darwin papers].

50 C. Darwin to Frankland, 3 May, 6 June 1876, 22 November, 2 December 1878, 4 January, 8 February 1879 [RFA, OU mf 01.04.0391, 01.02.1021, 01.08.0485, 01.02.0959, 01.08.0483, 01.04.1291].

51 Hermann Fick's essay *Ueber den Einfluss der Naturwissenschaft auf das Recht*

was sent to Darwin shortly after publication in 1872, receiving acknowledgement in a letter of 26 July: Hélène Fick, geb. Ihlee, *Heinrich Fick, Ein Lebensbild*, Zurich, 1897, pp. 278–313; Darwin's reply is in facsimile between pp. 314 and 315.

52 Maggie Frankland, *Diary*, 26 April 1876 [MBA].

53 Frankland was, of course, present at Darwin's funeral at Westminster Abbey (Admission Card [RFA, OU mf 01.02.0698]).

54 Frankland to Maggie Frankland, 1 September 1881 [JBA, OU mf 02.02.1657].

55 *Sketches*, 2nd ed., pp.40–51.

56 *Ibid.*, 1st ed., 1900, p. 57.

57 Copy letter, Frankland to F. Galton, April 1874 [AFA].

58 *Ibid.*

59 Frankland to T. H. Huxley, 16 February 1869 [Imperial College Archives, Huxley papers, 16/251].

60 Fred Frankland, *Diary,* 19 December 1869 (note 11 [MS. 799]).

61 Frankland to J. Tyndall (note 10).

62 Frankland to Sophie Frankland, 1 August 1863 [*Sketches*, 2nd ed., p.334].

63 Frankland to Sophie Colenso, 4 September 1881 [PCA 03.01.0616].

64 *Ibid.*, 2nd ed., p.51.

65 Frankland to Sophie Frankland, 24 April 1859 (*ibid.*, p.388).

66 *Gentleman's Magazine*, 1867, **222**, 674–5.

67 Frankland to Ellen Frankland, 31 July 1881, in *Sketches*, 2nd ed., p. 360.

68 A. D. White, *A history of the warfare of science with theology in Christendom*, Macmillan, London, 1896.

69 *Ibid.*, pp.xi and xii.

70 T. H. Huxley to C. Kingsley, 23 September 1860, reproduced in L. Huxley (note 7), vol. i, p.320.

71 W. J. Russell to Frankland, 22 March [1855] [RFA, OU mf 01.03.0262].

72 See *J. Chem. Soc.*, 1905, **87**, 591–7.

73 *Ibid.*, 20 October 1872.

74 Maggie Frankland, *Diary*, entry for 3 January 1880 [MBA].

75 Sophie Frankland to Maggie Frankland, 27 March 1875 [JBA, OU mf 02.01.0356].

CHAPTER 12

Family: years of crisis

On 18 January 1870 Edward Frankland celebrated his 45th birthday. There was a good dinner including turkey.[1] His two daughters and wife combined forces to produce a pair of slippers, and his two sons gave him a pen-tray, and afterwards they drank each other's health and made music. Frankland had delighted in musical activity since his Lancaster days, and late in 1868 had sent Maggie to learn singing from the celebrated García[2] ('it costs a deal of money and I hope you will make good use of it'[3]), encouraging her to persevere despite his occasional severity.[4] The *maestro*, whose pupils included Jenny Lind, taught at the Royal Academy of Music and became a family friend.[5] However on this occasion the girls played the piano and Frankland sang (there was no Tyndall to make deprecating comments).

They had celebrated birthdays like this for years past. Earlier in the day Sophie had taken her daughters Maggie and Sophie to a lecture by a Professor Humphry at the Royal Institution on the architecture of the human body.[6] It was a typical day in the life of the Frankland family at that time: plenty of cultural improvement for the young ladies, fun and music-making at home, good food and doubtless sincere demonstrations of family affection. On other days they would go to the opera or theatre and in the long winter evenings curl up in front of the fire with a novel by Dickens or Scott. Sometimes their father would read aloud to them and (in his absence) the boys would give much-practised impersonations of his friends, Tyndall and Spencer being specially good subjects. From the surviving diaries of three of the four children one receives an impression of a typical Victorian family, comfortably off and with all the values of the upwardly mobile middle class of that period.

Right at the beginning of a new decade they might have felt that this would not only be a typical day but that 1870 might turn out to be a typical year. In fact it was much more like the years immediately preceding it than those that followed. For the decade beginning in 1870 was going to witness profound and often painful changes to that family, with new and powerful centrifugal forces that would almost blow it apart.

But not just yet. Holidays were still taken together, in Scotland in 1870 and Switzerland in 1871. Dinner parties were given and, as they grew older, the girls were allowed to be involved. If justification were needed for this exercise in hospitality it was provided by one big change that was universally welcome. It had been known that the Royal College of Chemistry would be moving to South Kensington, the other side of London from Haverstock Hill. A change of residence was indicated and after house-hunting for some months they discovered a suitable property at 14, Lancaster Gate, just across Kensington Gardens from the new College site. Maggie describes her first reactions (with more than a hint of Victorian snobbery):

> I went with Papa to see our new house. The locality is delightful & the house is very large; but most frightfully damaged by the ex-tenants who must have been a dreadful vulgar lot. [7]

Percy, on the other hand, sounds just like an estate agent, even to an idiosyncratic use of capital letters:

> On the Basement there are the following rooms: Kitchen, Housekeeper's room, servant's hall [sic], Butler's pantry & the usual offices. On Ground floor, Dining room, breakfast room & library. On the first floor, enormous drawing [room], & a Boudoire . . .

and so on to the fourth floor; he adds: 'Speaking tubes laid on throughout'. [8]

Removal day began inauspiciously with the arrival of 'a wretched covered van with only one miserable horse', but reinforcements appeared and within a few days they were settled in, [9] assisted by the Helm grandparents from Lancashire who stayed a number of weeks. New furniture soon arrived, strolls were taken in Kensington Gardens, concerts enjoyed and lectures attended (though not always understood) at the Royal Institution. Things were getting back to normal. With three servants the family of six had time for such leisures, and there was still plenty of room in the house. [10]

(a) (b) (c) (d)

Fig. 12.1 Frankland's surviving children by his first marriage, in early adult life. (a) Margaret Nannie Frankland (Maggie) (1853–1941), later Margaret West [taken June 1871]. (b) Frederick William Frankland (1854–1916), first Government Actuary for New Zealand [n.d.]. (c) Sophia Jeanette Frankland (Sophie) (1855–1936), later Sophie Colenso [taken November 1873]. (d) Percy Faraday Frankland (1858–1946), first Professor of Chemistry at Birmingham [taken June 1879]. (J. Bucknall Archives.)

Yet now there were signs that the even tenor of their lives could not go on for ever. First was the fact of Frankland's increasingly frequent absences from home. These were occasioned chiefly by the work of the Rivers Commission and later by the demands made by his Presidency of the Chemical Society. Even in the 1870s his range of outside chemical activities was approaching Mancunian proportions. In one year (1878) his consultancies ranged from examination of rennet for a provision merchant in Bath[11] to experiments on crêpe for Courtauld & Co.[12] Then there were always the monthly dinners of the X-Club, scientific meetings to attend and a whole range of other consultancy business. The pace of life was visibly increasing and his temper getting shorter with the mounting pressures. At home the children became aware of it and the sensitive Fred records many rows and fits of passion. One example is probably more than sufficient. In early 1871 Frankland had returned from a trip to the north and was expecting a visit from Herbert Spencer the next day:

> In the evening there was a regular domestic row. Papa had made some alteration in the drawing room furniture by no means improving the look of the room. I was myself struck with the thing immediately on entering the room. It seems that Mamma, Maggy and Sophy had remonstrated with Papa on the subject & that Papa, unaccustomed to that sort of thing, had left the room in one of his passions. He came into the study, where I was sitting, & was followed by Mamma, Maggy & Sophy, who were crying bitterly & humbly asking pardon for an alleged fault, a fault which, even if they had been guilty of it, was far less than that which Papa is daily committing, nay far less than what he committed in the drawing room this very evening in my hearing when I chanced to go upstairs. He was then answering Mamma's friendly remonstrance by openly accusing her of stupidity before her son & daughters. How then can he complain? Nevertheless he was implacable.[13]

Maggie wrote in 1873:

> Really dear Papa you must be overwhelmed with work. Oh how I wish you could give yourself a little rest.[14]

In some of his trips Frankland was accompanied by his wife Sophie, as when he received an Honorary DCL from Oxford University in 1870. But joint ventures away became more difficult as a further shadow was slowly cast over the family. Sophie Frankland was far from well. That in itself compounded the stress caused by her husband's frequent absence and forced additional burdens upon their children, especially the two daughters. She had experienced a series of chest complaints since her early days of marriage. By 1870, despite the usual round of social visits, she was noticed to be taking less exercise than usual on holiday. Then, in November, she confessed to her children that she had coughed a good deal of blood the previous night; it just had to be one on which 'Papa did not come home'.[15] She was a courageous person and continued to battle against what was in fact tuberculosis ('consumption') and for the next few months was out and about in the usual way. However she often complained of weakness and loss of appetite. They had just returned from their holiday in Switzerland in the following September when they noticed her moving extremely slowly while walking in Kensington Gardens, and the following day she collapsed with expectoration of several millilitres of fresh blood. Frankland 'seemed very frightened' and summoned the doctor, who apparently reassured him with the statement that it was 'nothing serious'.[16] Nevertheless the invalid was confined to bed and the household responsibilities fell upon Maggie. Frankland himself was then away for a fortnight at Lancaster. Sophie improved a little and was able to leave the house in a cab or bath-chair. By August 1872 she was sufficiently unwell for physicians to be called. Again Frankland was away, this time in Norway. One afternoon she read some of his old letters to her, one of which she burnt.[17]

On Frankland's eventual return it was decided that convalescence in the Isle of Wight was indicated and so she, and her daughters, were despatched to Ventnor in September. Fred was already on the Island, having been sent by his father when the boy began to display symptoms of nervous breakdown. For the next couple of months the three oldest children took turns to be with their mother on the Island, a situation that might have continued indefinitely had not Frankland lobbed a metaphorical bombshell in their direction. He wrote proposing that he let their house for six months and urged Sophie to remain in Ventnor. It is not clear why he should do so but, despite her almost permanent 'low spirits', Sophie was sufficiently stung to send vigorous protests back on three successive days with the result that the project was abandoned and all returned home in time for Christmas.

The following year (1873) proved to be her last. There was no real progress in her condition so, next summer, a change of climate was again suggested, and on 28 August they set off for Switzerland.

Their destination (uncertain when they started) proved to be the high altitude resort of Davos. Now a fashionable meeting place for European politicians it was then a tiny cluster of buildings, to some of which convalescent and bronchitic patients were coming in increasing numbers. They arrived there on 6 September, and after a sorrowful parting from her husband Sophie was left in the care of the devoted Maggie. For the next two months the patient appeared to benefit from the clear air and unfiltered sunshine, but recovery was far from certain. Maggie's *Diary* was uncharacteristically brief, with many entries 'Don't remember'. In a moving series of letters Maggie and her mother maintained contact with the family in England and these documents have fortunately survived. They have their own story to tell about Victorian attitudes to illness and death, about religious values in adversity, about the strength of family bonds, about social rituals of convalescence and about contemporary medical attitudes to tubercular disease.[18]

On 20 December Frankland arrived for Christmas, by sledge. He seemed strangely ill at ease, for his daughter recalls his concentration on thermometric observations and tells little of his personal relationships. On 5 January he came to take his leave of Sophie:

> When Papa kissed her she said 'Lebewohl, lebewohl, lebewohl,' with short intervals in between and in such a voice as if she knew it was the last time on this earth that she would behold him.[19]

Three days later Sophie Frankland died.

In all the surviving documents Frankland's German wife rarely appears except as a figure in the background. Her brother Heinrich remarked that Sophie 'in her inborn great modesty was valued as the willing Cinderella of the family, and it turned out that she was not inferior in gifts'[20] to her artistically talented twin sister Marie. In England she is first seen as a stranger in a foreign land trying hard to adapt, then as a woman who made great efforts to accommodate herself to her husband's driving ambition, and finally as a devoted wife and mother. Victorian mores alone cannot account for her insubstantial appearance; after all her contemporaries Mrs Busk or Lady Lubbock showed none of the timid reticence displayed by Sophie Frankland. The explanation is much more likely to be found in her perennial ill-health, her sense of being an alien and perhaps the dominating personality of her husband. Yet when that is said she retained a great deal of intellectual independence, did not share in his antipathy to church-going, and on several occasions stood up to him and won the argument. Her children adored her, and when Frankland took Maggie to see her grave many years later he simply said 'She was a good mother to you', adding with peculiar emphasis 'she *loved* me'.[21] He spoke the truth.

The household at Lancaster Gate resumed its functions, though now domestic arrangements were in the hands of Maggie or her grandmother Margaret Helm who spent many months of 1874 in the Frankland home. The niceties of mourning were observed and Frankland with at least some of his children once more attended Dr Sadler's chapel in Rosslyn Hill. Gradually the usual round of visiting was resumed and Frankland, never at his best in times of emotional stress, threw himself again into his work.

His water analysis laboratory was now receiving a steady stream of orders (Chapter 14). Amongst the enquiries was one from a London barrister-at-law, Charles E. Grenside, regarding the water supply to his house at Wimbledon. No less than three analyses were undertaken in the first few months of 1874, and appropriate advice was no doubt given.[22] In fact the Grenside family was to play a prominent part in the future fortunes of the Franklands, but not until a remarkable encounter in the Alps.

That summer Frankland and the family took an extended holiday in Switzerland. It was in part a pilgrimage to Sophie's grave in Zurich and to visit her relatives at Cassel and other places *en route*. While Fred, Maggie and Percy remained at Zurich, Frankland took his younger daughter Sophie on a fortnight's excursion to Pontressina in the Italian Alps. There they encountered members of the Grenside family, including Charles' wife Ellen and their 26-year old daughter of the same name. This is how Ellen Grenside snr. describes the encounter:

> There has also arrived here Professor Frankland, of the Royal College of Chemistry, whom I had been corresponding with from Wimbledon about our well water although I had never seen him. He is a most clever and delightful man, and has kindly given us a great deal of good advice about it. He is a genial companion. On finding that Ellen was most anxious to explore a glacier, he determined that she should do it, & has got up a party to go, with a very experienced guide. . . Dr. Frankland had made us get strong alpine sticks, & have large projecting hob nails put into our boots. We had to jump over crevasses & walk on narrow ledges, & peep down almost bottomless holes of deep blue ice, with water swirling below, – Dr. F could explain to us all the wonders that

we saw, and while walking on the upper crust told us that we had a depth of 300 ft. of snow beneath our feet. . . . He gave us a rest of ten minutes to eat something, and supplied wine to us all round. After seeing all these wonders, I thought we must turn back & Charles came with us, & our second guide, but Ellen went on still further with the rest of the party, where they climbed up what is called the 'ice fall' where it descends in stupendous masses from the top of the mountain. The guide had to carve every step in the ice, and hand them up on to a narrow ledge, but Dr. F was very careful Ellen said & if a gentleman put out his hand to help, he would not let him but made the guide take her.[23]

Whether this meeting with the young Ellen was entirely un-premeditated must remain in doubt. Quite possibly it was a chance encounter. But for the widowed Frankland it marked the beginning of a relationship that rapidly developed into a courtship. At first the family were unaware of anything unusual. When Frankland took Maggie one Sunday to visit the Grensides at Wimbledon they were just 'a family whom Papa met in Pontressina'. Only the parents were there, but the visit was for 'some hours'.[24] However a fortnight later 'Mr., Mrs., and Miss Grenside came to an early dinner'.[25] The two families met at a photographic exhibition in Pall Mall (for which Frankland got them all tickets)[26] and at a concert at the Royal Albert Hall.[27] At this last event Maggie's *Diary* suddenly relapses into German, a phenomenon apparently associated with periods of secret grief. Had she realised that something was afoot? She had already dutifully accompanied her father on two further Sunday visits to Wimbledon and may well have had suspicions. [28]

With the coming of 1875 Frankland was often in the company of Ellen Grenside, and on 24 February his proposal of marriage was accepted; he had unsuccessfully tried twice before in the last three months.[29] The following evening he took his daughters to see Henry Irving as Hamlet, and then made his announcement. Maggie confided her reactions to her *Diary*, in German:

> When we got home Papa told us he had been engaged since the previous evening. The dear Lord gave me strength to bear it. Afterwards much was said and we did not get to bed until 1.30.[30]

She was still writing in German when the wedding took place in May 1875:

> We travelled to Wimbledon in two coaches, each drawn by two horses, Dr. Tyndall and Papa in one coach and Sophie, Percy and I in the other. We found Mrs. G. in a very sad mood and Mr. G. had been very

Fig. 12.2 Edward Frankland, 6 February 1874. (J. Bucknall Archives.)

Fig. 12.3 Ellen Grenside (1848–99), second wife of Edward Frankland.
(J. Bucknall Archives.)

ill and looked very bad. Ellen soon appeared in her newest suit and looked charming. She also had on a long white veil and wore petals in her hair and on her dress. The five bridesmaids (Emily, Miss Bowen, Miss Toynbee, Sophie and I) all had on light blue clothes and large white hats with blue trimmings. Only the close relatives of the Grensides were present. The church ceremony was very strange (for me). I floated as if in a dream. At breakfast there were only a few toasts. Afterwards Papa and Ellen travelled to the Isle of Wight and we went back at once with Dr. Tyndall to London. Our Grandparents shortly arrived to stay with us during Papa's absence.[31]

Frankland returned to Lancaster Gate with his bride three weeks later. Among the many letters of congratulation was one from his old Lancaster friend James Johnson.[32] Maggie and Sophie went to friends in the country, the former confessing :'right glad was I to get away for rest, peace & quiet; for we were perfectly worn out with work, worry and excitement'.[33] With these and other comments from the daughters it is clear that Frankland's second marriage looked as though it would disintegrate the family from the start. The sisters spent three weeks with the widow of Dean Dawes at Hereford, and then (after a brief return to London) nearly six months on the continent with their German relatives.

Meanwhile the family was already reeling under another blow that had been struck since Sophie's death. The absence of Fred Frankland from his father's wedding arose from no trivial mischance. He was at that time on his way to New Zealand. To the great distress of his sisters it was decided (by his father) that it was expedient for his health to live for a time in a better climate; conceivably Charles Grenside (who had travelled to Australia) was behind this particular suggestion. The compliant Fred agreed and left England on 25 January.

He was, by all accounts, a lad of unusual mental gifts, brilliant at mathematics and hard-working in all other subjects. He was sent to University College School, University College itself and then spent a year or two at the Royal College of Chemistry. He tended towards absent-mindedness and on occasion roused his father to impotent wrath. He would forget to deliver letters or to give messages; he would misunderstand his father's instructions and fail to meet his demands; he returned from a health trip to Leyland without permission; once he inadvertently locked his father in his study. Apart from this somewhat engaging vagueness his trouble was twofold. He had a sensitive nature that made him specially aware of other people's criticism, and he was a beaver for work and seemed unable to know when to stop. This is not a good combination of personal traits if your father happens to be Edward Frankland. Ambitious for his sons as well as for himself he frequently lamented the less than brilliant performance of his elder son. Fred's response was to work even harder, doubtless with paternal encouragement. The result was inevitable. 'Breakdown' is too vague a word to describe his condition, but he certainly went through a succession of mental crises where weariness and exhaustion compounded the problem. It would be probably unfair to suggest that a 'failed' scientist for a son would have been an intolerable embarrassment to Edward Frankland, but a spell abroad might well solve more problems than those of Fred's health.

He had to go. Kitted up by his father and provided with appropriate notes from the bank, he packed his bags and prepared for the long sea voyage to New Zealand. His *Diary* for the day of his departure is a revelation:

> Monday Jan. 25th. Was awoke [*sic*] by Rose about 7 o'clock. Came downstairs before any of the others. Early breakfast. Nothing of great importance discussed or talked over. A few hearty kisses (I ran upstairs to say goodbye to Papa) and then I was out of the house in the pouring rain. I ran as hard as I could to Praed Street, but just missed a train to Charing Cross. I reached Charing Cross just an instant too late for the train to Gravesend . . .[34]

Just that. Not a person to accompany him to the station, not a well-wisher on the quay, not a hired cab in torrential rain, not even a parent to join him at breakfast. Was the family really so uncaring? In fact the farewells had been made two days earlier – a kind of dress rehearsal – when the ship sailed from London to Gravesend, and Fred had returned home from there to spend his last night at home. Maggie recalled 'the parting this time, curiously enough, was not nearly so hard . . . but God grant not for very long'.[35] Fortunately she did not know that it was to be 13 more years before they would meet. It is hard to see this draconian action in any other light than that of a father who could not cope with a prolonged drain on his emotional energies and whose instinct was to dispose of the difficulty by whatever means was available.

Fred Frankland was to settle in New Zealand. In 1879 he married Miriam Symons, the first of his generation to marry. His mathematical skills were exercised in the Government Insurance

Service and he became Government Actuary and Statist. A friend and neighbour of the Prime Minister, Sir Harry Atkinson, he was largely instrumental in setting up the country's first National Insurance Scheme. He also wrote much on philosophy and religion; having been 'converted to Christianity in 1870' he did not share the scepticism of his father. But after 1875 he fades out of the Edward Frankland story, though he did maintain a correspondence and a surprisingly good – if distant – relationship with his father throughout his life. He died in 1916.[36]

With Sophie's death, Edward's remarriage and Fred's enforced exile it might be imagined that the troubles of the Frankland family were at an end. At the end of the eventful year of 1875 Maggie's reflections were more cheerful than might have been expected:

> How can I be thankful enough for the good fortune which has befallen us in the past year! It began with a difficult test, namely the departure of our dear brother Fred to New Zealand. God permitting, he will soon return to us, happy and well. The second test was our father's engagement and wedding. But this has turned out so well, and the Lord has been so gracious to us. The past year has brought me many lovely things. Our stay in Hereford with the good angels Miss Guthrie and Mrs. Dawes was most impressive. This three week break contributed greatly to my reconciliation with England, which I feel has helped me to bear the wedding of my dear father much more easily.[37]

After their long sojourn on the Continent, the sisters returned to England with their German cousins, Nanni and Luislie, daughters of Heinrich Fick and spent much of the spring of 1876 taking them to see relatives and friends. It was in April that they all met Darwin. On May 26 Sophie and Maggie were to have 'our dance' at home, perhaps a genuine gesture of friendship from Ellen to her stepdaughters, who were only a little younger than she was. Many friends were there and it was declared 'a great success'.[38]

Amongst the guests was a Mr Frank Colenso. Younger son of the controversial bishop, he had graduated in mathematics at Cambridge and was now in London to plead on behalf of an oppressed Zulu tribe led by Chief Langalibalele, and also to read for his law examination, having rooms overlooking Temple Gardens. J. W. Colenso, Bishop of Natal, had published the first part of *The Pentateuch critically examined* in 1862, and its unorthodox conclusions led to an end of his friendship with F. D. Maurice and to attempts by the English bishops to depose him, though an appeal by Colenso to the Privy Council was upheld three years later.[39] In the heat of controversy he was, as we have seen, the first guest to be welcomed to the X–Club. He and his family were also befriended by Charles and Mary Lyell who opened their home to them in Harley Street, and by Mary's sister Katherine who had married Charles Lyell's brother Henry. Katherine Lyell, a botanist and friend of Hooker, maintained a correspondence with Colenso's wife Frances for 30 years.[40]

Frank Colenso was no stranger to the Frankland household, probably having been introduced by the Lyells. Three weeks after the dance he wrote with consummate artlessness to Frankland 'asking whether he might see more of Sophie with a view to asking her to share his home in Natal'. Her sister recalls with fine restraint: 'Papa had a talk with Sophie after which I was told and then Ellen spoke with us. It was one of the most exciting mornings I have ever lived through.'[41] The engagement followed rapidly, on 2 July.[42]

All went well with the betrothed couple until November when Frank learned that he had failed his law examinations. These were passed shortly afterwards, he was called to the Bar and at the end of January, 1877, Frank displayed his gown and wig to the family before taking leave of Sophie as he returned to Natal. Despite his engagement he had decided to take up a legal career in South Africa and for three years practised as an advocate and attorney at the Supreme Court of Natal.

Understandably his departure put the relationship with Sophie under great strain, and with it created further problems in the Frankland household, often coming up in conversation. Letters to Frankland from both Frank and his father at the end of 1878[43] did little to mend matters and early in 1879 Sophie broke off the engagement.[44] Then, in October the hapless Colenso turned up unannounced on their doorstep and wanted to resume the relationship. 'It was a terribly painful day'.[45] Thereafter he was frequently to be noticed in the vicinity and Sophie was inclined to offer some encouragement. Frankland was adamantly against it, and there were 'disastrous and painful scenes with Papa and Ellen which made us dreadfully unhappy'.[46] The sisters would often see Frank at church, where Frankland forbade them to speak to him, and verbal violence at home escalated to such proportions that Maggie exclaimed 'Oh when will these scenes cease?'[47] Her rhetorical question was not answered until the following May when Frank, abandoning his promising legal career in Natal,[48] came to

(a)   (b)

Fig. 12.4 (a) Dorothea and (b) Helga, daughters of Ellen and Edward
Frankland [taken August 1880]. (J. Bucknall Archives.)

London with the offer of a job at £400 p.a. Frankland demanded a
week to reflect on 'whether he can consent to Sophie's marrying on
so small an income'.[49] By 23 May 1880 it was settled, a re-engagement
announced and on November 13 they were married. For Frank a
job at the Norwich Union Insurance Company opened up far better
prospects than before, and the couple soon moved to Norwich.[50]

During all the controversy over Sophie's engagement tensions
at home were increasing. To be sure there was the usual round of
social activity and Maggie and her sister were now doing voluntary
work at a local hospital, together with various kinds of teaching
activity. Meanwhile Frankland was busy starting another family,
and two more daughters were born, Dorothy in 1876 and Helga in
1878. Judging from Maggie's *Diary* (which obviously had its own
particular bias) further tensions arose over matters of religion.
Despite their father's views the children had all abandoned
Unitarianism for more orthodox Christianity, and in 1877 Maggie
was confirmed on her 24th birthday. She described the experience
as 'a turning point in my life' and was accompanied by her

brother and sister. Neither Frankland nor Ellen felt it necessary to be present.[51]

Maggie, too, had her taste of romance. A young doctor, Samuel West, was a friend of the Grensides, and Frankland had met him at Pontressina. Several times since then he had encountered Maggie at Wimbledon or at her own home. He was clearly attracted by the charming and intelligent young lady, and in the middle of 1877 proposed to her – and was refused. A note in her own hand over 40 years later says that she was then 'foolish enough to be in love with someone else'.[52] Such is the course of young love, but Sophie's problems seem genuinely to have taken up much of her sister's thought. The prospect of Sophie's marriage was the last straw for Maggie and she resolved to continue no longer under her father's roof. After one family row Maggie told Ellen of her intentions:

> I told her I must live with Percy as I could not contemplate living at home without either Percy or Sophie. God grant that I may be allowed to leave my home. It has been a very unhappy one in itself; but the love of us brothers & sisters for each other has been strengthened & made more full of comfort through the trials we have passed through.[53]

However within days her plans were modified and she looked forward to joining Sophie in her new home, perhaps with Percy as well. It seems that she was heading for a mental crisis as she also records (unusually for her) great weakness and malaise. In the event she weathered the storm, though after the wedding of her sister was clearly in great distress. Her moment of liberation came at Norwich when they were all together at the Colenso home: 'Frank put the plan of my living at Norwich before Papa who consented with very little demur'.[54] This did not actually happen, and within the year she was married to Samuel West;[55] it is pleasant to record that Ellen and Edward were 'very kind about everything'.[56]

The one remaining member of the first family unit was Percy. The laconic entries in his *Diary* contrast amusingly with the prolix compositions of Fred and Maggie. Yet his path through the 1870s was far from untroubled. Like Fred he went to University College School, and in 1875 he entered the Royal School of Mines where his brilliant performance secured him the offer of a Brackenbury scholarship of £130 to study medicine at St. Bartholomew's Hospital. However his father had other ideas. On a cliff-top walk in Norfolk Percy confessed to Maggie the reason for his 'low spirits':

> Papa had unsettled his mind & determination by trying to persuade
> him to study for a chemist [*sic*], a subject which Percy dislikes. He
> seems decidedly inclined for the medical profession.[57]

The issue was resolved by *force majeure*. Next October his birthday
present from his father was a copy of the latter's *Experimental
Researches*, and a gift of £130, precisely the amount of the
Scholarship.[58] Twelve days later he moved to Würzburg to study
organic chemistry under his father's old friend and distant relative
by marriage, Johannes Wislicenus. With two brothers abroad and
plenty of problems of their own the growing sense of alienation by
Sophie and Maggie can be well understood.

Percy was of course home during the summer, and in 1879 he
made sure that he spent a good deal of that time with a friend of his
sisters and of the Grensides, Grace Toynbee. The young lady in
question was daughter of Joseph Toynbee (1815–66), the first ear,
nose and throat surgeon in London and a notable resident of
Wimbledon.[59] Her brothers included Paget Toynbee (a Dante
scholar) and Arnold Toynbee (the social philosopher). Three
months after Percy's triumphant return from Würzburg with his
PhD he announced his engagement to Grace amidst much rejoicing.[60]

It is interesting to notice Frankland's own view of marriage at
this time. It emerges in the unlikely context of a correspondence
between himself and his son Fred. The latter seemed to suffer from
some kind of sexual incapacity and, from New Zealand, sought
advice from his father. Apparently Frankland had told his son that
'marriage is one of the best things worth living for' and 'the only
way of satisfying the sexual passion which is not open to the gravest
objections'. He regarded the complete isolation of couples (i.e.
monogamy) as indispensable to the eradication of the Victorian
scourge of syphilis. A further aspect of Frankland's sexual philosophy
was 'the limitation of families according to the means of the
parents'.[61] Like his father, Fred found it impossible to agree with
Darwin's opposition to birth control.[62] It seems that religious
attitudes to such matters remained, for Frankland may have
provided an additional reason for rejecting what he thought was
orthodox Christianity. As we have seen (p. 334) he vehemently
disputed (1) that marriages are made in heaven, and (2) that
Providence regulates the number of children.[63] By that time his
own marriage to Ellen had generated new problems, but his
contempt for 'religious' objections to family limitation was unchanged.

Soon the troubled decade of the 1870s was behind the Frankland family. All the children were married or about to become so (Percy's wedding was in June 1882[64]). There were still ominous clouds on the horizon as the health of the Helm family deteriorated. And before too long further family squabbles would arise on a scale and with a ferocity that was without precedent. But all that was unknown. The old house at Lancaster Gate had fulfilled its function and a new residence was in sight. For Edward and Ellen Frankland it was time for a fresh start.

Notes

1 Percy Frankland, *Diary*, entry for 18 January 1870 [RFA OU mf 01.03.1130].
2 Manuel García (1805–1906) belonged to a long line of distinguished musicians.
3 Frankland to Maggie Frankland, n.d. [JBA, OU mf 02.01.0183]; he signed his letter 'Ever your devoted Trovatore'.
4 Frankland to Maggie Frankland, 7 June 1869 [JBA, OU mf 02.01.0185].
5 García was sometimes a guest at dinner (Maggie Frankland, *Diary*, 28 May 1876, and 17 March 1877 [MBA]), was presented with a chemistry book by Frankland, probably his *Experimental Researches* (M. García to Frankland, n.d. [JBA, OU mf 02.02.1546]), and acknowledged Maggie's congratulations on his 99th birthday (M. García to Maggie West, 9 March 1904 [JBA, OU mf 02.02.1545]).
6 Maggie Frankland, *Diary*, entry for 18 January 1870 (note 5).
7 *Ibid.*, 12 February 1870.
8 Percy Frankland, *Diary*, entry for 21 March 1870 (note 1).
9 Maggie Frankland, *Diary*, entry for 11 – 19 April 1870 (note 5).
10 1871 Census [PRO, RG 10/25].
11 W. Titley & Sons to Frankland, 30 March and 11 April 1878 [RFA, OU mf 01.03.0810, 0812].
12 Frankland to W. Davison (S. Courtauld & Co.) 18 and 22 October, 17 November 1877, 26 January 1878; accounts, Frankland to S. Courtauld & Co., 6 and 8 February 1878 [Essex County Record Office, D/F 3/2/105].
13 Fred Frankland, *Diary*, entry for 18 February 1871 [Alexander Turnbull Library, Wellington, New Zealand, MS. 801].
14 Maggie Frankland to family, 5 December 1873 [PCA, OU mf 03.01.0460].
15 Maggie Frankland, *Diary*, entry for 24/25 November 1870 (note 5).
16 *Ibid.*, 22 September 1871.
17 *Ibid.*, 15 August 1872.
18 Over 50 letters have survived from mother or daughter to the family back in England during this period [PCA, OU mf 03.01.0227 on].
19 Maggie Frankland, *Diary*, entry for 5 January 1874 (note 5).
20 H. Fick, in Hélène Fick, geb. Ihlee, *Heinrich Fick, Ein Lebensbild*, Zurich, 1897, p.62.

21 Maggie West (*née* Frankland), *Diary*, entry for 7 July 1897 (note 5).
22 Frankland, Water Analysis Notebook no.1, entries 4063, 4074 and 4095 (12 February to 14 May 1874) [RFA, OU mf 01.09].
23 E. Grenside to F. Grenside, 9 September 1874 [RFA, OU mf 01.05.0051].
24 Maggie Frankland, *Diary*, entry for 11 October 1874 (note 5).
25 *Ibid.*, 25 October 1874.
26 *Ibid.*, 31 October, 1874.
27 *Ibid.*, 28 November 1874.
28 *Ibid.*, 8 and 22 November 1874.
29 *Ibid.*, 24 February 1875.
30 *Ibid.*, 25 February 1875.
31 *Ibid.*, 11 May 1875.
32 It was rather belated: J. Johnson to Frankland, 3 August [1875] [RFA, OU mf 01.03.0727].
33 Maggie Frankland, *Diary*, entry for 11 June 1875 (note 5).
34 Fred Frankland, *Diary*, entry for 25 January 1875 (in bound volume of memoranda) [AFA].
35 Maggie Frankland, *Diary*, entry for 25 January 1875 (note 5).
36 On Fred Frankland see G. L. Scholefield (ed.), *A dictionary of New Zealand biography*, Dept. of Internal Affairs, Wellington, 1940, vol.1, pp.279–80; R. W. W., *Trans. Actuarial Soc. America*, 1916, **18**, 382–3.
37 Maggie Frankland, undated note at end of *Diary* for 1875 (note 5).
38 *Ibid.*, 26 May 1876.
39 P. O. G. White, 'The Colenso controversy', *Theology*, 1962, **65**, 402–8.
40 W. Rees (ed.), *Colenso letters from Natal*, Shuter & Shuter, Pietermaritzburg, 1958, pp.86–7; both Lyell wives were daughters of Leonard Horner.
41 Maggie Frankland, *Diary*, entry for 14 June 1875 (note 5).
42 *Ibid.*, 2 July 1875.
43 *Ibid.*, 26 December 1878; the letters have not apparently survived and their contents are unknown.
44 *Ibid.*, 6 February 1879.
45 *Ibid.*, 13 October 1879.
46 *Ibid.*, 17 November 1879.
47 *Ibid.*, 21 December 1879.
48 His mother considered he 'would probably have been a judge before long': F. S. Colenso to K. Lyell, 8 February 1880, in Rees, p. 348 (note 40).
49 Maggie Frankland, *Diary*, entry for 7/8 May 1880 (note 5).
50 'Insurance celebrities', no. 28, *Insurance Sun*, 1887, **28**, 49–50.
51 Maggie Frankland, *Diary*, entry for 25 March 1877 (note 5).
52 *Ibid.*, 27/28 June 1877.
53 *Ibid.*, 5 July 1880.
54 *Ibid.*, 6 February 1881.
55 *Ibid.*, 22 December 1881.
56 *Ibid.*, 7 October 1881.
57 *Ibid.*, 28 September 1877.
58 *Ibid.*, 3 October 1878.

59 He is commemorated today in Wimbledon by a drinking-fountain in the High Street and by the Village Institute: 'M. G.', *Viewpoints in and around Wimbledon*, F. Cole, Wimbledon, 1926, pp.21, 31–2, 40.

60 Maggie Frankland, *Diary*, entry for 2 January 1881 (note 5).

61 Fred Frankland to Frankland, 20 September 1878 [RFA, OU mf 01.02.1114].

62 Fred Frankland to Frankland, 22 February 1879 [RFA, OU mf 01.02.1126].

63 *Sketches*, 1st ed., 1901, p.57. The passage was excised from the 2nd edition.

64 Maggie Frankland, *Diary*, entry for 17 June 1882 (note 5) .

CHAPTER 13

The analysis of water supply

1 Something wrong with the water?

When Frankland assumed the mantle of Hofmann in 1865 and took a temporary appointment at the Royal College of Chemistry, he received not only his predecessor's position as examiner for chemistry to the Department of Science and Art but also his job as official analyst for the London water supply, accountable to the Registrar General of Births, Marriages and Deaths. It was to be a momentous appointment. For the country it meant the vigorous prosecution of a sustained campaign for clean drinking water, and for Frankland a commitment that was to last until the year of his death, with engagement in controversies that would make him fervent friends and bitter enemies. Above all it would bring him into the public arena as never before and in a way that neither fundamental research nor popular exposition could begin to achieve. An obituarist was to describe him as 'the greatest living authority on water supply'.[1]

The importance of domestic water supply had increased far beyond anything known before the nineteenth century by the associated processes of population growth and urbanisation. It is neatly illustrated by the case of London.[2] Between 1800 and 1900 the country as a whole experienced an increase in population by a factor of 3, serious enough in itself, but the population of London rose twice as fast, from 1.1m to 6.6m. Other cities, especially in the industrialised north, showed comparable growth-rates, though their geographical expansion did not nearly keep pace with

Table 13.1. *Companies providing London's water supply*

Company	Date of origin	Source of water supply	Location relative to the R. Thames
New River	1619	R. Lea (Essex)	north
Chelsea	1723	R. Thames	north
Lambeth	1785	R. Thames	*south*
W. Middlesex	1806	R. Thames	north
E. London	1807	R. Lea (Essex)	north
Kent	1809	Wells in chalk hills near Crayford	*south*
Grand Junction	1811	R. Thames	north
Southwark	1834	R. Thames	*south*

demographic growth. There was a vast increase in numbers of slum dwellings and in some areas the population density reached unprecedented levels, with people living, almost literally, on top of each other.

In addition to all the attendant discomfort and squalor there were major problems of water supply. This may have been from shallow wells in the vicinity or from standpipes or domestic systems drawing water from a local stream or river. The latter were often perfectly adequate in rural areas. However in large towns, and in the neighbourhood of heavy industry, the water was usually so polluted as to be undrinkable. The case of London was not perhaps the most extreme but it was certainly the one that attracted the greatest notoriety when things went wrong. The capital was supplied by eight privately-owned companies, all but one serving river water to its customers:

A contemporary description of the effluents feeding the Thames in the nineteenth century left little to the imagination. There was a maze of underground tunnels conveying 31 billion gallons per annum to the river in streams 2 to 5 feet high:

> Through these secret channels rolled the refuse of London, in a black, murky flood, here and there changing its temperature and its colour, as chemical dye-works, sugar-bakers, tallow-melters, and slaughterers added their tributary streams to this pestiferous rolling river.[3]

Offensive to the eye and to the nose these effluents became the subjects of much complaint. In 1827 public indignation was raised by the quality of water taken in by one supplier, the Grand Junction

Company, at a point on the Thames known as 'The Dolphin', immediately adjacent to the outlet of a large and ancient sewer, the Ranelagh. A pamphlet war broke out and the Government (not for the last time) sought a quick technical 'fix' by appointing a Royal Commission, this one on 'Water supply'. It arrived at no clear conclusions because the real nature of the problem remained obscure. In the same year (1828) sand filters were introduced by the Chelsea Water Company to improve their product's appearance. Another thirty years were to elapse before another outcry based chiefly on aesthetic considerations. In 1858 Michael Faraday took a boat down the Thames at a time when the river was enveloped in a foul-smelling miasma and was itself so polluted that the bottom half of a card dropped in edgeways was invisible before the top had entered the water. Again there was public fury, in the columns of *The Times* and elsewhere, against London's 'Great Stink'.

During those 30 years another and more sinister element had entered into the debate. In 1828 there had been a general belief that foul water was unhealthy as well as unpleasant, and medical men tended towards a vague association of water pollution and disease. Thinking on such matters was greatly concentrated by a new and deadly phenomenon that appeared in the British Isles three years later. In 1830 an apparently new disease struck the citizens of Moscow, virulent and incurable. Fears that it might be carried by ship were realised when it claimed a victim in Sunderland in October 1831, after the docking of a ship from Riga. Cholera had arrived in Britain. Thereafter it spread rapidly and by the following June reached the Manchester area, where in nearby Salford lived a young mother from Garstang and her six-year old son. His name was Edward Frankland. The family prudently fled the plague and returned home to rural north Lancashire.

By early 1833 the epidemic was over, having claimed 21 882 lives in England alone.[4] Its precise cause, however, was uncertain. The remedies adopted by each citizen of Salford illustrate the prevailing folklore: 'a Burgundy pitch plaister on his chest and a little bag of camphor round his neck'.[5] Then, in 1848/9, the disease struck again and claimed 62 000 lives, the worst cholera epidemic in British history. Indignation was now fuelled by fear as well as disgust. Two reports to the General Board of Health by the sanitarian reformer Edwin Chadwick were quickly followed by a study by A. H. Hassall on *Microscopical examination of the water supplied to the inhabitants of London and suburban districts*. The latter came replete with horrifying illustrations of the micro-organisms that had revealed themselves to Hassall's microscope. Their precise epidemiological rôle was still unclear but they served as powerful tokens of the crucial fact discovered by a London anaesthetist, John Snow, in the previous year: that water supplies were a cause of cholera. Within the years 1849–52 there was fierce debate as to the politics of water supply (should it be a public monopoly?), the precise rôle (if any) of the millions of microbes swarming in London's water, and the significance of dissolved and suspended chemicals. With this last point, of course, went a critique of the part that chemical analysis should play in any evaluations of public water supply. Baffled by the chorus of conflicting voices the Home Secretary commissioned a chemical report from Hofmann, Graham and Miller in 1851 which, though sanctioning the existing *status quo*, was quite indefinite in relating the detailed analytical results to what might be termed 'desirable' water.[6] Meanwhile the two diseases of typhoid and typhus, not differentiated until 1869, were adding their toll of victims. By 1865, when Frankland was called upon to address the problem, public concern had been roused by a whole series of events (Table 13.2).

The cholera outbreak of 1848/9 produced several knee-jerk reactions from Government and water companies. Those that drew water from the Thames were required to remove their intakes upstream of the outlets of the London sewers (though there remained the problem of 800 000 people discharging sewage into the river further upstream). A Metropolis Water Act (1852) made filtration mandatory and also the installation of covered reservoirs. Lack of understanding of the precise value of filtration, other than aesthetic, was combined with a vagueness of specification that left engineers such wide latitudes for action that a Report from the General Board of Health could draw attention merely to the wide diversity of engineering technology and water quality from the eight companies. The time was ripe for concerted action based upon reliable scientific data. The chemists were to come into their own. At such a moment Edward Frankland entered the fray.

2 Appointment as official analyst

The appointment of Edward Frankland to the post of official analyst to the London water supply raises two important questions:

Table 13.2. *Water supplies and public alarm [courtesy Royal Society of Chemistry (from note 8(a)).]*

1828	'Dolphin' scandal
1829	
1830	
1831	CCCCCCCCCCCCCCCCCCCCCCCCCCCCCCCC
1832	
1833	
1834	
1835	
1836	
1837	TTTTTTTTTTTTTTTTTTTT
1838	
1839	
1840	
1841	
1842	Chadwick's *Sanitary Condition of the Labouring Population of Great Britain*
1843	
1844	
1845	
1846	
1847	TTTTTTTTTTTTTTTTTT
1848	CCCCCCCCCCCCCCCCCCCCCCCCCCCCCCCCCCCCCCCCCCCCCCCCCCCCCCCCCCCCCCCCCC
1849	
1850	Hassall's *Microscopical examination* [of London water]
1851	Hofmann's 'Chemical Report' on London water
1852	
1853	CCCCCCCCCCCCCCCCCCCCCCCCCCCCCC
1854	
1855	TTTTTTTTTTTTTTTTTTTTTT
1856	
1857	
1858	London's 'Great Stink'
1859	
1860	
1861	
1862	C = 1000 cholera deaths in Great Britain
1863	T = 1000 typhoid/typhus deaths in England & Wales
1864	
1865	
1866	CCCCC
1867	
1868	TTTT

why should a chemical appointment be deemed so important, and why should it be Edward Frankland?[7]

First, it may be far from obvious as to why chemistry was so relevant to a question that seems rather to be a matter for medicine in general and epidemiology in particular.[8] This is given extra poignancy by virtue of the problematic relationship between the results of chemical analysis and the aetiology of disease. For most of the nineteenth century there was no universal agreement as to the medical significance of certain chemicals in water. Yet the fact was that the public increasingly looked to the chemist rather than the doctor for guidance about the use of water supplies. Where medical men were involved in public discussion it was often because of their 'sideline' engagement in chemical analysis.

One explanation lies in the long history of water analysis by chemists. Frankland was heir to a tradition that stretched back to before the eighteenth century and included analysis of natural waters, whether from spas, wells or streams.[9] By the mid-eighteenth century crude titrimetric methods had been developed for determining the acidity or alkalinity of natural waters[10] and water analysis techniques formed the subject of a book in 1799 by Kirwan[11] and featured significantly in the first comprehensive textbook on analysis by Pfaff in 1821.[12] There were sound economic reasons for this interest, not least the need to determine 'hardness' of water, for a high concentration of calcium and magnesium salts was wasteful of soap when washing and their bicarbonates had the additional hazard of blocking steam-pipes by depositing insoluble carbonates.

Partly through the influence of Liebig, chemical analysis in general became the subject of numerous courses in Europe, not least in Britain. Beginning perhaps with the work of Frederick Accum, Richard Phillips and others in the early 1800s, it has a continuous history in London up to and beyond the establishment of the Royal College of Chemistry in 1845.[13] Gradually its usage spread in industry, especially in application to natural waters. As early as 1846 water analysis was playing a prominent part in the Museum of Economic Geology (where Phillips worked).[14] Elsewhere various consultants appeared, as Thomas Richardson of Newcastle who, in 1844, announced himself as 'assayer and professional chemist'[15]; his services included:

Qualitative examination of water   10s 6d to 21s
Quantitative examination of water   £2/2s to £5/5s

Nor was Richardson the sole purveyor of such services, even in Newcastle. In brewing and in agriculture water analysis was being taken seriously, as it most certainly was on the railways where hard water in locomotive boilers could have disastrous effects. So by 1860 the same Thomas Richardson is analysing the water of 'Lartington Pond' with a view to employing it as a source for the Stockton and Darlington Railway's extension to Tebay.[16]

There were, however, other reasons. If, as we have seen, science in mid-Victorian Britain was virtually equateable with chemistry, then chemistry had the image and authority of science as a whole. A scientific fix therefore meant a chemical fix. This authority depended as much upon the manifest usefulness of chemistry in the Industrial Revolution as upon its fashionable image associated with, above all, Faraday and Davy. And with this burgeoning of chemical activity in general came a pressure for chemists to secure in practice what Richardson had claimed for himself, the status of a 'professional'. They had to wait until 1877 before the movement for professionalization came to a head, but long before that there were territorial claims against both engineers and medical men. The very existence of disputes about who should analyse water (or anything else) gave coherence to the chemists' collective strategy to stand firm in their claim to unique rights to perform chemical operations.

All this is merely background to a rising tide of chemical employment in connection with urban water supplies. Nowhere was this more obvious than in the efforts of the individual water companies to establish the virtues of their product. How else but by chemical analyses? So they began to employ individuals to make and interpret their analyses. Many, but not all, of these were chemists. Some are included in Table 13.3.

However chemists were also employed on the other side of the debate, by reformists like Chadwick (whose champions were Angus Smith, Playfair and Hofmann) or by potential rivals to the existing companies who wanted to suggest alternative sources of supply; their numbers included Stenhouse, Thomas Clark and Miller. It was this reformist faction that found in Edward Frankland its most vociferous spokesman. Such was the ambiguity of chemical evidence

Table 13.3. *Chemists working for the London Water Companies*

1820s and 1830s	G. Pearson	1751–1828	physician	St George's Hosp.
	R. Phillips	1778–1851	chemist	Mus. Econ. Geol.
	J. Gardner	1804–1880	Prof. Chemistry	Apothecaries' Co.
1840s and 1850s	A. Aikin	1773–1854	chemist	Guy's Hospital
	W. T. Brande	1788–1866	Prof. Chemistry	Royal Institution
	J. T. Cooper	1790–1854	chemist	Russell Institution
	A. S. Taylor	1806–1880	medical jurist	Guy's Hospital
1860s and 1870s	H. Letheby	1816–1876	Medical Officer	City of London
1880s and 1890s	W. Odling	1829–1921	Prof. Chemistry	Oxford
	W. Crookes	1832–1919	Editor	*Chemical News*
	C. M. Tidy	1843–1892	Medical Officer	Islington
	J. Dewar	1842–1923	Prof. Chemistry	Royal Institution

at one time that chemists could cheerfully testify on behalf of whatever cause they were paid to defend. Nevertheless the level of ambiguity decreased with the passage of time and it is foolish and unhistorical to ignore the genuine advance in chemical knowledge in the second half of the century.

As if to stress the need for value-free science the Government offered places on its commissions to chemists who would (it was hoped) act with impartiality and fairness. The problem was that many Commission members had, at other times, served one or other of the contending parties. Thus the Royal Commission on Metropolitan Water Supply of 1828 had Brande as a member. The 1851 'Chemical report' was, as we have seen, a work by Hofmann, Miller and Graham. A General Board of Health Report of 1855 was served by Blyth and Hofmann, while the agricultural chemists J. B. Lawes and J. T. Way were members of the Royal Commission on the Sewage of Towns (1857–1861).

Thus when Frankland took Hofmann's place as analyst for the London water supply there was abundant precedent for this kind of work to be performed by a chemist and to be taken seriously by the public. The question 'why Frankland?' is answered most simply by the circumstances of his succession to Hofmann at the Royal College of Chemistry. It was just part of the package that the German chemist was anxious to hand over as expeditiously as possible. There was no other obvious candidate and considerable need for continuity.

This may be literally all there is to say at the level of administration and legality. But it leaves untouched the larger questions as to why he was so suited to the task and why he, more than any others, was to dominate the field of water analysis for the next 30 years. Unquestionably the answers lie deep in his own personal history.[17] To take the most obvious point first, he had, by 1865, an extraordinary degree of manual dexterity and the ability to perform with extreme accuracy the most intricate tasks in the laboratory. For this, as Armstrong pointed out, he should have been grateful for his years as a pharmacist's apprentice in Lancaster where neatness and precision were indispensable. He then had the good fortune to work in a laboratory, at the Museum of Economic Geology, where water analysis was just being developed. So, on a brief return to his native Lancaster in 1846, the topic of a lecture to the Mechanics' Institute was 'chemical analysis'. The next year one of his patrons, Dr Christopher Johnson, reminded him of the excellence of a piece of water analysis performed on his behalf.[18] Subsequent experience at Marburg had confirmed the critical importance of analysis for other fields of chemistry. And those fallow years at Manchester had, as we have seen, their compensations in industrial consultancies, not least among which were water analyses for a canal company (1851),[19] domestic consumers (1852),[20] and the Lancaster and Carlisle Railway (1855).[21]

Yet it is tempting to look beyond his education and his subsequent work opportunities to less tangible influences that were probably more powerful for being hidden. The life-threatening experience of cholera at Salford left an indelible impression on his mind. So, in a different way, did the experience of bathing in the sparkling waters of the Wyre or the unpolluted Lakes of Cumberland and Westmorland. They figure frequently in his data for potable water. His lifelong passions of fishing and sailing may owe their intensity to a feeling of identification with the waters not fouled by urban pollution. Is it too fanciful to ask whether his relentless battle with the purveyors of contaminated water reflected a longing for an ideal which he had first encountered (or thought he had) in his remote Lancastrian past?

Hofmann's legacy of the post of analyst to the Registrar General was at first almost a sinecure, requiring only monthly analyses of the water provided by the eight London companies. When, in 1872, the Registrar General's office was absorbed into the new Local Government Board, Frankland became an officer of that board and continued his analyses as before. Although he asked for a salary of £150 Frankland had to settle for the £100 negotiated by his friend William Farr, statistician to the Registrar General.[22] This, of course, was over and above the incomes from his post at the Royal College of Chemistry and his examinership for the Department of Science and Art. Although the 'official' status of his appointment was unclear Frankland continued this advisory service to the Government to the end of his days.

In addition to this work Frankland soon discovered a further outlet for his skills in water analysis. Cholera struck again in 1866 and, sure enough, the Government promptly established yet another enquiry: the Royal Commission on Water Supply (1867). Two chemists were appointed as consultants, Frankland and Odling. They reported with vigour, though not unanimity, on water in Wales and Cumberland and on the state of the Thames. Meanwhile another Royal Commission (on Rivers Pollution) was grinding to a halt because of disagreement between the chemical member, Thomas Way, and his two other colleagues. Accepting the inevitable the Government disbanded the contentious trio and, with much optimism, established a new Commission in 1868. To this they appointed Sir William Denison (an engineer) as chairman, together with J. C. Morton (an agriculturist) and Edward Frankland. Denison died in 1871 and Frankland became the dominant partner and chief spokesman. He took over the laboratory of Thomas Way in Westminster, where he was joined by his assistant H. E. Armstrong.[23] The Commission lasted until 1874 during which time Frankland received an annual salary of £800 and a laboratory allowance of £700. Once again Frankland was to merit his old *sobriquet* 'accomplished chemical pluralist'. The Commission's brief was far wider than merely the waters of the metropolis and Frankland found himself drawn in to analysis on a nation-wide scale, including (for example) advising on the water supply to the royal residences at Windsor,[24] Sandringham[25] and Balmoral.[26] As Hamlin has aptly said, 'regardless of whether he had set out to become a water expert, Frankland had become one'.[27] Even more pertinent are the comments of his former colleague and assistant H. E. Armstrong:

> The appointment was to change the entire tenor of his life. In the course of six years, in six reports to the House of Parliament, the Rivers Commission surveyed the whole problem of water supply. The final report, dealing with our domestic water supplies, was specially

valuable. The recommendations passed into the hands of the water engineers and soon we became the most favoured of peoples in our water supplies. The example passed to America and then gradually to the European continent. Frankland thus laid foundations of infinite importance.[28]

Given a secure institutional base Frankland was able rapidly to bring his formidable laboratory skills to the solution of totally new problems.

3 New methods of analysis

Water analysis had traditionally been concerned with the determination of dissolved inorganic matter (sulphates, chlorides, bicarbonates, nitrates *etc.*), together with rough estimates of organic material. Where the problem was fitness or otherwise of a given sample for drinking (potability) the chemist could easily measure its hardness and pronounce judgement on the basis of the received wisdom of the time, *i.e.* that hard water was or was not good for diet. With questions of disease, however, he was on far shakier ground. This is not the place to discuss the various competing theories of contagion, but it can be remarked that almost everyone linked it in some way to organic substances in the water. Crucial to the arguments were the germ particles revealed in 1850 by Hassall. When Frankland assumed Hofmann's water analysis duties 15 years later there were three alternative views about the rôle of germs. They might be:

(a) *The direct cause of disease.* William Budd had announced in 1849 the discovery of a water-borne cholera fungus, but very few apart from Budd regarded this as the real villain. Microbiology had yet to be born.

(b) *Indicators of disease.* Here the ultimate cause was deemed to lie in the decaying organic matter on which the microbes fed (Hassall). Hence absence of germs probably meant absence of pathogenic substances.

(c) *Merely disgusting.* That is, their importance was neither causal nor indicative but aesthetic. Angus Smith described the microscopic residents of the Thames as 'larger, fatter and uglier' than any he had seen before. That, in itself, was sufficient to render the supply unacceptable.

Clearly the situation was highly unsatisfactory for an analytical chemist. What on earth was he supposed to be estimating? Gradually a consensus emerged that what mattered was the organic content of the waters, together with those inorganic substances that could have been formed from the decomposition of organic matter. At this time the following procedures were in common use:

(a) Determine the total solids present by evaporation with a known weight of sodium carbonate;
(b) Ignite the residue from (a), loss in weight indicating organic matter + other volatile substances (as ammonia);
(c) Determine total organic matter by oxidation with potassium permanganate (where loss of colour indicates oxidation is complete);
(d) Estimate nitrite + nitrate by reduction with tin(II) chloride;
(e) Determine ammonia by liberating from ammonium salts with alkali and estimating colorimetrically with Nessler's solution.

Frankland's first strategy was to continue with these methods, but he quickly discovered grave defects in each. Moreover, the Hofmann and Blyth report of 1856 had stressed the desirability of determining the proportion of nitrogen in the organic constituents but no satisfactory procedure had been devised. His collaborator recalls:

> Thorough in everything he did, he soon became dissatisfied with the methods, especially the determination of organic impurities indicative of sewage contamination. He decided to revise them all. At the close of the summer session of 1866, he did me the great honour to propose that I should carry out the work for him. The method of combustion analysis *in vacuo* we devised was made known, the following year, in a lecture by Frankland at the Royal Institution; the work generally was described at the Chemical Society in February 1868.[29]

In this paper to the Chemical Society[30] the authors refer briefly to the deficiencies of existing practice. Thus, of the five procedures sketched above, they point out that (a) will lose ammonia from both ammonium salts and urea, (b) will give low values for urea because some of it will be converted into salts (as $NaCNO$), (c) is predicated on unwarranted assumptions concerning the amount of permanganate needed to oxidise a given amount of organic material, (d) is not specific to nitrites and nitrates since carbohydrates interfere, and

(e) ammonia is also liberated from urea under these conditions. The frequent reference to urea is significant for this substance, an end-product of animal metabolism of nitrogen, was a major constituent of raw sewage.

The simplest part of the Frankland/Armstrong reformation was to conduct the preliminary evaporation in the absence of sodium carbonate, thereby eliminating the possibility of expulsion of ammonia. They abandoned the ignition and permanganate procedures (except for very rough qualitative purposes). The biggest problem remained, apparently intractable: 'No process has yet been devised by which the amount of organic matter in water can be even approximately estimated'.[31] After nearly two years' work they found an answer.

The new method was a development of Liebig's classical technique for estimating C and N in organic compounds. The essential feature of this variation was oxidation of the evaporated residue with lead chromate in an evacuated tube, the resultant gases (CO_2 and N_2) being estimated volumetrically. To avoid interference from carbonates the preliminary evaporation was first conducted with 'sulphurous acid', thereby expelling CO_2, and also nitrogen from nitrates and nitrites (a 'remarkable reaction' which 'could scarcely have been predicted'). The evacuation was effected, before and after combustion, by the mercury fall-pump described by Sprengel in 1865.[32]

In this connection Armstrong throws some interesting light on the professor/assistant relationship:

> The experience was invaluable, as I was thrown much on my own resources, though sufficiently aided at critical times. . . .The task was not entirely simple. I recollect, when he suggested that we should make the attempt, that he referred me to the account given by Graham of the use he had made of the Sprengel mercury pump in studying the diffusion of oxygen through india-rubber. A few days later, seeing that I was in difficulties at the blow-pipe table, he came to my aid and made the first Sprengel fall-tube for me We were the first after Graham to use the pump. If the traditions of the early Frankland school had been kept alive and the few students with hands [sic] had been regularly tutored in the methods of gas analysis, the vacuum combustion process would now be a preferred process, I believe.[33]

In fact it was not to be. While this work was in progress an alternative attack on the problem had been developed by Frankland's former student J. A. Wanklyn (now Professor of Chemistry at the

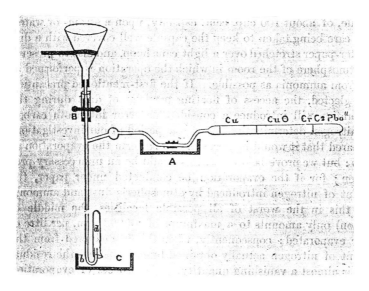

Fig. 13.1 Frankland's apparatus for determining N and C in residues from water. The residue is mixed with solid oxidants in the combustion tube on the right, heated in a furnace. The Sprengel pump (on the left) evacuates the tube by the dropping of mercury. Gaseous products from the ignition collect in the tube d, and are subsequently analysed eudiometrically. (E. Frankland, *Experimental researches*, p. 594.)

London Institution) and two of his collaborators.[34] This was based on the assumption that the dangerous materials in polluted water are albuminoid in nature, *i.e.* to be proteins which always contain nitrogen. By treating the water with, first, alkali to expel ammonia from ammonium salts and urea, and, second alkaline potassium permanganate, the organic material will be quantitatively oxidised to ammonia which may then be estimated colorimetrically using Nessler's solution. The method was announced to the Royal Commission on Water Supply on 20 June 1867 and to the Chemical Society on the evening of the same day.

Without doubt the Wanklyn method was simpler than that of Frankland and Armstrong. The important question was whether it was as accurate. A letter from Frankland, then on holiday with his parents in Leyland, asked Armstrong to compare their results with those of Wanklyn as a matter of urgency.[35] The superiority of the rival process was vigorously denied by Frankland and his colleague, who pointed out that the oxidation is never complete, even for albumen.[36] Wanklyn's later admission, that only two thirds of the organic material is so oxidised,[37] brought a further denial from his rivals that the fraction of ammonia evolved is at all constant.[38]

There then followed a series of public altercations of such bitterness that one must suspect more than professional disagreement over purely technical matters. Wanklyn published his own account which included several unsavoury innuendoes. Odling (Secretary of the Chemical Society) might have 'abused his office and used my unpublished manuscript' in order to use the process himself, ahead of publication; the results of Frankland and Odling were 'spurious' (a pretty serious charge); and the correspondence between Frankland's expectations and results must be described as 'sinister'.[39] In fact it was just now that Wanklyn was involved in the notorious 'black-balling' incidents in the Chemical Society even though he was then a member of its Council.[40] In these escapades a group of young men led a 'rebellion' in which candidates for Fellowship proposed by the Council were rejected by a vote of the membership. As one of Wanklyn's former colleagues admitted, 'Unhappily nature had endowed (or cursed) him with a spirit of pugnacity which seemed almost inevitably to involve him sooner or later in personal quarrels with those with whom, or for whom, he worked'.[41]

On January 16 1868 Frankland read his paper to the Chemical Society but Wanklyn and Chapman attacked his numbers and methods and praised their own. Acrimony continued until 10.35 p.m. and the discussion was adjourned until the next meeting on 6 February. Herbert McLeod recorded his impressions:

> Wanklyn was speaking about water analysis and was followed by Chapman. Dugald Campbell afterwards contested Wanklyn's results and said that the bringing forward of unfinished experiments was an insult to the Society. Wanklyn contradicted him and said that his statements were false. The President called them to order. Frankland replied and disposed of Wanklyn and Chapman's objections.[42]

Following Frankland's criticism Wanklyn asked the Council of the Chemical Society to institute an official enquiry into water analysis; the motion received no seconder and therefore lapsed. Two years later he was expelled from the Society (for non-payment of subscription?),[43] and after helping to found the Society of Public Analysts in 1874 became similarly estranged and announced his secession. With evident relief that Society's journal remarked 'we are happy to confirm the accuracy of his statement'.[44]

That Wanklyn's objection to the Frankland/Armstrong process

was not devoid of personal animosity may be gleaned by the unprecedented question asked him when he testified to the Royal Commission on Water Supply as to whether he was on good terms with Dr Frankland, and his own complaint that 'at that period, as is pretty well known, there was a strong prejudice against myself and my colleagues'.[45] He had been a guest at the Frankland household in January 1867 and gave an affirmative reply to the Commissioners' very personal question a few months later as to whether he was 'in communication with Dr. Frankland'.[46] But the situation was already changing. It is likely that his growing hostility to Frankland, when the latter discovered a rival process to his own, was reinforced by other events. It came at a time when relationships at the Chemical Society and elsewhere had reached a record low for other quite unconnected (and largely unknown) reasons, and Wanklyn was deeply involved. Such was his personality that the poison of hate remained to ruin his life into the distant future. Despite a genuine talent for analysis, recognised by Liebig amongst many others, Wanklyn succeeded in alienating most of his colleagues and expended much of his creative energy in strife and controversy.

In retrospect it is clear that the Frankland technique had several advantages. First, it was extremely accurate. To read Frankland's papers on the subject is to be transported into a world of painstaking attention to detail, of careful standardisation of reagents and calibration of apparatus, of blank control experiments, of rigorous procedures for quantitative transfer, of flawless vacuum techniques, even of meticulous instructions for 'reading' Nessler tubes. It is a world long forgotten today, except perhaps by those with long memories stretching back into a pre-electronic era. Secondly, the method was capable of amendment and adjustment as experience demanded. Those (including Frankland's critics) who saw this as a sign of inadequate technique completely missed the point that *all* scientific procedures need constant updating, though often only in minor matters of detail. It was, as Odling said, 'a new and very refined process of water analysis'.[47]

However it also had a number of disadvantages. First and foremost among these was the sheer difficulty of the techniques involved. For a Frankland or an Armstrong this was not a problem, nor was it likely to be so for the numerous assistants whom they had trained. But for lesser mortals it presented serious difficulty, though this can be exaggerated. The American chemist J. W.

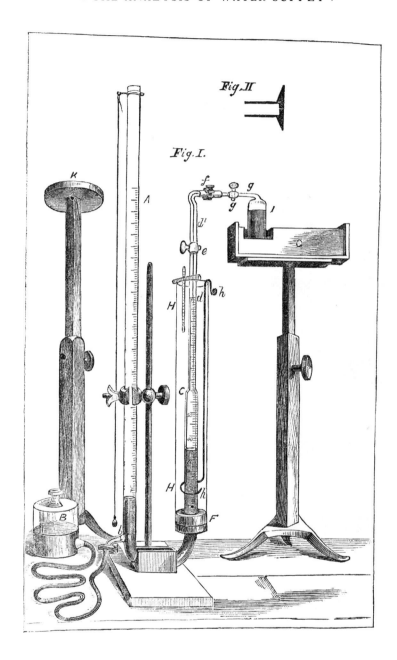

Fig. 13.2 Frankland's apparatus for 'analysis of gases incident to water-analysis':
CO_2, NO and (by difference) N_2. (E. Frankland, *Experimental Researches*,
p. 594).

Mallet, writing many years later and after some simplifying improvements, described it as 'a process of great delicacy, and quite satisfactory in its details with proper precaution and in trained hands'. He concluded:

> The Frankland process is quite within the reach of the manipulative skill of any fairly-trained chemist, but it requires practice, and probably pretty constant practice. It cannot be taken up off-hand, and even tolerable results obtained at once. From the hands of a person without proper laboratory training, its results are utterly valueless. It is hence better adapted to regular use in the examination of many samples of water in a large public laboratory than to an occasional use by a private individual.[48]

Two further and related problems were the time taken for a complete analysis (up to two days) and the financial cost (in terms of time, apparatus and chemicals). Finally there was the critical question as to what exactly was being measured. One analyst put his finger on the problem thus:

> Most chemists are, we believe, convinced that, assuming the organic matter to be once inside the combustion tube free from an admixture of nitrates, its carbon and nitrogen can be estimated with an extraordinary degree of accuracy, by means of Frankland and Armstrong's process. The real questions, however, are, firstly: does the process enable us to get the organic matter dissolved in a litre of water into our combustion tube undiminished in quantity, and freed from the large excess of nitrates, with which it is often associated, and, secondly, can we make accurate allowance for any ammonia which may be present in the water?[49]

By comparison with the Frankland/Armstrong process the alternative introduced by Wanklyn and his colleagues was both faster, taking about three hours, and cheaper. On the other hand it suffered from the defect that albumen is not quantitatively decomposed into ammonia under the test conditions, and, moreover, that urea, supposedly eliminated as ammonia by preliminary treatment with sodium carbonate, does not give 100% conversion. For these reasons Wanklyn introduced his correction factor of two-thirds. This stratagem, together with his failure to conduct his own tests of the rival process, led to much acrimony in the pages of *Chemical News*.[50] The results by this method were almost always lower than those obtained by Frankland and raised the critical question as to whether albuminoid substances were in fact the right subjects for analysis.

So what happened? By 1872 it was claimed that most leading analysts had rejected Frankland's process in favour of Wanklyn's, their number including Thomas Way, Angus Smith, W. A. Miller, A. Voelcker and Henry Letheby. Crookes, the editor of *Chemical News* remarked 'we scarcely know a single chemist of reputation who approves of Dr. Frankland's water analysis'.[51] Four years later Frankland himself conceded the popularity of the ammonia process.[52] Nevertheless it is important not to exaggerate the monopolistic position claimed for the latter. Thus the Nottingham Public Analyst, E. B. Truman, assured Frankland in 1874 that he was using his method for 'hundreds of analyses',[53] J. W. Thomas of Cardiff wrote that, though he had temporarily gone over to Wanklyn's method in order to keep fees down, he hoped to return to the Frankland method as soon as possible,[54] and Odling told Frankland 'you may always refer to me as a chemist habitually using your process of water analysis', even though they might occasionally disagree over the conclusions.[55] An undated note in Frankland's handwriting discloses that his process was in use by Bischof, Campbell Brown, H. Brown, Donkin, Hill, Moulting, Odling, O'Sullivan, Russell, Tate, Thomas, Truman and Williamson.[56] A copy-letter to Sir Hugh Owen, also undated but probably in the early 1890s, names Hill, Mills, Odling and Tidy as users of his own process which, however, is 'tedious, laborious and costly'. But the ammonia process, though 'utterly untrustworthy', is used by 'the profession at large'.[57]

As for Frankland himself, he continued to use his own method for at least two decades, sometimes supplementing it by the Wanklyn method, modifying it with the growth of experience and performing in all many thousands of analyses. In 1876 he delivered a lecture to the Chemical Society 'On some points in the analysis of potable waters'[58] in which he paid tribute to improvements progressively introduced by his assistant W. Thorp, described an invention by Bischof to reduce the time of evaporation by two-thirds and elaborated his critique of the ammonia-process. The hapless inventor of the latter was present, and Frankland's children described his reactions in a letter to Fred, by now in New Zealand. He replied 'Wanklyn's confusion and chagrin I can easily picture to myself'.[59] Family loyalty evidently ran high.

It was not long before Frankland was discovering that water analysis made immense demands on his time, in his capacity as a

River Commissioner, as analyst to the Local Government Board or as an independent consultant. Once again he came to rely heavily on his loyal band of assistants. One of them wrote as early as 1869:

> We have an enormous lot of work at present in the Laboratory . . . [Frankland] seems awfully busy and can hardly find time to come to the Laboratory once a week and he does not do much more at the College.[60]

In fact Frankland was to find that much travel was involved, partly to procure the necessary samples but also to consult with water authorities and others on the proper courses to take once analysis had been completed. And that meant advice founded on certain principles of interpretation.

4 The birth of a concept: previous sewage contamination

Since the ultimate cause of water-borne diseases was still a matter of speculation the chemists could not immediately identify offending substances as they could, for instance, in a case of lead poisoning. This placed them in an acute dilemma. By the 1870s they had grown accustomed to speaking with authority on many matters of public health, though this authority had taken many years to win. They were used to writing reports, testifying before commissions, defending or attacking a brief in court. It was also the time when they were becoming something like a profession, and that meant a heightened awareness of their collective place in society. There certainly was a connection between 'water analysis and the hegemony of chemistry', as Hamlin entitles one of his chapters. The difficulty was that at this critical time for chemistry in Britain the science was faced with a problem insoluble *in its own terms*. With hindsight we know full well that the pathogens were not simple chemicals but micro-organisms. But, as we have seen, the importance of germs was then highly problematic and the public demanded certainty, or the nearest approximation to it, from those employed to ward off epidemic disease.

Several approaches were possible. One, the 'minimalist' position, was to search the water for known toxins and if these were absent to declare the water safe. At the opposite position, the 'maximalist', any hints gatherable by a chemist that the water *might* (not *did*) contain nasty things inaccessible to chemical enquiry would cause that water to be branded as unsafe or even dangerous. Many chemists, of course, occupied the middle ground between these extremes.

In these terms Frankland was a maximalist. Recognising that nothing that he, as a chemist, could find in water could be unequivocally identified as a cause of cholera, typhoid *etc.*, he took the view that the deadly agents, whatever they were, were almost certainly distributed into the system by disposal of sewage. Therefore any clear traces of sewage detritus were clear warnings of danger. Occasionally this would be more a matter for a microscopist than a chemist, though the water would not always be polluted to so gross an extent. But was it possible that sewage could leave *chemical* traces? Frankland believed it was.

The key element was nitrogen. Substances produced from living organisms, whether from sewage or elsewhere, would contain the elements C, H, O and N, of which the carbon might be partially degraded by natural processes to CO_2, while the hydrogen and oxygen would end up as water. The nitrogen, however, could still be found in residual organic matter together with some carbon (of which albumen was but one possibility out of many). That would be *actual sewage contamination*. Results of analysis by the Frankland/Armstrong method would give the amount of organic nitrogen (and also the C/N ratio which came to be useful later on when he regarded lower values as increasingly indicative of sewage).

But nitrogen might also be present as inorganic nitrites, nitrates or ammonia. These would be the very end-products of sewage degradation, the mere 'skeletons' as it were of the organic material, but nevertheless indicators that it had once been present. And so, with great boldness, Frankland introduced into the chemistry of water analysis an entirely new concept, that of *previous sewage contamination*. He did so in a preliminary lecture at the Royal Institution in March 1867 and in his paper to the Chemical Society in the following January:

> In view of the opinions now very generally entertained with regard to the propagation of certain forms of disease by means of spores or germs contained in excrementitious matters, the search for nitrates, nitrites, and ammonia is second only in importance to that for actual sewage contamination, because, although these substances are in themselves innocuous, unless present in excessive quantity, yet, when contained in a water in more than a certain proportion, they betray previous contamination by sewage or by manured land. The nitrogenous

organic matters contained in sewage or manure, undergo slow oxidation and conversion into mineral compounds when mixed with water; their carbon is converted into carbonic anhydride, and their hydrogen into water. These mineral products can no longer be identified in the aërated waters of a river, spring, or lake; but the nitrogen is transformed into ammonia, nitrous acid, and nitric acid; the two latter combine with the bases contained in most waters, and, together with the ammonia, constitute a record of the sewage or other analogous contamination from which the water has suffered. With certain corrections, mentioned below, the determination of the nitrogen contained in these mineral compounds proclaims the previous history of the water as regards its contact with decomposing nitrogenous organic matters. We propose to employ this determination for the expression of the previous sewage contamination of a water, in terms of average filtered London sewage which, if thus oxidized, would yield a like amount of nitrogen in the form of ammonia, nitrites, and nitrates. For this purpose average filtered London sewage may be assumed to contain 10 parts of combined nitrogen in 100,000 parts, as deduced from the numerous analyses of Hofmann and Witt, and of Way and Odling.[61]

This concept, introduced in 1867/8, was to be a guiding light for Frankland's water analysis for the next six years. Table 13.4 shows how it was presented in an early Report to the Registrar General (February 1867) (p. 384).

5 Opposition

The concept of previous sewage contamination proved to be intensely controversial. Some of the criticisms, especially at first, were founded on genuine misunderstandings, while others came because of its impact on vested interests. No less a body than the Royal Commission on Water Supply found itself in difficulty with the concept and it is fortunate that a detailed statement of Frankland's reactions survives in the form of a 17-page letter to the Registrar General to whom he was responsible for the monthly analyses of London water.[62] The document offers a convenient way into the argument.

After dealing with the Commissioners' misconceptions about his phrase 'total solid impurity' he addresses their equally lamentable failure to understand the meaning of 'previous sewage contamination'. With blunt candour he begins: 'I regret that the Royal Commissioners

Table 13.4. Analysis of Metropolitan Waters in February 1867

Companies	Date & Place of Collection	Total Solid Impurity	Organic Carbon	Nitrogen, as Nitrates and Nitrites	Ammonia	Total Combined Nitrogen	Previous Sewage Contamination (estimated)	Total Hardness
Thames								
Chelsea	1st February Cab-rank, Horse Guards	28.58	.433	.337	.004	.371	2,420	16.2
West Middlesex	1st February Great Portland Street	28.68	.340	.356	.006	.412	2,630	16.2
Southwark & Vauxhall	1st February Barclay's Brewery	29.08	.293	.357	.005	.361	2,630	16.8
Grand Junction	1st February Royal College of Chemistry	29.44	.417	.322	.004	.325	2,270	17.1
Lambeth	1st February Cab-rank, Westminster Road	29.36	.423	.341	.005	.356	2,470	16.0
Other Sources								
New River	15th February Cab-rank, Tottenham Court Road	29.72	.272	.350	.003	.396	2,540	18.5
East London	16th February Cab-rank, Shoreditch	33.56	.293	.357	.004	.392	2,620	18.8
Kent	15th February Waterworks, Deptford	39.84	.088	.421	.008	.428	3,300	23.1
South Essex*	16th February Mr. Whiffin's, Romford	38.32	.143	.844	.007	.850	7,520	21.1

*This company does not supply London.

should have founded their strictures almost entirely upon mere statements of opinion, unsupported by facts.'

First there was the charge that *'previous sewage contamination' excluded the effects of animal manure*, to which Frankland complained 'I have repeatedly stated in my reports to you and elsewhere that the term is an abbreviation for "previous sewage and manure contamination".' That was easily dealt with.

The second reservation of the Commissioners was rather more serious. It was that *nitrates in water might come from sources other than sewage or manure*. Frankland referred to the 'overwhelming number of analytical facts showing that waters of known purity draining from almost every variety of uncultivated and unmanured soils do not exhibit previous sewage contamination' but felt they had been ignored in the face of three relatively trivial observations. One was the discovery by Boussingault of small amounts of nitrates in the soil, to which Frankland responded that traces of nitrates are known to exist in rain water, are corrected for in water analysis and that was sufficient explanation for the observations by the French chemist. He added:

> The Royal Commissioners appear to have lost sight of the fact that such a correction is always made in my reports to you and that no water is pronounced guilty of previous sewage contamination unless it contains more nitrogen in the shape of nitrates, nitrites and ammonia than that contained in rain water under these forms.

Another subversive observation had been made by Angus Smith: the formation of nitrates from putrid yeast. Since the animal/vegetable status of yeast was in doubt, since it was often inhabited by grubs, and since 'it can scarcely be contended that yeast or any material closely allied to it can play any important part in the contribution of nitrates to Thames water', the facts were dismissed as irrelevant. A third difficulty under this heading was an observation by W. A. Miller that 'the waters from Watford, which come from chalk wells, where there can be no suspicion of previous sewage contamination, always contain nitrates'. But that was to ignore the possibility of seepage into those wells of human waste from the local population, for previous sewage contamination must surely include 'all kinds of decomposed animal matter whether contained in sewage, privies, cesspools or the solid manure applied to land'. He then continued with a fourth apparent problem of his own discovery. This was the fact that Lawes and Gilbert had been making experiments at Rothamsted on wheat cultivation, with and without application of manure, and they supplied Frankland with samples of water draining from the different plots:

> On submitting these samples to analysis I was surprised to find that they all contained large quantities of nitrates, and, what was still more remarkable, the drainage water from a plot which had received no manure for 25 years contained a larger proportion of nitrates than the water from the adjoining plot which had been dressed with 14 tons of farmyard dung per acre every year during the same time.

The difficulty was resolved by a visit to the site where he discovered that the plots were parallel strips of a large field and that nothing existed to prevent a complete underground diffusion of the water leaking from each plot, so 'the water flowing from each pipe represents rather the drainage of the whole field'. As a matter of practice Frankland considered that, though nitrates may be derived from the nitrogenous salts applied to land in artificial manures, 'the amount of these salts so used is far too small to require consideration even if we were to assume that the plants to which they are applied as a costly manure make no use of them'.

A third problem was the frequent *discrepancy between Frankland's condemnation of a water and its appearance and reputation.* He stoutly denied any evidence 'of the innocence of any sample of water which has been found guilty of previous sewage or manure contamination on analytical grounds', continuing:

> I have now examined upwards of 1000 samples from all parts of the United Kingdom and have not yet met with a single case of clear analytical guilt which has not been sustained on further investigation. It is true that my verdict has repeatedly been met with vehement protestations of innocence, but further investigation always proved that these could not be sustained. The other day a gentleman brought to me two samples of well water for examination. I reported both as exhibiting great previous sewage contamination; he protested that it was impossible as the waters were bright and sparkling and possessed a high reputation; a week later he informed me that the source of contamination had been discovered, – one of the wells was situated close to a large cesspool; the other received the drainage from a dog kennel.

Related to this was a fourth consideration: the *self-healing properties of river water.* Did not nature effect a removal of the pathogens in the ordinary course of river flow? This had certainly been the view of Letheby who had informed the Commissioners that sewage added to a river up to a dilution of 5% would lose its character as sewage at the end of the seventh mile.[63] In his address to the same Commissioners Frankland had compared the fate of two eggs dropped into the Thames at Oxford, one broken and one unbroken. The former would mix its contents with water immediately but the latter might well survive intact at London. Its vitality would be undestroyed. Could not germs be similar?[64] The Commissioners were not convinced but Frankland had done some tests: 'I find that percolation through 5 feet of gravelly soil removes much more organic impurity from sewage water than does a flow of 50 miles in a river at a rate of one mile per hour.'

A fifth difficulty was that of *negative evidence*: 'the absence of nitrates is not absolutely conclusive evidence of immunity from this pollution':

> Nitrates are removed from water by the action of aquatic plants and they also disappear when the water putrefies. It is probably by the first of these processes that in summer they are more or less completely removed from the East London Co.'s water which is often stored in reservoirs for a long time before use. I have also found that the addition of fresh sewage to previously contaminated water destroys more or less completely the evidence of the previous contamination. Thus the evidence of the previous contamination of the London water supply is sought for in vain in the discharge from the metropolitan sewers, – the nitrates present in London water are absent from the sewage which barely contains a trace of nitrates. It is to this circumstance that the reduction of the amount of nitrates in some streams after the addition of fresh sewage is due.

Does this therefore mean that the whole concept is suspect? By no means:

> Shall we reject this ordeal because polluted waters may pass through it without conviction? Do we reject criminal prosecution as a safeguard to society because the evidence notoriously fails in some instances to convict the guilty? In neither case can it be said that the guilty and convicted are unfairly treated because a certain number of guilty but unconvicted escape.

Which brings us to a sixth difficulty: Frankland's approach stressed the *severe limitations of a chemical method*. It was well-known that Frankland on other grounds had condemned sources that did not display signs of previous sewage contamination. The example of the East London Company above is a case in point. He wrote:

> At the risk of being tedious I cannot close this letter without calling
> your attention to what chemical analysis can and cannot tell us
> respecting potable water. The following sentence in the report of the
> Royal Commissioners seems to render it necessary that this should be
> clearly stated as the Commissioners themselves entertain opinions
> respecting the powers of chemical analysis which the latter is not at
> present able to satisfy. They say (page CI) in the present state of
> chemical science, analysis fails to discover, in properly <u>filtered
> Thames water</u>, anything positively deleterious to health. Now this
> statement applied to filtered Thames water is equally applicable to
> London sewage or to the dejections of persons suffering from cholera
> or typhoid fever. Chemical analysis has failed to discover anything
> positively deleterious to health in these matters in their most
> concentrated condition and it is not therefore likely that analytical
> research will be more successful when applied to them after dilution
> with the water of the Thames.

For the Victorians it was hard to imagine why water that
contained no dangerous chemicals should be condemned. It may be
so for us, – until we remember germs. The distinction between
actual sewage contamination and *previous* sewage contamination is
quite a subtle one. Moreover, as has been repeatedly emphasised,
this was a time when chemists were struggling towards professionalization. The last thing they needed to be told was that their
work was of restricted value, and that their tests had great
limitations. And government officials, accustomed to prompt and
unambiguous replies from chemical witnesses, could simply not
understand the apparent coyness with which Frankland was giving
his answers. How could he say that 'chemical analysis is utterly
powerless to detect any matter positively injurious to health in any
of the forms of animal refuse which go to contaminate water'? The
trouble was that they were looking in the wrong direction:

> The use of chemical analysis in the investigation of water for sanitary
> purposes lies almost exclusively in a totally different direction. It is
> only by bringing to light the previous history of potable water that
> chemistry can help the custodians of public health. Even the discovery
> of actual sewage in water has only a sanitary significance on the same
> ground, it only shows that human excrements are in the water, it
> throws no light upon the noxious or innocuous character of these
> excrements; that they may be comparatively innocuous is proved by
> the fact that thousands of our fellow countrymen habitually drink their
> own sewage which has percolated through a few feet of earth; that they
> may be noxious is rendered very probable by the decimation of the
> same people when a case of cholera or typhoid fever is imported

amongst them . . . It is for the physiologist, not the chemist, to say what
influence the admission of excrementitious matters into drinking
water has upon the health of the community. If his verdict is that they
have none, then water analysis for sanitary purposes becomes useless;
but if the accumulated experience of 30 years leads him to the contrary
conclusion, then chemical analysis can aid him by showing, in a great
majority of cases, the previous history of the water as regards its
companionship with excrementitious matters.

All these arguments were circulating by 1870. There was no lack of
spokesmen for the opposition. Here is an opinion from the
Chemical News of 1872:

> In truth, if we determine the amount of nitrates in 100,000 parts of
> water and deduct from that the amount of nitrates present in average
> rain water, we get a number which means nothing in particular, and
> which certainly is no index to the goodness or badness of the water. It
> is high time that the Registrar-General of Great Britain should cease
> stultifying the national statistics by so ridiculous a return.[65]

That dismissive put-down came from an article on the Registrar
General's monthly reports on London waters and was a characteristic
attack on Frankland by his old adversary Wanklyn. He was, of
course, still embittered by the invention of the rival C/N process
and had been, or was shortly to be, embroiled in controversy with
Frankland over the structure of certain organo-sodium derivatives[66]
and the action of sodium on ethyl acetate.[67] The substance of his
objection now was the 'most extraordinary psychological fact' of
Frankland's failure to take adequate account of the relative
magnitudes of vegetable action on dissolved nitrates and previous
sewage contamination. Wanklyn continued to be a thorn in
Frankland's side for many years to come, even arguing that
Frankland had stolen his results during the years in the Manchester
laboratory and that, from about 1865, Frankland's best research
was over.[68]

More serious opposition was already in the air. The water
companies, now very sensitive to changes in public opinion, were
alarmed at the implications of previous sewage contamination.
Many of their fears were focused by the medically trained Henry
Letheby, retained by the East London Water Company to defend
their product. As early as 1867 Letheby stood Frankland's
argument on its head and argued that the presence of nitrates in
water was good news, because that meant that the decomposition of
the sewage was complete.[69] This, of course, was merely to confuse

actual with *previous* sewage contamination. Addressing the Association of Metropolitan Medical Officers in 1869 he took it upon himself to attack 'the persistent use of certain expressions of an improper kind', referring of course to 'previous sewage contamination'. Although he complained about 'undue excitement' created by Frankland's monthly reports, he did not address the underlying scientific issues. Frankland gave a confident reply and received some support from representatives of both medical and engineering professions. Letheby's power to contradict Frankland reached a maximum in 1873 when his own analyses were included in the monthly reports of the Local Government Board. Frankland's view was understandably jaundiced:

> Letheby is at the present moment a most serious obstacle in the way of sanitary reforms. You will always find him on the side of joint stock companies and against the public – companies pay well, the public does not pay, – *Voila!*[70]

Letheby died in 1876 and was succeeded by his former pupil C. M. Tidy. He became, as Hamlin notes, Frankland's principal antagonist. In conjunction with Crookes and Odling (also retained by the water companies) he produced in quasi-official form their own analyses of London's water. Frankland duly protested.[71] Crookes had opposed Frankland since 1869 when the Rivers Commission had condemned a process of the Native Guano Company with which Crookes was closely associated; it proposed to purify sewage by addition of coagulants to precipitate out suspended matter that could then be sold as a fertiliser. Odling, a former ally on matters of water analysis, was enticed temporarily to forsake the peace of his Oxford department for incursions into the world of water controversy and, at the behest of the water companies, to undermine the credibility of Frankland's methodology. It was now the 1880s and the trio of chemists proclaimed – against Frankland – the value of chemical water analysis over all alternatives. It was a strange irony that at that very time microbiology was about to take over the task.

Tidy conceded advantage to Frankland in terms of C/N technique but attacked the view that rivers do not naturally cleanse themselves.[72] At a meeting of the Chemical Society he was routed by Frankland who was giving a paper 'on the spontaneous oxidation of organic matter in water'[73] and had previously sought moral support from his friends Huxley and Tyndall. The question at issue was whether running water can be safely used for dietetic purposes a few hours after it has been mixed with sewage. Dr. Tidy contends that it can, on the ground that the dead organic matter of the sewage will be burnt up by spontaneous oxidation, and the living germs will burst their envelopes by endosmic action and die![74]

He wanted Tyndall to say 'even a dozen words about germs', in which case 'we should not leave Tidy a leg to stand upon'. Tyndall replied:

> The levity with which certain of our medical men throw off hypotheses regarding germs is extraordinary. They shut their busy eyes to the most conclusive evidence, and give themselves over to the strangest delusions.[75]

In the event the discussion was animated, the meeting continued until 11.30 p.m. (a record), Tidy's theories were savaged by Huxley, and victory for Frankland was 'complete and crushing'.[76]

Tidy lived to fight another day. In fact he remained on curiously good terms with Frankland. Tidy was at least once his guest at the Athenaeum,[77] and the two families would sometimes dine together.[78] Frankland agreed with his daughter Sophie that his old antagonist indeed gave sumptuous dinners, adding that such was possible by 'devoting yourself body and soul to private companies, – gas, water, etc.'. Ruefully (but surely tongue in cheek) he reflected on the humbler destiny awaiting those merely in 'the service of the public'![79] Abel was probably right when he told Frankland that, despite a dearth of published work, Tidy 'ranks with us in the eyes of the Public as an Analyst and a consultative authority'.[80]

6 Later analytical work

As we have seen, Frankland's task as official Government analyst was to prove a lifelong commitment from 1865. The audacious publication of rival analyses under the guise of equally 'official' results did little to harm his authority and, apart from the letter to Owen, he seems to have let the matter pass. In February 1882 a new laboratory was established at Grove House, Pembridge Square, near Notting Hill Gate, and water analysis moved there from South Kensington. Most of the fittings were paid for by Frankland, together with the salary of an assistant; his son Percy, who had joined him a short time before, received half the fees for private water analyses.[81] Throughout the 1870s and 1880s his laboratory poured out a ceaseless stream of analytical results, whether in the service of the Government or (as was more often the case) for private individuals or companies. Alone of his scientific notebooks those relating to water analysis have survived (from 2 January 1868 to 7 July 1899),[82] and with them dozens of letters on the same subject, as well as a pocket notebook from 1891 to 1898.[83] It is clear that water analysis was to be the centre of gravity of his scientific interest in the last 20 years of his life.

From the eleven thousand analyses recorded in the books it is possible to draw some clear conclusions. The first is the enormous variety of clients bringing water analyses to Frankland. Many were water suppliers outside London. Overseas calls for help, advice and actual analyses, came from Shanghai, Mecca, Buenos Aires, India and many places in Europe. To take one year as an example, we might select 1876 when all the Rivers Commission work had finished. Of the 186 analyses recorded for that year clients included:

- water companies at Hitchin, Newcastle and Gateshead, Wakefield, Llandudno, Newtown, Chesterfield, Colne Valley, Lambeth, Bournemouth, Broadstairs and High Wycombe
- Rivers Balder, Lune, Tees
- gas companies at Llandudno and Brentford
- brick-works at Wakefield
- colliery at Wath, near Wakefield
- breweries at Leeds, Maidenhead and London
- copper mine in S. Africa
- hospitals at Homerton and Bristol
- asylums at Derby, Banstead and in the West Riding of Yorkshire
- schools at Ilford and Ashford
- the *Arethusa* at Greenhithe
- homes of the landed gentry in Cumberland, Ascot, Bracknell, Burton-on-Trent (not forgetting Buckingham Palace)

The country was traversed from Dawlish and Hastings in the south to Campbeltown and Berwick in the north, from Aberystwyth in the west to Beccles in the east. At this stage the overseas work had hardly begun. It is interesting to note clusters of sites, *e.g.* at Wakefield This was a common pattern and possibly reflects the rôle of personal recommendation in securing more business.

The other interesting thing to emerge from the water analysis notebooks is the use of 'previous sewage contamination'. Far from being abandoned in the mid-1870s the concept was still being routinely employed. The column with this heading was regularly filled up until 25 June 1881 (an entry for Wolverton railway works).[84] Thereafter it was either left blank or else used for some other purpose. It is, of course, exactly at this time that chemical debates on water pollution were becoming (unlike the water they discussed) increasingly sterile. Perhaps Frankland had had enough of that particular controversy. More probably he saw that analysis of the total organic residue in water told enough, for that technique continued with unabated interest. It is also possible that he may have been guided by the results of further work at Rothamsted which now suggested that added nitrogen to agricultural land remained in the soil as vegetable, not animal, matter. Hence its eventual appearance in effluent water may not necessarily imply animal origin. This was suggested to him by Lawes, who stated his warm admiration of the 6th Report of the River Commissioners, adding that 'probably no one has had the Report before him more permanently than myself'.[85] He later cited Frankland's discovery of seasonal variation in the nitrogen washed away from unmanured plots.[86]

Of all the cases of water analysis undertaken in Frankland's later years none can have been more spectacular than that of the water from the great Islamic shrine of Mecca. One day in August, 1881, two bottles of water were delivered to him from William Thistleton-Dyer, at that time Assistant Director of Kew Gardens. He had received them from the British Consul at Jeddah with the information that they came from Hagar's Well in Mecca. A label indicated that the water was holy and was sent all over the Islamic world as gifts. Recognising the prevalence of disease in the area he added 'It may be the heaven-sent messenger of all sorts of zymotic diseases'.[87] Would Frankland analyse it? He did so, was horrified by the result and promptly wrote to the *Times*.[88] Most alarming of all was the discovery that the nitrate concentration was *six times that in strong London sewage*. For all the abandonment of previous sewage contamination in his reports in England, and recently from his laboratory notebooks, it leapt back into his consciousness, into one of his official forms[89] and into his subsequent correspondence.

A letter from Thistleton-Dyer (who had seen the *Times* letter) breathed the caution of a civil service official. Was he sure the sample was genuine? He could write to the consul at Jeddah who happened to be back in London between tours of duty.[90] Frankland evidently took the advice for only three days later a letter from the Consul (a Mr James Zohrab) was on its way. It brought plentiful assurances that the sample was genuine: 'It was brought to me last January by a Mohammaden gentleman in whom I have implicit confidence'. He added the important information that the source was not a spring but a *well*, supplied by filtration, and that at times of pilgrimage thousands of pilgrims daily washed in its vicinity.[91]

This was just the kind of situation to appeal to Frankland: it needed action, it needed human concern, above all it needed science. He did not waste time. Within a day or so he was writing, not to Thistleton-Dyer, but to his boss, Sir Joseph Hooker. Giving details of his analysis, of the immense 'previous animal contamination' (a variant on his usual phrase), of the authenticity of the sample and of the conditions described by Zohrab, he concluded the well was 'a most potent source of cholera poison'. In the whole of his career he had never encountered a sample of drinking water 'making even a distant approach to the degree of pollution exhibited by the water of Hagar's Well'. His conclusion was simple: 'In the interests of the health of Europe and Asia efforts ought to be made to guard the water from this abominable and dangerous pollution'.[92] Shortly afterwards Hooker's subordinate reported: 'I have forwarded the very important document you have sent us', together with an official letter, to the Secretary of State for India.[93]

There matters rested. Or rather they did so for ten more years. Then, in September 1891, Frankland received a telegram announcing 11 000 deaths in Mecca from cholera. At the same time a letter reached him from Constantinople. It was a very lengthy composition in French from a Colonel Bonkowski-Bey, chemist to the Sultan. He and Frankland had met previously in Vienna and Budapest (presumably at International Hygiene Congresses). Now, prompted by a further outbreak of cholera at Mecca, he had performed his own analysis of the well water. He quoted Frankland's old figures correctly except that the value of previous sewage contamination was 1000 times too low (possibly through misreading a decimal point). His own analysis is much less alarming than Frankland's, but contains no organic carbon or nitrogen determination. Bonkowski-Bey concluded that his figures disproved the genuineness of the 1881 sample.[94]

Frankland was on a Scottish holiday when the letter arrived but on his return to London he sent a reply. He confirmed his belief in the authenticity of his sample and stressed the differences between their two sets of results, particularly the fact that his nitrate estimate of 1881 was 15 to 30 times larger than that of his correspondent. With courtesy but more than a hint of impatience he asked 'Do you not think, dear Colonel, that in the fact of the fearful mortality from cholera at Mecca . . . a strict inquiry ought to be made into its causes on the spot?' He went on to suggest seven questions that he could ask at Mecca, including location of drains, nature of soil, depth of well *etc.* The issues were starkly simple:

> Which is the water which poisons the pilgrims at Mecca? A careful investigation on the spot would certainly answer this question as it has done in scores of cases in this country.[95]

He did not believe in quarantine or disinfection, but in eliminating the source of the plague. That should be the objective of Colonel Bonkowski-Bey.

Nearly two more years passed until the issue surfaced in public consciousness. In June of that year reports reached Europe of a new and fearful epidemic of cholera at Mecca. In one day 60 persons died, but the Sultan did not take the action recommended by the international community, fearing foreign interference in Turkey. By the end of the month the daily death-toll had reached 999, though within a week it had dropped to 260 (with many more at Jeddah). By 8 July it was reported that in the previous month 7 000 persons had died in Mecca, but the Egyptian Quarantine Board considered the real numbers were twice that of the official figures. At Mecca corpses were piled by the roadside as there were simply not enough grave-diggers. Of the 60 000 pilgrims in Mecca that summer over 10% were victims of cholera.[96] By now the European pubic was thoroughly alarmed and on 22 July the *British Medical Journal* wrote to Frankland for a copy of his analyses of Mecca water.[97] On the very same day Frankland composed a second letter to the *Times*. The expectations of an enquiry last time had not been fulfilled. The last hope was a question by Roscoe in the House of Commons. That might cause the Foreign Secretary to put pressure on the Turkish Government to abate the risk.[98] On 24 July Roscoe asked whether it was not time for 'International representation' on the matter, but the First Lord of the Treasury, Sir Edward Grey, assured him that an Ottoman Board of Health had been set up and advised waiting for their response. On the following day Roscoe again rose in the House, demanding immediate publication of that body's report, but Grey was unable to suggest how long that would take or how long the Board had been constituted. But of course 'no effort will be spared' by HM Government, *etc.*[99]What, if anything, the Government actually did we do not know, though by 8 September the disease had spread to Constantinople and alarm of an impending epidemic had reached Rumania.[100] Quite apart from the scale of human tragedy and the international dimensions of the problem, the episode illustrates much of Frankland's underlying approach to water analysis.

Frankland's analytical work was not limited to actual analyses performed by himself or his army of faithful assistants. Ever the polemicist he also took delight in the many legal actions that took place over questions of water supply or purity. He was no stranger to litigation and continued to be involved in lawsuits relating to other matters (including carbon batteries[101] and smokeless powders[102]). On one occasion he was advising Counsel on the pollution of oyster-beds in the Isle of Wight's River Medina by sewage from Newport and Cowes. This time Frankland was concerned to prove *minimum* sewage contamination, while the opposition was represented by Voelcker and Letheby (an interesting inversion of rôles). His suggested questions were cleverly designed to show that their opponents could not prove the organic origins of various products in the river, and that their sampling techniques might have been defective, involving agitation of the mud bed. He regarded Letheby as the more formidable foe.[103] Another contest was between the Eastbourne Water Works Company and its consumers, who felt disinclined to pay their bills on account of the poor quality of water supplied.[104]

In addition Frankland was called on to give evidence before Parliamentary committees. He must have been gratified to read the following in a printed sheet[105] to be gummed on to the official record (p.233) of the report on the Southwark and Vauxhall Company's Bill:

> With regard to the answers which I gave to Questions 2256 and 2508 and 2533 to 2542 in my evidence before the House of Commons Committee on the 5th of May last, I desire to put on record that I in no way impeach the honesty, or, as I said in reply to Question 2577, the bonâ fides of Mr. James Mansergh, Mr. George Henry Hill, Sir

EDWARD FRANKLAND, or any other gentleman connected with the case, and I desire to apologise if, in the heat of cross-examination, any words may have escaped me which can bear such an interpretation.
(Signed) ALEX^R. R. BINNIE

6th July, 1898

Nor was the excitement limited to Parliament. Frankland had frequently to adjudicate in controversies involving local Vestries, for parish pump politics (a very literal phrase) could generate immense heat. As he remarked to Crookes after one bruising experience 'Mr Cassal should be more cautious when he reports to a highly inflammable vestry'.[106]

Since 1871 Frankland had included bacteriological work in his investigations of drinking water. In May 1892 he took this aspect of analysis much more seriously, for it was now clear to him that among the millions of bacteria swarming in a glass of water there might be some that had lethal effects on human beings. The nearest he got to a modern understanding (and it was surprisingly near) was a statement to the effect that 'the virulence of pathogenic microbes is due to the chemical poisons they excrete'.[107]

At one stage Frankland had argued for filtration through spongy iron filters, as advocated by Bischof,[108] but by now he was convinced that such microbes could be at least partially removed by sand filtration[109] and by Clark's process for water softening.[110] This, of course, was music in the ears of the water companies but not at all pleasing to the hawks of the Local Government Board. Convinced of the value of a concerted effort by chemistry and microbiology, Frankland began to include data from bacteriology in his reports, but in August 1898 was astounded to receive a request from the Local Government Board to discontinue the practice forthwith. He responded with alacrity[111] and the Board observed that this extra work had been paid for, not by them, but by the water companies[112] Clearly the Board was worried that its bulldog was losing his bark, or had even defected to the enemy. But they need not have worried. Within a few months he was dead.

7 To what end?

It remains to ask two questions: why did Frankland act as he did, and what were the effects of his actions? In relation to the first of these, it might well be thought that Frankland acted as a disinterested scientist, dispensing knowledge and advice impartially to all and simply following his scientific instinct. This view of value-free science is distinctly unfashionable today, and the only writer to deal in any detail with his work in water analysis has argued a very different case. Christopher Hamlin has performed a great service to our understanding of this aspect of Victorian science and his splendid work has relieved this chapter of the necessity to repeat many of the details that he has provided, particularly from Government reports and papers. For him Frankland is anything but a disinterested scientist pursuing solely the cause of truth.

Hamlin asserts that Frankland was developing a strategy to make analysis a basis for social action and this involved the 'practice of colouring facts'. In more detail he defines the principles of Frankland's strategy in the following way:

1. that both negative and positive results of water analysis were untrustworthy (analytical nihilism);
2. that the onset of water-borne zymotic disease was un-predictable;
3. that decision-making bodies, having great and misplaced faith in the abilities of chemistry, would not take decisive action in the face of uncertainty;
4. that therefore, in attempting to protect the public, the scientist could not rely on normal democratic processes, but would have to pre-digest information.[113]

However, these generalisations require considerable qualification. The first proposition as it stands enshrines an obvious ambiguity. If it means 'results of chemical analyses *on their own* cannot predict epidemics' then it is true but, in the light of knowledge at the time, unremarkable. Indeed it is what Frankland repeatedly stated. One needed a great deal of supplementary information, such as the proximity of sewer outlets and (later) the results of microscopic examination. However it might mean that chemical analyses *could not reliably signal dangers* in water (even though that did not extend to epidemic prediction), either because there was no clear correlation between such dangers and chemical data, or because the latter were just inaccurate. Such 'analytical nihilism' is contradicted by all we know of Frankland. Though negative results did not necessarily acquit water, positive results inevitably condemned it (until very late in his life); and the accuracy of his methods was confirmed even by his opponents.[114] If, on the other hand, it suggests that chemical analyses *taken together with other data* are untrustworthy it is manifestly untrue, for the use of data from chemical analysis, microbiological examination and topographical enquiries was a central plank of Frankland's argument. The question is not whether *we* think such strategy to be valid but whether *he* did. Of course mistakes were made, as in any science, but they did not affect the issue in principle.

The second proposition is certainly correct for at least the 1860s to the 1890s. The third contains a number of ideas that need unpacking. Who, for example, were the 'decision-making bodies'? As Frankland knew all too well, the ultimate decision-makers (Government) did not even have small faith in chemistry. Farr had told him in 1865 that 'it was impossible to get the Government to see the importance of the water analysis',[115] and subsequent experience had not contradicted that. To some extent the Local Government Board, and to a greater extent the water companies, might be said to have exercised 'great faith' in chemistry (and chemists would have much enjoyed it). But if their faith was 'misplaced' that is only from a very modern, Whiggish standpoint. There is no evidence that Frankland saw it thus, even after he recognised that one had to use chemistry and bacteriology in tandem. Faith in chemistry, as such, was not misplaced. But it remains a matter of common experience that decision-makers (whoever they might be) do want as much certainty as possible (whether in chemistry or anything else) before they take decisions.

Proposition 4 is so convoluted that it is hard to know what to make of it. What are 'normal democratic processes' in this context? Votes? Delegation of powers? And what have they to do with the transmission of chemical ideas? And what scientist, then or now, does *not* 'pre-digest information' in order to make it accessible to his or her intended audience? That Frankland made every effort to present his ideas effectively was a fundamental feature of his style, as has been repeatedly demonstrated in the last few chapters. But it is pure speculation to imply that the clear presentation of data reflected an intention to dress up known uncertainties as empirical facts in order to please the authorities. Hamlin makes much of Frankland's method of expressing previous sewage contamination as parts per 10^5, specifically contrived to give large numbers which might impress (or depress) members of the public. One could equally well argue that this was done to make the results accessible to lay-persons because such people are normally happier with whole numbers than with decimal fractions. But either way it is idle to speculate without the most solid evidence. The figure was simply the proportion of the water that had at one stage been in the same condition of London sewage; it was generally a 3- or 4-figure number.

That, however, is not the whole burden of Hamlin's charge. He says 'previous sewage contamination was an analytical construct designed to serve a political function',[116] as was Frankland's depiction of the germ. There are two contentious points here. The first concerns the use of the value-laden term 'political'. If the mere *identification* of 'safe water' was a political function then disputes about the word become merely a matter of semantics; it is hard to see any investigative process to which it could not be applied. More obviously political was the *delivery* of that water to the British public, accompanied by dire warnings as to the possible harm in water contaminated by sewage *etc*. To be sure, Frankland endeavoured to identify safe and unsafe sources of domestic water in the expectation that the companies and local government would take the appropriate action. But that is a fairly weak sense of 'political function'.

Even that is not Hamlin's main point, for he argues that the concept of previous sewage contamination was value-laden, that it was *designed* to achieve a certain end, carefully planned so as to achieve maximum social effect. The phraseology implies that non-scientific issues ('political' ones) obtruded into a programme of scientific research. It implies also that we can be sure of Frankland's actual intentions, and can read behind his rhetoric a hidden strategy that was somehow more 'real'. That may be possible through inspecting private correspondence, diaries and the like. The difficulty in Frankland's case is that nothing whatever has come to light that even remotely suggests that such concepts were ever seen as anything more than shrewd scientific hypotheses designed to indicate possible hazards in water.

But let us not be easily deterred. If documentary evidence of his own words is not forthcoming we can instead examine what other people said about him. Here there is no shortage of material, much of which is brilliantly exposed in Hamlin's book. The difficulty is that nearly all the data come from those with an obvious axe to grind: Wanklyn with a deep personal grudge; Crookes who fell foul of the River Commissioners over his manure company; above all the water companies' chemists who were actually paid to deny Frankland's allegations and therefore to discredit their author. The fact that Tidy, Letheby and Odling complained that 'Frankland was trifling with public sensibilities in a manner inappropriate to his official status' at the very time when they were setting up a rival analytical establishment tells us more about the determination of the water companies than about the true motives of their opponent.

Finally, if all else fails, we may take refuge in inferential conclusions based upon an individual's behaviour. We may point out the large political consequences of Frankland's advocacy (to which we come in a moment) and deny that they could have come about by anything other than a settled programme of political action. That again has its dangers, not least the simple fallacy of asserting that, because a process has certain consequences, it was therefore necessarily designed to that end.

There is an alternative scenario to that of political schemer. It takes into account three elements of Frankland's life that Hamlin tends to underplay or even ignore. The first is his dedication to effective scientific communication; that alone could be sufficient explanation for the manner in which he presented his analytical results. The second is his track record of chemical research. A paper by Hamlin[117] speaks of Frankland's 'early career as London's official water analyst, 1865–1876'; this a perfectly legitimate description for its own purposes. Yet at the beginning of that period Frankland had already been engaged in analysis for 20 years, he was at the peak of his career as a chemist and had already shown himself as one of Britain's most distinguished chemical researchers. Any person's work must be seen in the context of what he or she has done before. Consideration of his research record renders it unlikely that he would resort to cheap fudges or misrepresentations. Apart from anything else he had his already towering reputation to consider. The third underestimated element in Frankland's life is the pressure of professionalization. It was during the late 1860s and 1870s that this pressure was at a maximum and chemists were more glaringly in the public eye than ever. Accuracy, integrity and all the other virtues were now at a premium. Frankland, who was to lead that movement, had more to lose than any by misleading the public. The movement came to a head in 1877 with the establishment of the Institute of Chemistry.[118]

The alternative scenario is therefore this. Frankland's work in water analysis was undertaken at the behest of the Registrar General on behalf of the public. The water examiner had to be impartial, and had to use all information available, whether from his own speciality or not, to arrive at the best answer. That is why Frankland, risking the wrath of fellow-chemists who did not have his responsibility, acknowledged the dangers inherent in untreated sewage. To say that Frankland's rules 'permitted one to arrive at an opinion of a water's quality by whatever route and on whatever basis one chose' is to forsake critical analysis for caricature, though to add that one could 'defend it by whatever means seemed most effective'[119] is to state exactly what happened, namely the exercise of sound common sense.

However a biographer does no one a service by implying his subject has no faults. If Frankland was not a man to subordinate his scientific enquiries to political expediency (and he emphatically was not) he can be criticised on several other grounds. Thus he was prepared to tailor his reports at the request of a client, though only within certain limits. This happened over a report on the waters of Constantinople where various potentially inflammatory phrases were suitably toned down.[120] He was often criticised for farming out the laboratory work to students and assistants,[121] and towards the end of his life, his son Percy (then having a major row with his father) threatened to expose his practice of leaving blank analytical report forms with his signature at the bottom; if the public had got to know 'the consequences might have been very serious'.[122] In these and other ways Frankland could be faulted even by the standards of his own times.

So what did he accomplish by his 34 years of 'official' water analysis? Whereas Hamlin has asserted 'Frankland shamelessly used his position as a forum for social change'[123] it would be much more accurate to say 'Frankland courageously used his position as a means for improving public health' (courage was certainly required for he was often in a small minority, sometimes almost alone). In any event he undoubtedly had a profound effect on the health of the nation.

The public pressure generated by his relentless quest for water that was likely to be pollution-free led to the closure of multitudes of shallow wells and springs, to the abandonment of hundreds of contaminated sources at home and abroad, to proper control of reservoirs and (in his later years) to the greater use of filtration and other purification techniques. Numerous pieces of hygiene legislation resulted from government reaction to his disclosures, including the Metropolitan Water Act (1871) creating the post of Water Examiner, and the Public Health Act (1875) establishing posts of Medical Officer of Health and Public Analyst.

Perhaps three final points should be made about Frankland's methodology in all his water analysis. Though much misunderstood by many opponents and one or two historians, his insistence that chemistry was relevant to the *history* of a water specimen was central to his whole strategy. That science could draw conclusions about the past from the present was a basic thesis of geology and may have been suggested by his own work on glaciers and by the interest of his friends Tyndall and Huxley (see Chapter 14). The methodology is closely similar. He even referred to nitrates as 'skeletons' of previous sewage contamination, calling to mind the inferential language of palaeontology.

A second feature was the language and thought-forms of jurisprudence. The frequent references to water being 'guilty' or 'acquitted', to 'convictions' and even to 'ordeals' suggest vividly a court of law.[124] There is no doubt that, as official Government analyst, he regarded himself as something very like prosecuting counsel. But there was one vital difference from legal practice, and he often referred to it: 'my motto, unlike that in criminal cases, has always been assume water to be guilty until it is proved innocent.'[125]

Finally there is a piece of advice, again repeated several times, which moulded his whole method of presentation. He applied it to himself, increasingly as the years passed. It was the antithesis of the attitude displayed by that beleaguered official in Constantinople: 'I do not want the public to have a voice in anything but the result'.[126] For Frankland policy was determined by a lifetime of effective communication to people of all kinds. His motto was 'Take the public into your confidence'.[127] If ever that approach was necessary it was in the vexed questions of water supply and water analysis in Victorian England.

DEPARTMENT OF SCIENCE AND ART.

ROYAL SCHOOL OF MINES.

LECTURES
TO WORKING MEN.

The following Courses of Six Lectures each will be delivered in the Evening during the Session 1880—81.

1. On **Chemistry** ················· By EDWARD FRANKLAND, D.C.L., F.R.S.
2. On **Applied Mechanics** ········ By T. M. GOODEVE, M.A.
3. On **Mineralogy** ··············· By WARINGTON W. SMYTH, M.A., F.R.S.

THE FIRST COURSE,

ON

WATER,

WHICH WILL BE DELIVERED IN

The Chemical Theatre of the Science Schools at

THE SOUTH KENSINGTON MUSEUM,

Will commence at EIGHT O'CLOCK, on MONDAY, the 8th of November, 1880.

LECTURE I.—*Monday, November 8th,* 1880.
Solid Water.—Properties of Ice.

LECTURE II.—*Monday, November 15th,* 1880.
Liquid Water.—Its solvent powers and influence on climate.

LECTURE III.—*Monday, November 22nd,* 1880.
Gaseous Water.—Properties of Steam.

LECTURE IV.—*Monday, November 29th,* 1880.
Chemical constituents of Water.

LECTURE V.—*Monday, December 6th,* 1880.
Water as it occurs in Nature.

LECTURE VI.—*Monday, December 13th,* 1880.
Drinking Water.—Its Pollution and Purification.

Tickets for this Course may be obtained at the Museum of Practical Geology, Jermyn Street, and at the South Kensington Museum, by Working Men only, on MONDAY EVENING, the 1ST NOVEMBER, 1880, from **SIX** to **TEN** o'clock, p.m.—Registration Fee for the Course of Six Lectures, 6d. Each applicant is requested to bring his Name, Address, and Occupation, written on a piece of paper, for which the ticket will be exchanged.

F & T 700 10—80

Fig. 13.3 Science for the masses. (J. Bucknall Archives.)

Notes

1 [J. R. Japp], *Proc. Inst. Civil Eng.*, 1900, **139**, 343–9 (347).

2 See, *e.g.*, (a) Anne Hardy, 'Water and the search for public health in London in the eighteenth and nineteenth centuries', *Med. Hist.*, 1984, **28**, 250–82, and (b) W. Luckin, *Pollution and control: a social history of the Thames in the nineteenth century*, Hilger, Bristol, 1986.

3 E. Walford, *Old and new London*, Cassell, London, n.d. [c.1880], vol. v, p.238.

4 Cholera graveyards from this epidemic may be seen in several cities today, notably at York just outside the city wall near the railway station.

5 Frankland, *Sketches*, 2nd ed., p.3.

6 T. Graham, W. A. Miller and A. W. Hofmann, *J. Chem. Soc.*, 1851, **4**, 375–413.

7 C. Hamlin, 'Edward Frankland's early career as London's official water analyst, 1865–1876: the context of previous sewage contamination', *Bull. Hist. Medicine*, 1982, **56**, 56–76.

8 (a) For a general survey see C. A. Russell, 'Taking the waters: chemistry and domestic water supply in Victorian Britain', in M. Fetizon and W. J. Thomas (eds.) *The role of oxygen in improving chemical processes*, Proceedings of the 6th BOC Priestley Conference (Paris 1992), Royal Society of Chemistry, Cambridge, 1993, pp.174–87; (b) on the social history of water analysis at this time see C. Hamlin, *A science of impurity: water analysis in the nineteenth century*, Hilger, Bristol, 1990.

9 See *e.g.*, A. Debus, 'Solution analyses prior to Robert Boyle', *Chymia*, 1962, **8**, 41–61; N. G. Coley, 'Physicians and the chemical analysis of mineral waters in 18th century England', *Med. Hist.*, 1982, **16**, 123–44, and 'The presentation and uses of artificial mineral waters, *ca*. 1680–1825', *Ambix*, 1984, **31**, 32–48.

10 F. Szabadváry, *History of analytical chemistry*, trans. G. Svehla, Pergamon Press, Oxford, 1966, pp.201, 212.

11 R. Kirwan, *An essay on the analysis of mineral waters*, London, 1799.

12 C. H. Pfaff, *Handbuch der analytischen Chemie*, Altona, 1821.

13 W. A. Campbell, 'The analytical chemist in nineteenth century English social history', M.Litt. thesis, University of Durham, 1971; M. M. Mackin, 'Analytical chemistry in the British Isles in the nineteenth and early-twentieth centuries: aspects of the professional and educational developments', M.Sc. thesis, Faculty of Science, Queen's University, Belfast, 1984.

14 Chemical laboratory notebooks, 1839–51 [British Geological Survey Archives, Keyworth, Royal School of Mines, GSI/728].

15 Williams, *Directory of Newcastle*, 1844, advt. p.22.

16 T. Richardson to W. Bouch, 15 August 1860 [Newcastle-upon-Tyne Central Library, Letters and cuttings on the early history of railways, vol i, f.65].

17 *Lancastrian Chemist*, pp.141–61.

18 C. Johnson to Frankland, 9 April 1847 [RFA, OU mf 01.03.0095].

19 Frankland to B. P. Gregson, 25 October and 8 November 1851 [RFA, OU mf 01.06.0113 and 0118].

20 Frankland to G. Carruthers, 30 September 1852; Frankland to J. F. Foster, 3 November 1852 [RFA, OU mf 01.06.0267 and 0278].

21 Frankland to S. N. Borlam , 11 June 1855; S. N. Borlam to Frankland, 2 and 18 June 1855 [RFA, OU mf 01.08.0266, 0265 and 0264].

22 W. Farr to Frankland, 5 October 1865 [RFA, OU mf 01.03.0090]; it was raised to £150 after his work with the cholera outbreak of 1866, and stayed at that level until his death (undated note, almost certainly 1899 [RFA, OU mf 01.04.0545]).

23 H. E. Armstrong to J. J. Day, 4 April [1868] [Imperial College Archives, Armstrong papers, C-243]; the laboratory was at 111 Victoria Street.

24 Frankland to Sir T. M. Biddulph, 23 January 1874 [RFA, OU mf 01.07.0680].

25 Sir W. T. Knollys to Frankland, 16 and 22 January 1875; Frankland to Sir W. T. Knollys, 15 March and 19 January 1875 [RFA, OU mf 01.01.1181 and 1179; 1175 and 1177].

26 Frankland, Water analysis notebook, 2 September 1878 [RFA, OU mf 01.09].

27 Hamlin (note 8(b)), p.173.

28 H. E. Armstrong, First Frankland Memorial Oration of the Lancastrian Frankland Society, *Chem. & Ind.*, 1934, **53**, 459–66 (462).

29 *Ibid.*, p.462; the paper was read on 16 January and there was further discussion on 6 February 1868.

30 E. Frankland and H. E. Armstrong, *J. Chem. Soc.*, 1868, **21**, 77–108.

31 *Ibid.*, p.87.

32 H. Sprengel, *J. Chem. Soc.*, 1865, **18**, 9–21.

33 H. E. Armstrong, *Pre-Kensington history of the Royal College of Science and the university problem*, Royal College of Science, London, 1920, p.7.

34 E. T. Chapman, M. H. Smith and J. A. Wanklyn, *J. Chem. Soc.*, 1867, **20**, 445–54.

35 Frankland to H. E. Armstrong, 21 September 1867 [Royal Society Archives, MM.10.98].

36 Frankland and Armstrong (note 30), p.83.

37 J. A. Wanklyn, *J. Chem. Soc.*, 1867, **20**, 591–5.

38 Frankland and Armstrong (note 30), p.82.

39 J. A. Wanklyn and E. T. Chapman, *Water-Analysis: a practical treatise on the examination of potable water*, 4th ed. rewritten by J. A. Wanklyn, Turner & Co., London, 1876, pp.160–178; an anonymous reviewer, while 'strongly recommending' this 4th edition expressed 'deep regret' at the controversial matter thus introduced, castigating it as 'wholly out of place' and 'a very serious blemish in an otherwise most meritorious work' (*Analyst*, 1877, 1, 166).

40 See C. A. Russell, N. G. Coley and G. K. Roberts, *Chemists by profession*, Open University Press, 1977, pp.115–17, 138–9 *etc.* The unnamed instigator of the 'rebellion' turns out to be none other than Wanklyn's colleague, Chapman (H. McLeod, *Diary*, reprinted as *Chemistry and theology in mid-Victorian London*, ed. F. A. J. L. James, Mansell, London, 1987: entry for 30 March 1867).

41 B. Dyer, *The Society of Public Analysts and other analytical chemists: some reminiscences of its first fifty years*, Heffer, Cambridge, 1932, pp.12–13.

42 H. McLeod, *Diary* (note 40), entries for 16 January and 6 February 1868.

43 Minutes of the Chemical Society, 19 May 1870 [Royal Society of Chemistry Archives].

44 *Analyst*, 1877, 1, 157.

45 Wanklyn (note 39), pp. 160–1.

46 *Royal Commission on Water Supply*, P.P., 1868/9, XXXII, minutes of evidence, pp. 303 and 305 (Qu. 5421 and 5461, 20 June 1867).

47 W. Odling, testimonial for Armstrong, 1 December 1870 [Imperial College Archives, Armstrong papers, 1st series, 335].

48 J. W. Mallet, *National Board of Health Bulletin*, Washington D.C., Supplement no. 19, 27 May 1882, p.4.

49 Anon., *Analyst* (note 39).

50 *E.g.* editorial comments in September and October 1868: *Chem. News*, 1868, 18, 151, 153, 165.

51 Editorial 'Water analysis' , *Chem. News*, 1872, 25, 157.

52 Frankland, *J. Chem. Soc.*, 1876, 29, 825–51 (847).

53 E. B. Truman to Frankland, 25 June 1874 [RFA, OU mf 01.04.0409].

54 J. W. Thomas to Frankland, 22 December 1876 [RFA, OU mf 01.04.0415].

55 W. Odling to Frankland, '27 April' [RFA, OU mf 01.04.0418].

56 Frankland, undated memorandum [RFA, OU mf 01.04.0421].

57 Frankland to H. Owen, n.d. [RFA, OU mf 01.07.0848]. As far away as New Zealand, and as late as 1896, the Dunedin Public Analyst, A. G. Kidston-Hunter, was arguing the diagnostic importance of albuminoid nitrogen and ammonia in water, whilst earlier estimations of oxidisable organic matter probably employed permanganate oxidation; if so Frankland's techniques do not appear to have made much progress in that part of the Empire (R. J. Wilcock, 'Water chemistry', in P. P. Williams (ed.), *Chemistry in a young country*, New Zealand Institute of Chemistry, Christchurch, 1981, pp.195–205).

58 Frankland (note 52). It was clearly a significant occasion. Frankland had offered a Sprengel pump, together with mercury, to the Chemical Society, and the latter's Council agreed to purchase from the photographer the plates to illustrate the paper at the then astronomical sum of £43 [Royal Society of Chemistry Archives, Chemical Society Council Minutes, 3 February and 20 April 1876].

59 Fred Frankland to Maggie, Sophie and Percy Frankland, 24 May 1876 (the date is incorrectly given as 1874) [RFA, OU mf 01.04.0259].

60 J. J. Day to H. E. Armstrong, 27 March 1869 [Imperial College Archives, Armstrong papers, C-246].

61 Frankland and Armstrong (note 30), p.106.

62 Copy-letter, Frankland to Registrar General, 10 July 1869 [RFA, OU mf 01.04.1134]; the letter was printed in *Reports on the analysis of waters supplied by the Metropolitan water companies during 1869, 1870, and 1871, by Professor Frankland, P.P.* 1872, XLIX, pp.32–8.

63 H. Letheby, *Royal Commission on Water Supply*, P.P. 1868/9, XXXII, Minutes of evidence, p.235 (Qu. 3897, 29 May 1867).

64 Frankland, *ibid.*, p. 350 (Qu. 6372, 27 February 1868).

65 J. A. Wanklyn, *Chem. News*, 1872, **25**, 159–60 (159).

66 J. A. Wanklyn, *Ber.*, 1869, **2**, 64–5; *J. Chem. Soc.*, 1869, **22**, 199–202.

67 J. A. Wanklyn, *Chem. News*, 1872, **25**, 225–6.

68 J. A. Wanklyn, *Sanitary Record*, 1878, **8**, 174–5. Hamlin (note 8(b), p.190) attributes to Wanklyn the opinion that 'Frankland's great discoveries had ceased when he had left Frankland's lab.', though in fact Wanklyn wrote [my italics] that 'Dr. Frankland's latter work – *that dating during the last twelve or thirteen years* – is unmistakably inferior to that of the earlier period', *i.e.* since about 1865, not 1857 when Wanklyn left him. There is still a sting in the remark, though not so venomous as Hamlin seems to suggest.

69 Royal Commission on Water Supply, evidence, p.334 (note 63).

70 Frankland to Tyndall, 19 January 1871 [Royal Institution Archives, Tyndall papers, 9/E3.12].

71 Frankland to Owen (note 57).

72 C. M. Tidy, *J. Chem. Soc.*, 1880, **37**, 267–367.

73 Frankland, *J. Chem. Soc.*, 1880, **37**, 517–46.

74 Frankland to J. Tyndall, 14 May 1880 [Royal Institution Archives, Tyndall papers, 9/E6. 20].

75 Tyndall to Frankland, 16 May 1880 [RFA, OU mf 01.04.1661]; this passage was removed from the Royal Institution copy-letter [Royal Institution Archives, Tyndall papers, 9/E7.8.10].

76 Frankland to J. Tyndall, 21 May 1880 [Royal Institution Archives, Tyndall papers, 9/E6. 21].

77 Undated dinner list [RFA, OU mf 01.02.0916].

78 Maggie West (*née* Frankland), *Diary* [MBA].

79 Frankland to Sophie Colenso, 24 January 1883 [PCA, 03.01.0641].

80 F. A. Abel to Frankland, 2 May 1887 [RFA, OU mf 01.04.1693].

81 Frankland, undated notes [RFA, OU mf 01.04.0161].

82 Frankland, Water Analysis Notebooks [RFA, OU mf 01.09].

83 Frankland, Notebook [RFA, OU mf 01.04.0435 – 0461].

84 Hamlin (note 8(b)) claims Frankland had dropped it from his analytical returns in 1876 (p.160) or at the beginning of 1877 (p.233). Accuracy apart, this is largely a semantic point since the *concept* (under whatever name) was still in use in his private reports until at least 1892 (*e.g.* copy-letter Frankland to Local Government Board on Plumstead Well, 5 February 1892 [RFA, OU mf 01.07.0817]).

85 J. B. Lawes to Frankland, 21 December 1878 [RFA, OU mf 01.01.1281].

86 J. B. Lawes, cutting on 'nitrogenous manures', 1878 [Rothamsted Cuttings Book, vol. ii, p.18], reproduced in G. V. Dyke, *John Bennet Lawes: the record of his genius*, Research Studies Press, Taunton, 1991, p.183.

87 W. T. Thistleton-Dyer to Frankland, 26 August 1881 [RFA, OU mf 01.08.0571].

88 Copy-letter Frankland to *The Times*, 8 September 1881 [RFA, OU mf 01.01.0168]; published 9 September (p.9).

89 Analysis form, n.d. [RFA, OU mf 01.01.0173].

90 W. T. Thistleton-Dyer to Frankland, 26 October 1881 [RFA, OU mf 01.01.0180].

91 J. Zohrab to Frankland, 29 October 1881 [RFA, OU mf 01.01.0176].

92 Copy-letter Frankland to J. Hooker, October 1881 [RFA, OU mf 01.01.0164].

93 W. T. Thistleton-Dyer to Frankland, 18 November 1881 [RFA, OU mf 01.01.0183].

94 Bonkowski-Bey to Frankland, 8 September 1891 [RFA, OU mf 01.08.0581].

95 Frankland to Bonkowski-Bey, 31 October 1891 [RFA, OU mf 01.08.0574].

96 *The Times*, 10, 22 and 28 June, and 5, 8, 10 and 18 July 1893.

97 E. Hart to Frankland, 22 July 1893 [RFA, OU mf 01.01.0186].

98 Copy-letter Frankland to *The Times*, 22 July 1893 [RFA, OU mf 01.01.0187]; published 26 July (p.3).

99 *Hansard* for 24 and 25 July 1893 (pp.307, 483).

100 *The Times*, 8 September 1893, p.3 (the last reference that year to cholera in the region).

101 W. Crookes to Frankland, 11 May 1886, and Frankland to F. L. Rawson, 6 June 1887 [RFA, OU mf 01.04.1443 and 01.01.0030].

102 Frankland to I. C. Andersen, 29 November 1892 [JBA, OU mf 02.01.0476].

103 J. Eldridge to Frankland, 16 November 1871, and Frankland to J. Eldridge, 17 and 21 November 1871 [RFA, OU mf 01.07.0741, 01.07.0733 and 01.01.1237]; it is probable that Odling was also acting for the defence: W. Odling to Frankland, '18 October' [RFA, OU mf 01.07.0696].

104 Numerous papers 1895–8 [RFA, OU mf 01.04.0475–6 and 0480–0494].

105 Printed sheet and typed copy [RFA, OU mf 01.04.0474 and 01.03.0856].

106 Frankland to W. Crookes, 2 April 1899 [RFA, OU mf 01.03.0854].

107 Frankland to Local Government Board, 18 October 1898 [RFA, OU mf 01.04.0534].

108 A letter from Thomas Spencer to a 'Mr. Marshall' complains that he had anticipated Bischof by 20 years and that his filters were installed in numerous government offices, Buckingham Palace *etc.* He adds that 'Bischof is backed up by Dr. Frankland with whom I have been at feud, as the Scotch say, for several years on the subject of sewage irrigation': T. Spencer to 'Mr. Marshall', 18 April 1877 [Royal Society Archives, MC11/50]. Bischof's process lasted only for a few more years, being replaced by sand filtration.

109 Copy-letter Frankland to Sir A. R. Binnie, 28 May 1898 [RFA, OU mf 01.03.0844].

110 Copy-letter Frankland to East Surrey Water Co., 22 March 1898 [RFA, OU mf 01.04.0498].

111 Frankland to Local Government Board (note 107).

112 Local Government Board to Frankland, 30 November 1898 and 27 February 1899 [RFA, OU mf 01.04.0537 and 0539].

113 Hamlin (note 8(b)), p.161.

114 An amusing example is supplied by Crookes, after Frankland had challenged his analysis of Chelsea water :'I am sorry to say our man has been giving us New River water for some time past instead of Chelsea. The policeman

objected to his taking water from the Horse Guards standpipe, but sent him to one a little way off, saying it was the same water. He thinking it did not matter never told us. Your suggestion was therefore perfectly correct, and the possibility of your so accurately hitting the nail on the head is in itself one of the best proofs of the *extreme accuracy of water analysis.* I have made a clean breast of it in the forthcoming report, and can only hope that my friends (and, alas, opponents), will let us down easily. Such a mistake cannot arise in future' [my italics] (W. Crookes to Frankland, 11 August 1894 [RFA, OU mf 01.07.0753]).

115 Farr to Frankland (note 22).
116 Hamlin (note 8(b)), p.161.
117 Hamlin (note 7).
118 Hamlin's incorrect date ('the early 1880s') may well have led him to underestimate this factor in the earlier years: Hamlin (note 8(b)), p.179.
119 Hamlin (note 8(b)), p.207.
120 Among other things Frankland's correspondent feared misunderstanding of his sentence 'I should not despair of making this an excellent potable water' (by filtration), as 'the word despair has frightened everybody': Frankland to J. F. Bateman, 29 June and 10, 12 and 16 November 1874 [RFA, OU mf 01.01.1126, 1124, 1128, 1107]; J. F. Bateman to Frankland, 9 June, 10 and 29 October, 9, 11, 14 and 17 November 1874 [*ibid.*, 01.08.0547, 01.01.1115, 1113, 1114, 01.08.0535, 01.01.1109, 1103]; another case was a report on signal lights: Frankland to N. J. Holmes, 23 February 1872 [*ibid.*, 01.07.0772].
121 Frankland once wrote 'it is neither possible nor desirable that the principal of a laboratory should personally make the analytical determinations, but they should be made under his direction or supervision' (Frankland to A. Hill, *Analyst,* 1888, 13, 143; reproduced in Dyer (note 41), pp.96–7).
122 Percy Frankland to Frankland, 12 December 1889 [RFA, OU mf 01.04.0219].
123 Hamlin (note 8(b)), p.153.
124 *E.g.,* Frankland (note 62).
125 Frankland, undated memorandum (*c.*1898) [RFA, OU mf 01.03.0865].
126 J. F. Bateman to Frankland, 14 November 1874 (note 120).
127 Frankland, undated memorandum (*c.*1896) [RFA, OU mf 01.04.0427].

⋆ CHAPTER 14 ⋆

'The wildest parts' of nature

1 Science from the mountains

It is time now to retrace our steps and to record what might be called the more peripheral scientific work performed by Frankland, peripheral in the sense that it did not obviously relate to the main themes of his chemical research. While his most lasting fame will be for his work in valency and synthetic organic chemistry, and as we have seen (Chapter 13), he was chiefly renowned in his lifetime for water analysis, Frankland touched many other fields of scientific research. Of those to be discussed in this chapter, every one owed its inspiration, directly or indirectly, to a fascination with the high places of nature, the wilder parts of the natural world. As he put it 'the researches in physical chemistry were, as a rule, either executed during vacation rambles or were suggested by observations made during such holiday trips'.[1] It is as though from time to time Frankland was irresistibly drawn from the hum-drum world of laboratory analysis and academic politics into a closer, more direct, communion with nature, as has many another physical scientist before and since. In Frankland's case it was plain for all to see.

Ever since Frankland espied the dim outlines of the Cumbrian mountains from his school playground at Lancaster they, and their counterparts in Scotland and the Alps, cast a lasting spell upon him. Year after year he was lured from London to explore their mysteries, to wonder at their stupendous beauty and to attempt an assault on their more difficult and challenging peaks. Mountaineering became a sport that rivalled his other favourite pastimes of fishing and sailing. And yet it was much more than a sport. Like many Victorians he found a surrogate for worship in their remote grandeur, and like many of his scientific comrades he became acutely inquisitive into the natural phenomena they displayed.

Frankland's first visit to the Alps appears to have been in 1856 when he met Tyndall at the Finstermuntz Pass in the Tyrolese Alps (p. 188).[2] Their tour of the English Lakes in the spring of 1859 (p. 232)[3] was followed by an unforgettable experience in the Alps later the same year. This was the ascent of Mont Blanc and the spending of a night on its summit.

Mont Blanc, at 15 782 ft., is the highest peak in the Alps. Long regarded as unconquerable it was first successfully climbed in 1786 by a local doctor, Michel Paccard, and a guide, Jacques Balmat. Ascending the north west buttress of the mountain they avoided the lesser peak of the Dôme de Gaûter that had been the termination of all three previous attempts and instead struck up the formidable Valley of Snow, thence reaching the top after an unbelievable climb of 14 hours in one day.

A few others followed in their steps over the next half century but it was not until the 1850s that Alpine climbing became a fashionable pursuit for Victorians generally. It began to be particularly favoured by British academics on their long vacations, railway travel now suddenly bringing the Alps within their reach. Amongst their number was John Tyndall who, with Huxley and Hooker, began exploring the Oberland in 1856. The next year he ascended Mont Blanc by Paccard's route, together with his friend Thomas Archer Hirst and a guide. A second ascent the following year, with Auguste Balmat as guide, was undertaken in order to place a maximum/ minimum thermometer on its summit. That was to be the prologue to an adventure in which Frankland found himself involved in 1859.[4]

Tyndall had long wished to have scientific data about temperatures in high altitudes and had been encouraged by the Royal Society and others to conduct further experiments. Accordingly he now intended to ascend Mont Blanc for the third time, to recover a thermometer he had left on the summit the previous year, to spend more time there gathering information and to set up a series of thermometers attached to stout posts at intervals between the valley and the summit. Now Frankland was able to join him, having already had recent experience of his company on the lesser peaks of the English Lakeland.

They foregathered at Chamonix at the foot of the mountain, and spent a week exploring the other side of the valley, waiting for suitable weather and preparing themselves for the major assault. On 20 August they set out, with three guides (including another Balmat) and no less than 26 porters. They were prepared, if necessary, to spend up to a week on the mountain so needed plenty of supplies. Since it was widely known that they intended to stay on the top for some time they left with more than the usual expressions of goodwill and proceeded by the route now familiar to Tyndall. After seven hours' steady climbing they reached Grands Mulets; here was a small climbing hut and at this point 14 porters returned to Chamonix. The remainder spent the night at this desolate place, though the two scientists pitched their tent for themselves as the party was too large for the hut. They were well into the glaciers and the internal temperature of the tent was −2.8°C. Thanks to the continuous roar of avalanches uncomfortably nearby they had a sleepless night, but emerged before dawn to witness a dramatic multicoloured halo as the sun approached the rim of the horizon. At first light they set off, Frankland exhilarated by the prospect and the bracing air. They entered the long Valley of Snow and, unlike some of their predecessors, were fortunate to avoid almost any wet snow underfoot. On arrival at the Petit Plateau they were enjoined by the guides to keep total silence lest noise should dislodge part of the glacier above them and induce an avalanche; the precaution was probably unnecessary. Three further hours of climbing brought them to the Grand Plateau, an almost level snow platform about $1\frac{1}{2}$ miles in diameter. Here a 30-minute halt for refreshment was permitted before the serious business of climbing a range of ice precipices was attempted. After this it was a mile or two of 'monotonous and wearisome plodding' through knee-deep snow in the declivity known as the Corridor. A 300-feet climb to the Mur de la Coté brought them within sight of the final cone and after a further hour of stiff climbing they reached the summit. To Frankland 'the scene was unutterably grand . . . a panorama which I could never forget'. Seven more porters were now despatched back down the mountain, some with letters home written on the journey. Then came the first indications of potential disaster.

Tyndall, unlike Frankland. had been unwell throughout the climb, suffering from headache and general lassitude. Within half an hour of reaching the summit the rest of the party complained of similar symptoms. Altitude sickness was beginning to take its toll, so the tent was erected as a matter of urgency. Fortunately Frankland was relatively unaffected at first but gradually he too developed symptoms, though less markedly than anyone else. He wrote: 'I . . . soon found myself tacitly installed in the office of cook to the whole party, who now began to cry out loudly for tea but who, with the exception of Balmat, seemed utterly unable to afford any assistance in making it'. At which point a further disaster stared them in the face: the guides had brought only enough charcoal fuel to melt an urnful of snow.

Mercifully Frankland had insisted as a back-up precaution that the porters carried a large spirit lamp and a gallon of alcohol. Placing hard-packed snow and some tea in a small copper pan he was able to provide a warming drink for everyone, though it took two hours to do it. It was their only dietary requirement, neither alcoholic drink nor solid food being in the least attractive to any. By 9 p.m. Frankland himself was succumbing to the general torpor but with much difficulty he managed to fire a rocket from the summit as a pre-arranged sign to the people of Chamonix that all was (comparatively) well. Thereafter he and the whole party settled down for the night in a seven-foot diameter bell-tent, on a rubber ground sheet, with the 14 bodies arranged spokewise towards the centre. In other circumstances their dilemma would have been comical:

> the pile of legs in the centre, armed with heavily-nailed boots, produced a pressure on the lowermost pair that could not be borne for more than a few minutes; consequently there was a continual transference of the undermost pair of legs from the base to the summit of the pile, causing not merely a disturbance to the equilibrium of the pile itself, but also of the tempers of the individuals owning the legs.

Though several porters had to leave the tent during the night to be sick, by 8 a.m. most were much better, including Tyndall who busied himself with setting up the first thermometric post and then measuring the heat of the sun's rays. Frankland then conducted an experiment of burning stearin candles for an hour in order to ascertain whether they lost more or less weight than when burning in Chamonix. The experiment was conducted within the tent, from which all except himself were excluded to minimise moving currents of air.

It was now clear that further time on the summit was neither necessary nor prudent, so 22 hours after their arrival they began the descent, erecting thermometric posts on the way and pausing at the Grand Plateau to satisfy their returning appetites with small amounts of food. They reached Chamonix an hour after dusk, having completed an expedition that Frankland claimed was 'the first and only one in which a night has been spent on the summit of the mountain'. Many had not thought it possible, so the achievement itself was a modest piece of research into the limits of human endurance. Yet for some strange reason it has not entered the literature of Alpine climbing, was not included in Tyndall's *The glaciers of the Alps*, and was sceptically received by many, of whom a resident of Davos may be typical:

> The Director simply said 'impossible, one would freeze', & in spite of all we said he persisted in disbelieving us but admitted the possibility of sleeping a few 1000 feet below the summit; but as to such a feat as sleeping at the top, he totally ignored, saying that if such a thing had been done it would have been published in all the papers.[5]

Frankland afterwards said that the ascent of Mont Blanc furnished him with 'experimental work extending over nine years'.[6] The experiments formed the major scientific 'spin-off' from the expedition, and were all concerned with combustion phenomena.

1.1 Combustion rates and pressure

It all began with the examination of candles burning on the summit. The general conclusion to which he came was that pressure changes have little effect on the rate of combustion. Each candle was allowed to burn for an hour at Chamonix and an hour on the mountain-top, being weighed before and after each experiment. The differences were not large enough to be significant, especially as the possibility existed that temperature, rather than pressure, might be a critical factor. The temperature on the summit was much lower than that in the valley. A controlled experiment at ground level later confirmed the general impression that pressure was not very important, only a 15% drop in combustion rate being detected at quite low pressures.

This was just about the time of a revived interest in combustion generally. Of the various discussions one of the most important was a communication by an artillery officer in India, Quartermaster

Mitchell.[7] He examined the variation of burning times for shell-fuses at different altitudes (and therefore pressures). Unlike Frankland's candles Mitchell's fuses took much longer to burn at high altitudes than at low levels. Frankland's response was the research on time-fuses he had mentioned to Stokes[8] (p. 211) and was entirely typical of the man.

First he used his considerable network of contacts to acquire the raw materials, in this case six-inch fuses from F. A. Abel at Woolwich Arsenal. Then he constructed an apparatus that embodied many familiar features, a large iron cylinder with an air-pump at one end and at the other a six-foot long gas-pipe connected to a mercury pressure-gauge and capable of receiving the fuse. An iron ball was suspended by a thread at the far end of the fuse so when combustion eventually reached that end the thread would burn and the ball would drop on to an iron plate below. It seems to be a rare case in Frankland's work where the critical signal was an acoustic one.

He concluded that *each diminution of one inch of barometrical pressure causes a retardation of one second in a six-inch or thirty-seconds fuse*. The military implications of this discovery, which closely agreed with that of Mitchell, were fairly obvious, especially in mountainous terrain. Scientifically the differences between burning of fuses and candles could be explained thus. The latter receive their supply of fuel by diffusion up the wick of high-boiling hydrocarbons, a process unlikely to be pressure-dependent. On the other hand the passage of combustion along a fuse depends on ignition at any one point being triggered by hot gases from adjacent material that has just burnt. If the external pressure is reduced these gases will be more readily dispersed and 'the number of ignited gaseous particles in contact at any one moment with the still unignited disc of composition will be diminished'.[9]

1.2 The origins of illuminating power

That settled, there remained another puzzle from the Mont Blanc observations on candles. Whatever their rate of burning might have been on the summit there was no denying a considerable reduction in their *luminosity*. This would immediately have roused Frankland's interest because during the past few years he had performed many experiments on illuminating power for gas producers, manufacturers of candles and others. Further experiments would need to be conducted, and so they were, on his return to England.

Using another ingenious apparatus he observed combustion of candles in an enclosed space, at varied pressures. He concluded that within a fairly narrow range of pressures diminution of illuminating power was proportional to decrease in external pressure. An almost linear relationship was established for pressures from atmospheric down to 4.6 inches of mercury (when the photometer failed to respond). If, however, pressure was increased above that of the atmosphere a similar relation held true until it was doubled, after which there was wide divergence from linearity, possibly due to experimental error.

The cause of this behaviour was intriguing. Two explanations suggested themselves. First, the loss of illuminating–power at low pressures might be attributable to 'imperfect combustion'. This was unlikely, in view of the known tendency of a flame to become smoky with unburned carbon on compression (not decompression). It was decisively ruled out by Frankland's analysis of gaseous products which proved 'no escape of unconsumed combustible gas' at low or ordinary pressures. The alternative explanation was in terms of the different temperatures of the two candle-flames, that at high altitudes being possibly much cooler. However it was not difficult to show that there was little variation in 'the heat of the flame', thereby confirming an assertion of Humphry Davy.[10]

Frankland concluded that the variations depend on the relative ease of access of oxygen to the interior of the flame. Citing the well-known effect to be seen on opening the air-hole of a Bunsen burner Frankland argued that *at low pressures* the increased mobility of the reacting gases and the increased volume (and therefore surface area) of the flame render more complete the admixture of oxygen with the burning gases. At the heart of his argument lay the supposition that the light emitted depended on 'the unconsumed amount of carbon separated within the flame', and plentiful access of oxygen would ensure this was kept to a minimum.

The results of these two sets of experiments he communicated to the Royal Society in June, 1861.[11] Yet in fact, like so much in scientific research, they raised more problems than they solved and initiated another set of experiments. The work falls into three further parts.

First came some investigations relating combustion to temperature.[12] This paper was apparently occasioned by investigations into
gas explosions in London in which he observed a differential
diffusion of gases as coal-gas escapes from a confined space. Hence
the inflammability of any given sample would depend on its exact
composition which, in turn, would be governed by the relative rates
of diffusion of its components. It was therefore important to
discover the ignition temperatures of constituents of coal gas. He
showed that they rose in the order: $CS_2 < H_2 < CO < C_2H_4 < CH_4$.
Added carbon disulphide would reduce the ignition points of the
other components but not of ethylene. Hydrogen may be ignited by
sparks from flint and steel, so Davy lamps need to be used with
circumspection where (as in coal-gas) gases other than methane are
also present. By its emphasis on diffusion of gases this paper is
closely related to his speculations about combustion of candles, and
thus to the experiments on Mont Blanc.

Next, in 1867, Frankland returned to the problem of why some
gas flames have high luminosity but others do not. His explanation
of the Mont Blanc experiments was, as we have seen, in terms of
Davy's theory that flames owe their brightness to particles of
incandescent carbon resulting from incomplete combustion in the
centre of the flame. The immediate occasion was a series of lectures
at the Royal Institution on 'Coal gas'. Here of all places the
memory of Humphry Davy was sacred, yet it was Davy's theory
that was found to be defective. Frankland was able to show that the
black deposit obtained by insertion of a cold surface into a flame
also contains combined hydrogen (removal by chlorine at high
temperature). He cited the brilliant flames obtained by burning
arsenic, phosphorus and carbon disulphide for none of which can
incandescent solid particles be responsible. He concluded that
luminosity must be due to dense but transparent hydrocarbon
vapours.[13]

His final paper on the subject was read to the Royal Society in
1868 on the combustion of hydrogen and carbon monoxide under
great pressure.[14] Combustions of the two gases in oxygen were
conducted in another iron pressure vessel, furnished with a suitable
glass window. Pressures of up to 20 atmospheres were used, and as
these were approached, hydrogen and carbon monoxide each
emitted a brilliant light (whose spectrum was continuous). Similarly,
passage of electric sparks through hydrogen, oxygen, chlorine and

sulphur dioxide and (above all) mercury vapour gave out light whose brilliance increased with the density of the gas.

During the next few years he continued to work at the subject but did not publish. Thirty years later he wrote in a private letter to Sir George Stokes:

> At that time, I attempted to solve the problem in a different way. I endeavoured to ascertain the effect of pressure upon the luminosity of (a) incandescent solids (lime, Pt, and Ir, in the oxyhydrogen flame) and (b) incandescent gaseous matter (SH_2 in oxygen). For several years I spent a good deal of time upon this problem; but I met with unforeseen and very formidable difficulties which were, in fact, insurmountable in the case of (a) but which were so far overcome in the case of (b) as to show that the luminosity of a flame of SH_2 in oxygen follows approximately the same law as that which governs gas and candle flames under varying atmospheric pressures. Of course the light of incandescent solids in a gaseous medium might follow the same law and then the results would contribute nothing to the solution of the problem. The light of an electrically ignited solid I found to be greatly increased by the rarefaction of the medium, but then this was obviously due to the cooling action of the medium.[15]

Interest in the subject had been rekindled in 1889 when Stokes wrote to Frankland about 'two patches of bluish light' obtained when beams of sunlight, focused by a lens, were passed through a candle flame. 'The cause of the patches of light was evidently suspended matter in a state of very fine division'; they also 'showed all the polarisation phenomena of very fine solid or liquid particles in suspension.' This disproves Frankland's conjectures in his 1868 paper 'and pretty well establishes the correctness of the common view'.[16] A further letter on the same day reported no dispersion with a Bunsen burner flame, but evidence for it if a bead of salt was introduced.[17]

Frankland, writing from Norway, thanked Stokes for his letters of 1 August and referred to a paper by Heumann[18] that threw doubt on his own simple view that no solid particles exist in the flame:

> Ever since the conclusive experiments of Heumann (I am not sure that I spell his name correctly) which I have carefully verified, I have abandoned my too sweeping statement that there are no solid or liquid particles in gas and candle flames, and I have since that time yearly shown to my class at South Kensington the shadows cast by these particles when the flames are placed in the path of an electric beam. By a modification of Heumann's experiment I have also made these particles roll themselves into balls large enough to be visible to the

> naked eye. Nevertheless I still hold that a very large proportion of the
> light of candle and gas flames is not due to solid or liquid particles.[19]

He maintained this view on several grounds: flames from phosphine, arsenic, hydrogen sulphide *etc.*, where solids are not likely to be present; addition of extra oxygen can increase luminosity (would be expected to burn up solid particles); and a large excess of soot gives a decrease in light emitted.

A reply from Stokes a few days later explains his experiments further, as Frankland seems not to have grasped the point. The flame was 'used as a screen on which to receive the image of the sun' and the polarisation was observed not from the flame but from scattered light coming from the image of the sun.[20] A further letter the following day indicated that Stokes had been unable to locate in the University Library the paper by Heumann (presumably because he was searching in the *Berichte* instead of *Annalen*). He was continuing experiments with a candle flame, including observing it under a bell-jar, and speculated that an intermediate in the combustion might be marsh gas.[21]

Frankland duly returned thanks for Stokes' letter, admitting that he now understood the nature of his experiment. 'No doubt, as you say, the temperature has much to do with the luminosity of flame'.[22] The final letter from Stokes in this exchange withdrew his proposal about a methane intermediate and suggested acetylene instead. He had still not seen the Heumann paper but reckoned that the hypothesis in it was probably correct.[23] That was the end of their correspondence for another two years. Then it was reopened in 1891 by a letter to Frankland from Stokes.

Writing from Ireland the Cambridge physicist wrote to Frankland enclosing a paper he had just written on flame[24] and reverting to his previous opinion that the intermediate in combustion is methane, not acetylene.[25] Frankland, also on holiday, replied from Scotland, expressing pleasure at receiving Stokes' letter and indicating an intention to resume work on the subject when he returned home.[26] Stokes repeated his point about methane in a further letter, this time from Wales.[27] He indicated that he had been unaware of a paper by Burch which anticipated his own conclusions and that Thomson, Rayleigh and Tait had all seemed unaware of it.[28] A further missive (from Cambridge) reported that two others had since mentioned Burch's paper to him.[29] Yet another letter conveyed the joyful news that he had at last been able to track down the

Heumann paper. Calculation on his own experiments with a lens revealed that the sunlight must have been condensed by a factor of 1000 to 2000, thus differing greatly from an experiment described by Heumann where no lens was used. He candidly admitted 'so far I grant there is not much to choose between the common theory and yours', but challenged his correspondent to explain 'the extreme thinness of the stratum containing suspended matter if that stratum is nothing but the outer surface of a space containing some gaseous hydrocarbon to which the luminosity is mainly due'.[30]

Frankland's reply repeated his reference to Heumann's work and his demonstrations at South Kensington. It was in this letter that he summarised his unpublished work in the 1870s.[31]

A final exchange between the two men took place four years later. It was initiated by speculation by Stokes ('walking up from dinner at the Athenaeum') that the brightness of a phosphine flame might be due to solid particles of phosphoric acid or anhydride.[32] Frankland's reply was brief and to the point: phosphorus pentoxide and phosphoric acid are volatile below flame temperatures so cannot be 'ignited solids' in flames of phosphine.[33]

At the end of this long correspondence it is hard to avoid the conclusion that Stokes was ill at ease with the chemistry involved and that Frankland was out of his depth in the physics. Neither man seems to have made the subject a matter of priority but Stokes at least seems to have found it a pleasant matter for reflection on holiday. In the absence of more sophisticated techniques in spectroscopy, it is difficult to see that further substantive progress could then have been made.

1.3 Muscular power and nutrition

It is appropriate to begin with Frankland's own words:[34]

> In August 1865 I sat at the window of an hotel in Geneva, anxiously watching the weather. It appeared hopelessly bad; and so I telegraphed to my brother-in-law, Professor Fick, Weather too unsettled for Faulhorn. I return to Zürich. We had arranged, in concert with Professor Wislicenus, to submit to a crucial test the theory which assigns the source of muscular power to the oxidation and destruction of the muscles themselves. We intended to confine ourselves to a non-nitrogenous diet, and to ascend the Faulhorn, taking strict account of the greatest possible muscular oxidation by determining the

amount of nitrogen expelled from the body of each person before, during, and after the ascent of the mountain. The above telegram, or rather its cause, deprived me of the honour of taking part in what I regard as one of the most important chemico-physiological experiments ever made. A fortnight later that experiment was carried out by Dr. A. Fick, Professor of Physiology in the University of Zürich, and Dr. Wislicenus, Professor of Chemistry in the same University.

Any mountaineer or athlete must sometimes wonder where his or her energy comes from. By the middle of the nineteenth century there was no doubt among the scientific community that it was of chemical origin, its source being oxidation of organic matter in the body to (eventually) carbon dioxide, water *etc.* In that respect it was similar to the development of body heat. There had been a growing belief in such ideas from Lavoisier, through Berzelius and up to Liebig whose chemical views of life processes were of very great influence. The big question was the nature of the chemical fuel consumed in exercise, and whether it was the same as that used to maintain body temperature. Liebig had shown that the latter was non-nitrogenous material, what we should call carbohydrates and fats. During the 1860s, however, several authors had suggested a different fuel for development of muscular power. Arguing on the basis that muscle is rich in albumen[35], and that some was probably consumed in violent exertion, they proposed that oxidation of albuminoid substances must be the source of that energy.[36] Such a view was, however, opposed by two German scientists, Adolph Fick and Johannes Wislicenus, both of whom were relatives of Frankland by marriage. They did so on three quite different grounds:[37]

1. *Argument by analogy*: if albumen is consumed it must effect structural replacement in muscle tissue, one can apply the same idea to a locomotive: 'This machine consists essentially of iron, steel, brass, &c.; it contains but little coal; therefore its action must consist of burning iron and steel, not of the burning of coal', – a true *reductio ad absurdum*.
2. *Argument from design*: making the strange concession that 'in the light of Darwin's theory, the employment of teleological arguments in a certain order of cases has once more become admissible', they point out the oddity of a design (the body) with two separate fuel systems, 'as uneconomically as a manufacturer who should put up a stove near a steam-engine'.
3. *Argument from experiment*: it had been several times shown

that muscular exertion causes a dramatic increase in the expiration of carbon dioxide but not in the excretion of urea.[38] However nitrogenous waste-products might leave the organism in some other form, and the evolution of CO_2 might be a mere side-effect.

The authors readily accepted that none of these arguments was conclusive and that what was needed was a quantitative correlation between energy expended by muscular action and amounts of nitrogen-excretion. They proposed to conduct the experiment on themselves. Expenditure of energy could be determined by ascending a mountain of known height. They artlessly confessed: 'We preferred the mountain to a *treadmill*, not merely because the ascent is a more entertaining employment, but chiefly for the reason that we had no suitable treadmill at our disposal'! The mountain chosen was the Faulhorn in the Bernese Oberland, giving an ascent of exactly 1956 m. (6417 ft.) and with a convenient hotel on the summit at which to spend the night. That was the context of the projected experiment in which Frankland much wanted to take part but from which he was deterred by the weather.

When the ascent did take place, on 29 August 1865, with the mountaineers on an albumen-free diet, the results seemed clear: the energy expended was about 1.75 times that expected on the basis of measured nitrogen excretion. They concluded, *since the muscle machine can undoubtedly be heated by non-nitrogenous fuel, this fuel is in all cases that best suited for it.*[39] However Frankland was not to be denied his place in this assault on nature if not the actual mountain. Asserting somewhat inconsequentially that 'the Faulhorn experiments and these subsequent experiments gave the death-blow to the old theory of a *vital force*',[40] he stressed the connection between the *actual energy* expended as heat and mechanical motion and the *potential energy* locked up in food and released by chemical change in the body. To determine the latter he resorted to a series of calorimetric experiments.[41] These were conducted in a Lewis Thomson calorimeter, a copper tube containing the sample mixed with potassium chlorate. A fuse was inserted consisting of a dry cotton thread, previously soaked in potassium chlorate solution. The closed end of the tube was submerged in water and as soon as the fuse had been lit a bell-jar was placed over the open end to collect gaseous products. The heat released by combustion was determined from the difference in water temperatures before and after the explosion.

Frankland first measured the energy developed by combustion of one gram of each of the following: beef-muscle, albumen, beef-fat, hippuric acid, uric acid and urea. However an end-product of this experiment *in vitro* was nitrogen gas, not as *in vivo* when almost exactly one third of the nitrogen from beef-muscle or albumen ends up as urea. On that basis he was able to show that Fick and Wislicenus over-estimated the energy derived from nitrogenous products. His conclusion was

> The combustible food and oxygen coexist in the blood which courses through the muscle; but when the muscle is at rest, there is no chemical action between them. A command is sent from the brain to the muscle, the nervous agent determines oxidation: the potential energy becomes actual energy, one portion assuming the form of motion, another appearing as heat. *Here is the source of animal heat, here the origin of muscular power!* Like the piston and cylinder of a steam-engine, the muscle itself is only a machine for the transformation of heat into motion; both are subject to wear and tear, and require renewal; but neither contributes in any important degree, by its own oxidation, to the actual production of the mechanical power which it exerts.

Frankland then turned to measurements on common constituents of food, determining the energy released from one gram of each by combustion in oxygen and oxidation in the body. He showed that ingested fats, though not essential for diet, nevertheless had a high calorific value. He then made the vital connection between calorific value and actual cost, showing that fats, because of their cheapness, were an important item of diet for the poor. It accorded with his experience in Lancashire where the food of agricultural labourers contained a high proportion of fat (notably dumplings and egg and bacon pies). He added a vivid recollection from his childhood:

> I well remember being profoundly impressed with the dinners of the navigators employed in the construction of the Lancaster and Preston railway; they consisted of slices of bread surmounted with thick blocks of fat bacon in which mere streaks of lean were visible. These labourers doubtless found that from fat bacon they obtained, at the minimum cost, the actual energy required for their arduous work.

That must have been indelibly stamped on his memory.

Frankland reported his calorimetric work to the Royal Institution on 8 June 1866, an event his loyal assistant McLeod described as 'a capital lecture'.[42] When his paper was read to the Royal Society in September Frankland sent a copy to Darwin who confessed 'I have liked it very much, though here and there there were bits I could not fully understand'.[43] A couple of years later Liebig commented to Wöhler on both the work of Fick and Wislicenus and Frankland's extension of it which 'sent to the grave my former theory'. He tried to accommodate the new work by supposing the muscular system to be a kind of reservoir of energy (*Kraftmagazin*) that might be derived from food with or without nitrogen. It would then gradually release it 'like the stretched spring in a clock'.[44] Other, less serious reactions were expressed in a piece of contemporary doggerel on the ascent of the Faulhorn by Fick and Wislicenus and on Frankland's further work:

> The result seemed to be, that the work they had done,
> And the waste in urea, were nowise as one;
> But more wonderful still; – ere the climbing begun,
> The urea was greatest – when work there was none!

> They and Frankland now say, though some critics look sour,
> That Liebig and Playfair have both had their hour;
> For the Doctors, they swear, got their muscular power
> From the starch and the fat they took care to devour.

> But I think that these Doctors, though plainly no fools,
> Have ignored both Digestion's and Logic's first rules;
> And, comparing to urine the use of their tools,
> They entirely forgot to examine their st----.

> So while thus Wislicenus and Fick make a bustle,
> And in victory's plumes Frankland's trying to rustle,
> Many think that they haven't the best of the tussle,
> But that Liebig and Playfair are right as to muscle.[45]

But on the whole reactions, positive or negative, were muted. Frankland's research was the very beginning of measurements of what came to be called 'calorific value' of foodstuffs, an important element in all modern studies of dietetics. Yet as one historian has rightly said:

> The omission of his name from even comprehensive treatises on nutrition indicates that he is not generally known as the first to study foods for the quantitative energy values which they yielded on combustion.[46]

That, as Frankland might have said, was the story of his life.

1.4 Glaciers

No Alpine visitor could be unaware of the existence of glaciers. In 1856 Tyndall and Frankland had entertained some ideas for an Alpine holiday, but on 1 August the former suggested a change of plan to enable him to obtain a better look at the glaciers. The suggestion was that Frankland and he should join Hooker and Huxley at Basle, neither of whom Frankland had then met.[47] A letter sent three days later[48] was the one seeking forgiveness from Mrs. Frankland for imperilling her husband (p. 188), though he also promised 'that in treading the glaciers I shall tie you with a rope to my girdle'.[49] This was Frankland's first Alpine experience. Their visit culminated in a traverse of the Hochjoch in a hailstorm. When the weather had cleared,

> at the top of the pass we found ourselves on the verge of a great *neve*, which lay between two ranges of summits, sloping down to the base of each range from a high and round centre: a wilder glacier scene I have scarcely witnessed.[50]

Frankland could hardly have been more impressed.

As always the mountains opened up new scientific questions, as well as athletic and aesthetic experiences. High on the agenda of many an alpinist with scientific inclinations was the origin, nature and function of the glaciers. The revolutionary suggestion that they (or their predecessors) had been important non-volcanic agents of geological change had been advanced by Louis Agassiz in 1837. They, rather than water, had been responsible for shaping the face of much of Europe. Underlying this most controversial proposal lay questions to which no obvious answer could be supplied. One of these was why Ice Ages had occurred in the first place, and another was the actual nature of a glacier and the mode by which it moved. Some progress with the latter question had been made between 1841 and 1851 by the Edinburgh geologist J. D. Forbes (1809–1868).[51] In brief Forbes believed that glaciers were viscous liquids, moving faster at the centre than at the edges. At about the same time an apparently similar view had been expressed by an obscure ecclesiastic, L. Rendu, later to become Bishop of Annecy. Forbes was assailed by the Cambridge mathematician William Hopkins and by John Tyndall. The former complained that Forbes' proposals violated the laws of mechanics while the latter criticised a lack of precision in his use of the term 'viscous' and became a strong opponent on matters of glaciology.[52]

It was Tyndall's hostility to Forbes that was, in a curiously roundabout way, to bring Frankland into the argument. Together with Huxley, Tyndall proposed an alternative mechanism for glacier flow that involved regelation, the power of pieces of ice in contact in freezing water to weld themselves together.[53] Controversy continued in growing bitterness with matters coming to a head in 1859 when the Royal Society met to consider the award of its Copley Medal, one of the contenders being Forbes and the other W. E. Weber (the German physicist famous for his work in electrodynamics). By now a caucus of London scientists, led by Huxley and Tyndall, was determined to deny the Medal to Forbes. The latter, for his part, was desperate that the award should not go to Tyndall (though in fact the latter was not in the running). Since his supporters included the Scottish physicists P. G. Tait and William Thomson it looked as though the conflict would become a classic case of north v. south, or more precisely Scotland v. London. That, however, would be to overlook the underlying differences which were far more ideological than geographic, for Tyndall, Huxley and Hooker were later to unite as members of the X-Club. Already past masters at the art of lobbying they threw their efforts into discrediting their enemy's candidature, and how better to do it than to gain the ear of their friend on the Council,[54] Edward Frankland? Accordingly Huxley addressed to him a long and urgent letter. It adds up to an accusation of Forbes for having selectively quoted from the work of Rendu by omitting all reference to the Bishop's comparison of a glacier to a viscous stream.[55] A few days later Huxley was writing in alarm that a leak had occurred and it was now rumoured outside Council that he 'intended to charge Prof. Forbes with having directly borrowed from M. Rendu's work', which – though Huxley denied it – was virtually what he had done.[56] A further letter of the morning of the Council meeting urged Frankland to 'trust in the Lord and keep your powder dry' and also to quote very selectively from the writings of Hopkins and Tyndall.[57] Within hours a message was winging its way back to Huxley with Frankland's assurance that 'the battle is over and the victory ours', and that had only been possible through Huxley's 'Herculean aid'.[58]

It was a squalid story of vindictiveness and intrigue and no one comes out of it very well. It was, unfortunately, also characteristic of Victorian science at this time. While the future X-Club members succeeded in denying to Forbes the coveted Medal, the modern view of glaciers is closer to that of Forbes than that of Tyndall.[59] Frankland, meanwhile, had become embroiled in glacier controversy not of his own making. He was soon to become involved in one for which he was directly responsible.

Frankland produced his own theory of glacial action. Greatly daring, he attempted to answer that other large question prompted by Agassiz: what brought the glaciers into existence in the first place? He gave his own explanation in a paper 'On the physical cause of the glacial epoch'.[60]

It was prompted by a holiday in Norway, with his wife and two eldest children, in 1864. For a few days the whole family walked and climbed together. But it was obvious that Frankland had other ideas, for Sophie writes of his 'long intended trip to the Jostedals glacier', which was not to be abandoned on account of 'a very bad cold in the chest'. Something of her anxiety and frustration can be read from her letter to her parents-in-law which indicates total uncertainty about Edward's movements and the length of his absence. She did not think he would be away much more than a week as the weather had now broken with torrential rain. 'Oh I dare not think of dear Edward – I hope he will return and give up any attempt'.[61] He did not give up, and returned after 16 days. He added a postscript to his wife's next letter to the Helms, reporting 'a most delightful trip through one of the wildest parts of Norway'.[62] In his diary of the trip he wrote 'this is the first glacier I have seen [in Norway] and is much grander than the Mer de Glace'.[63]

His excursion assured him of the glacial origin of much of the scenery, and he was convinced by the view that the glaciers had moved in a north-westerly direction, travelling over the high ground to the sea and carving out the fjords on their way. If they had not come from the polar regions, what was their origin? He argued that they were the result of a 'distillation'. Water had been evaporated from the ocean, had condensed in the cold regions of the upper atmosphere and been deposited on the high ground which thus acted as a 'receiver'. For this to happen the ocean had to be relatively warm, and his great conclusion was '*The sole cause of the phenomena of the glacial epoch was a higher temperature of the ocean than that which obtains at presen*t'. That rise in temperature would result from the internal heat of the earth, a topic of great current interest.[64]

Reception of this theory was generally glacial (if that term may be permitted), and by 1877 Frankland himself largely recanted: 'I can no longer entirely subscribe to' it.[65] It was incompatible with subsequent measurements of deep sea temperatures and, moreover, it aroused the wrath of the biologists. Hooker wrote to Darwin:

> I was very near printing an exposé of Frankland's glacial theory, which cost me a sleepless night to concoct, but F. begged me to wait till he had published at length. I never read anything so wrong, geologically or meteorologically.[66]

He reported that after Charles Lyell had 'spent Sunday afternoon and evening with us, very pleasantly full of chat; we discussed Frankland's new glacial theory which I think the most monstrous absurdity ever broached in science and the most ill-digested attempt'.[67]

Regarding the theme of Frankland's paper on glaciers Darwin replied:

> I had intended exploding on same subject; it is absurd and I am the more bound to say so, as a former warm sea would never suit the Algae, Crustaceans and Fish common to North and South. It amused me to see how coolly he assumed the whole world was cooler during glacial period.[68]

Darwin was never too keen on any glacial theory, and expressed great caution:

> Apropos to what Frankland quotes I shd be very much obliged if you wd ask Tyndall when you next see him whether he supposes if only $\frac{1}{2}$ the present amount of snow fall on the Alps, that the climate of Europe fell to that of Greenland, whether the glaciers wd not greatly advance? I see the importance of the fall of snow but does not Frankland exaggerate its importance? Frankland ought to look into my journal for the extraordinary flexure in the snow line in S. Chile.[69]

It was indeed ironic that the people with whom Frankland was soon to ally himself, in or out of the X-Club, should have rejected his theory so strongly. However there was still one further area in which mountain-experiences were to lead to scientific research.

1.5 Meteorology

In Victorian times – as now – the weather was a talking-point for Britons, and frequently occurs in the Frankland correspondence.

In addition he actually made modest contributions to the science of meteorology.[70] A report 'On the composition of air on Mont Blanc' addressed the question of variable proportions of oxygen and nitrogen. Samples from the mountain were examined at St. Bartholomew's Hospital laboratories. It was confirmed that the N_2/O_2 ratio was 'within the limits of variation noticed by other experimenters' but that the carbon dioxide concentration reached a maximum (0.111%) at Grands Mulets, *i.e.* about 11 000 feet.[71] A brief paper in 1861 speculated on the cause of 'ground-ice' in rapidly flowing streams, connecting it with the ability of materials to crystallise on rough surfaces such as pebbles in a river-bed.[72] In the hot dry autumn of 1865 William Farr had enquired whether it would be possible to obtain periodical chemical and microscopical analyses of the atmosphere;[73] Frankland's response is not known.

Frankland does not seem to have made serious efforts in meteorological enquiry until late 1873. During that year, when his wife Sophie had become progressively worse with tuberculosis ('consumption') it was resolved to send her, with her elder daughter Maggie, to Davos, a Swiss mountain village with a growing reputation as a health resort. Here Maggie was charged with thrice daily observations of temperature and atmospheric pressure. Soon after arrival she wrote:

> We have now commenced our meteorological observations. The doctor has placed several thermometers in favourable places for ascertaining the temperature 3 times a day in the shade and sun and the minimum temperature during the night. He has made out a table or 'Tabella' on which to write the observations, also the remarks on the day as to sunshine or rain etc. The table hangs outside our room.[74]

A week of her results was summarised in the table:[75]

Nov.	minimal temp.	Schatten Temp.			Sonnen Temp.		
		9	12	3	9	12	3
1	$-\frac{1}{2}$	+4	+5	+5	+4	$+13\frac{1}{2}$	$10\frac{1}{2}$
2	$-2\frac{1}{2}$	+4	$+6\frac{3}{4}$	+5	+6	$+9\frac{1}{2}$	+6
3	$-\frac{1}{2}$	$+1\frac{1}{4}$	$+7\frac{1}{2}$	+5	+21	+11	+6
4	$+\frac{1}{4}$	+4	+6	+4	+6	$+17\frac{1}{2}$	+6
5	$-2\frac{1}{2}$	0	+4	+5	$+19\frac{1}{2}$	$+28\frac{1}{2}$	+16
6	-4	$-\frac{1}{2}$	+6	$+3\frac{1}{2}$	+14	+9	$+3\frac{1}{2}$
7	-2	+3	+8	$+2\frac{1}{2}$	$+4\frac{1}{2}$	+12	$+17\frac{1}{2}$
8	$-2\frac{3}{4}$	$+1\frac{1}{2}$	+6	$+4\frac{1}{2}$	$+3\frac{1}{2}$	$+12\frac{1}{2}$	$+4\frac{1}{2}$

At a later date she said 'these observations are of course taken with the Reamur thermometer, but the minimal thermometer hangs outside the corridor window & Mama thinks it is not cold enough there'.[76]

She promised to send the results home at the end of the month.[77] She does not refer to these responsibilities in her diary; while caring for a terminally ill parent she had more important things on her mind. At Christmas, 1873, Edward Frankland came to spend two weeks at Davos. He had acquired a set of thermometers from the London instrument supplier Louis Casella, complete with certification from Kew.[78] Mounting his instruments a few feet above ground level he recorded hourly temperatures in both shade and sunshine. His daughter recorded his unsuccessful efforts to make water boil by using a lens to focus the sun's rays on to the kettle.[79] Attempts to compare the Davos figures with data obtained at the same time at Greenwich were frustrated when Frankland discovered that the English observations had been made by thermometers lying on the ground, whereas his were on posts (and therefore unsusceptible to heat transfer from the earth directly). Nevertheless his results led to a clear conclusion:

> They explain the cause of the extraordinary winter climate of that valley, and they illustrate the conditions which ought to surround a winter sanatorium for diseases of the chest. It is proved that the unconcentrated rays of the sun at Davos in midwinter are capable of producing, under favourable circumstances, a temperature of 221° Fahr., or 9° Fahr. above the boiling-point of water.[80]

The causes of this desirable situation were altitude, reflective snow or sheet of water, sheltered position, southerly aspect and above all clean air ('comparatively free from zymotic matter').

Frankland's last paper of substance in this subject was again prompted by travels to mountainous parts. It concerned accurate measurements of solar intensity. They were made in England, Germany, Switzerland, Italy, Norway and even by his son Fred on a voyage to New Zealand. The conclusion, hardly surprising, was that solar intensity varies widely even though the sun is at constant altitude, elevation and reflecting surfaces as snow having a marked effect. Perhaps the most important feature of the investigation was the emphasis on standardisation of conditions for measurement. In true Frankland fashion a device was described for eliminating unnecessary variables and enabling reproducible results to be obtained.[81]

Frankland's final work in meteorology was, in practical terms, perhaps the most important. In many ways it was a fitting consummation of all his previous labours to penetrate the mysteries of the atmosphere. It sought to establish chemical links with the two kinds of air with which Frankland and his family were familiar: the pure fresh air of the mountain tops and the polluted, fog-laden, sulphurous atmosphere of London. He had already given a short account of 'dry fog' in London, when oily particles in the air retard evaporation of water by coating the droplets; the result ('smog') is irritation of the respiratory system.[82] On 10 February 1882 he gave a lecture in the Royal Institution : 'Climate in town and country'. To a packed house he delivered an impassioned plea for the banning of coal as a fuel and its replacement by coke or anthracite. Most exceptionally the lecture continued ten minutes over the allotted time, but no one seemed to mind.[83] As with some of his work on water (Chapter 13) this lecture had an even greater importance as persuasive rhetoric than as a report of fresh scientific research. When it was printed[84] it impressed many. Sir James Paget, then Vice Chancellor of London University, read it in a train and wrote 'I thank you very much for it and would, I think, reform my stoves and chimneys if my home had more than a year to run'.[85] An American, G. A. Smyth, wrote from Burlington, Vermont, thanking him for his paper and hoping to use it in the USA where, however, there was massive indifference to such matters.[86] Frankland backed up his rhetoric in other ways. Thus he advocated a 'Smoke Abatement League' and proposed to speak on its behalf at a meeting of the Council for Smoke Abatement.[87]

However it was a long haul and he was not to see a successful outcome. Writing to his daughter Sophie in 1886 he complained that the fogs were still 'fearful in London', though at his home in Reigate, Surrey, the November day was bright and sunny.[88] Other campaigners rallied to the cause and on 16 March 1887 the *Morning Post* carried a letter from 'Smokeless Fire' advocating that innovation and citing Frankland and others in support. The next day the author wrote to Frankland, identifying herself as Henrietta Baden-Powell and querying the dangers of carbon monoxide fumes from coke stoves.[89] She was doubtless reassured by Frankland's response in the form of another letter to the *Morning Post*. Writing from his own experience over five years at his Reigate home he said there was no reason to fear the fumes, given adequate ventilation.

He added that he controls the space above his fire by an 'iron curtain' (the first appearance in print of that Churchillian phrase?). Efforts to filter out the fog in his London home in 1870 were futile. Cotton pads were inserted between perforated zinc plates placed between a raised sash window and the sill. Within a fortnight they were clogged up, even though the fog was not thick.[90]

Those who deem environmental concern to be a feature of science only in the late twentieth century would do well to ponder the investigations of Edward Frankland into the quality of the air we breathe.

Frankland often went back to the mountains, though never to an experience that could match the ascent of Mont Blanc. Years later he accompanied his family to Switzerland, and took two of his children on extensive tours on foot; on one day they accomplished 20 hours of mountain walking, including four on ice, 'but Maggie was by no means over-fatigued'.[91] Holidays were not always spent in the Alps: Scotland, the Lake District, Norway and other places accounted for many a long vacation. But one momentous summer involved a return to the Alps a few months after his wife Sophie had died (p. 348). He could hardly have imagined that there he would meet Ellen Grenside whom he eventually married. If the mountains provided Frankland with an *entrée* into new areas of science they may also have helped to introduce him to his future wife. Even more remote than the mountains, however, were the depths of space and Frankland responded also to the challenge of the sky.

2 Science from the sky

2.1 *Telescopes on earth*

Frankland displayed a practical interest in astronomy from his early years in London. In the 1860s he erected a small observatory in the garden at Haverstock Hill, grinding, polishing and silvering a 7″ mirror for a reflecting telescope which he mounted equatorially. This was a considerable feat of craftsmanship and, like many other amateur astronomers at the time, he exchanged a voluminous correspondence with others, exchanging news of technical problems and their solutions. It may be that he first determined to make his own telescope after one of his many visits to Hereford, for two of his correspondents came from that part of England. One was Henry

Cooper Key, a clergyman from the nearby Stretton Rectory, and the other was George H. With, for 25 years Master at the Blue Coat School, Hereford, and later a consulting chemist to the Hereford Society for Aiding Industries. George With (b.1827) was one of the earliest people in England to make silver-on-glass mirrors, a spare-time hobby that produced some very fine samples that still survive in use. Generous with his advice, he wrote freely to Frankland and also helped Cooper Key to develop a new method for grinding specula, which was improved on by With himself. Key later became a distinguished amateur observer, especially of the moon and nebulae. George With produced a considerable number of mirrors for sale.[92] They may well have met through the introduction of Richard Dawes.

Another prolific correspondent on such matters was James Nasmyth, an old friend from Manchester days and now retired to Penshurst, in Kent. He was now more interested in telescopes than steam hammers, and wrote many letters of advice to Frankland. It must be a possibility that it was Nasmyth who first encouraged Frankland to embark on practical astronomy. In any event a large number of letters from these three men between 1862 and 1864 has survived.[93]

The observatory afforded its owner great pleasure, though the poor quality of London air must have severely limited its usefulness. Little is heard of it after about 1865. However when Frankland moved to Reigate, he returned to practical astronomy, doubtless tempted by better prospects of observing through clean air. With advice from With he purchased a $12\frac{1}{4}''$ reflector, and had it mounted and installed by the Dublin firm of Howard Grubb.[94] This was a much more substantial affair than the previous equipment and was extensively used, mainly as a hobby and relaxation.

The growth of amateur astronomy in Britain in late Victorian times is well known, and Frankland was part of a considerable movement. In two areas of the subject he brought important issues before the wider scientific community.

2.2 Glaciers on the moon?

The ill-fated theory of Frankland about the origins of glacial action has already been encountered (p. 428). In his 1864 lecture to the

Royal Institution he concluded with some astronomical speculations by remarking that his hypothesis 'suggests the probability that other bodies belonging to our solar system have either already passed through a similar epoch or are destined still to encounter it'. He declared that observations through his 7″ reflecting telescope created an impression that the valleys of the lunar surface might well have originated through glacial action. The fact that no evidence exists for the presence of water on the moon does not eliminate the possibility that it might have done so in ages past though it had now been absorbed into the 'cavernous structure' of the interior caused by the cooling process. If so,

> the moon, then, becomes to us a prophetic picture of the ultimate fate which awaits our earth, when, deprived of an external; ocean . . . it shall revolve round the sun an arid and lifeless wilderness, each hemisphere alternately exposed to the protracted glare of a cloudless sun and plunged into the gloom of an arctic night.[95]

As with the rest of this controversial paper there was a mixed reception. Nasmyth regarded the lunar section as 'heretical' and Hirst condemned it for its 'wild speculations'.[96] Thirteen years later, in 1877, Frankland declared that this part of his glacial theory was still credible,[97] though in 1874 he had admitted to Galton that it was 'a rather wild theory founded on the views obtained by my speculum'.[98] There was a rather special reason for prolonging its useful life.

For some years the moon had been used as evidence against the alleged Biblical account of creation. Changes on its surface analogous to those on earth would help to diminish the special status of our planet, and might in any case suggest a slow uniformitarian series of changes instead of an 'instant' catastrophic transformation. Thus Frankland was able to write gleefully to Huxley some years later:

> The 'lunar politics' are splendid and altogether your lecture will frighten the parsons more than anything they have encountered for a long time.[99]

It is mildly amusing that science from the sky should have been invested with such theological significance.

2.3 Helium on the sun

One of the greatest achievements of Frankland's former teacher, R. W. Bunsen, was the establishment of chemical spectroscopy, in conjunction with G. R. Kirchhoff. With this technique caesium and rubidium were discovered in 1860/1, and in 1859 they had begun the examination of spectra from the heavenly bodies. They wondered whether the dark lines in solar and stellar spectra might arise from substances whose arc spectra might show corresponding bright lines.[100] If so it might be possible to identify the elements in sun and stars. A few years later W. A. Miller (Professor of Chemistry at King's College, London) and William Huggins (an amateur but extremely able astronomer) joined forces to examine the problem further. In 1864 Huggins discovered two green lines in the spectrum from the Great Nebula in Orion, attributing them to an unknown element 'nebulium'.[101]

Meanwhile Edward Frankland had been experimenting with the spectroscope and had used it for demonstrations at St. Bartholomew's Hospital, following the method employed by Tyndall at the Royal Institution.[102] In 1861 he published a short letter to Tyndall on the spectrum of lithium.[103] Another amateur astronomer, Norman Lockyer, had commenced his own observations of the sun. Employed as a clerk at the War Office Lockyer made the acquaintance of Frankland a year or two later and learned some chemistry from him at the Royal College of Chemistry. During an eclipse of the sun in August 1868 they worked together on the spectrum of one of the solar prominences. A bright line was observed that was unknown in the spectrum of any terrestrial element, and they concluded that here was a new element, only present on the sun. They gave it the name helium.[104]

Frankland wrote to Lockyer saying that he had 'written to Carpenter about Helium and asked him to put it right'.[105] Comparing their discovery with that of thallium by Crookes, where the substance was actually isolated on earth, they admitted that

> when Frankland & Lockyer, seeing in the spectrum of the yellow solar prominences a certain bright line not identifiable with that of any known Terrestrial flame, attribute this to a hypothetical substance which they propose to call Helium, it is obvious that their assumption rests on a far less secure foundation.[106]

Further papers reported evidence for a *gaseous* condition of the sun's photosphere, in contradiction to earlier speculation that the incandescence resulted from solid or liquid particles, but in harmony with Frankland's own work on combustion at high pressures.[107]

The two men worked together through 1869, mainly at Frankland's laboratory at the Royal College of Chemistry,[108] but also visiting Lockyer's observatory at his home in Wimbledon. Thus Maggie Frankland reports that on 4 April 1869 'Papa went to Mr. Norman Lockyer to see the sun. He told us that he saw a mountain of hydrogen upon it of immense height'.[109] The collaboration came to an end only when the heavy work of the Rivers Commission obliged Frankland to withdraw in the early 1870s, with great reluctance.[110]

Notes

1 Frankland, *Sketches*, 2nd ed., p.234.
2 J. Tyndall, *The glaciers of the Alps*, Everyman Ed., J. M. Dent, London, 1906, pp.28–30.
3 *Sketches*, 2nd ed., pp.375–91.
4 A full account by Frankland himself is given in *Experimental Researches*, pp.864–73, and repeated in his *Sketches*, pp.302–19. It is based upon a 13-page document that appears to have started off as a letter to his wife (dated 23 August 1859), but finished, unsigned, in the middle of a page; a further MS copy of this document also exists, in a different hand, but with even diagrams meticulously copied (by tracing?); the latter was perhaps intended for the printers of *Experimental Researches* [JBA, OU mf 02.02.1684 and 1822].
5 Maggie Frankland to family, 26 October 1873 [PCA, OU mf 03.01.0367].
6 Frankland, *Experimental Researches*, p.873.
7 J. Mitchell, *Proc. Roy. Soc.*, 1855, **7**, 316–18.
8 Frankland to G. G. Stokes, 14 January 1862 [Cambridge University Library, Stokes papers, Add. MSS 7656/F365].
9 Frankland, *Phil. Trans.*, 1861, **151**, 629–53.
10 H. Davy, *ibid.*, 1817, **107**, 75–85.
11 Frankland, 1861 (note 9).
12 Frankland, *J. Gas Lighting*, 1862, **11**, 320–1; *J. Chem. Soc.*, 1863, **16**, 398–403.
13 Frankland, *ibid.*, 1867, **16**, 172, 233–5.
14 Frankland, *Proc. Roy. Soc.*, 1868, **16**, 419–22.
15 Frankland to G. G. Stokes, 23 November 1891 [Cambridge University Library, Stokes papers, Add. MSS 7656/F375].
16 G. G. Stokes to Frankland, 1 August 1889 [RFA, OU mf 01.07.0511].
17 G. G. Stokes to Frankland, 1 August 1889 [RFA, OU mf 01.07.0513].
18 K. Heumann, *Annalen der Chemie*, 1876, **184**, 206–54.
19 Frankland to G. G. Stokes, 18 August 1889 [Cambridge University Library, Stokes papers, Add. MSS 7656/F372].
20 G. G. Stokes to Frankland, 28 August 1889 [RFA, OU mf 01.07.0515].
21 G. G. Stokes to Frankland, 29 August 1889 [RFA, OU mf 01.07.0504].
22 Frankland to G. G. Stokes, 30 August 1889 [Cambridge University Library, Stokes papers, Add. MSS 7656/F373].
23 G. G. Stokes to Frankland, 2 September 1889 [RFA, OU mf 01.07.0508].
24 G. G. Stokes, Letter to Professor Tait, 15 June 1891, *Proc. Roy. Soc.*

Edinburgh, 1891, **18**, 263–4; the research was prompted by 'some questions about lighthouses'.

25 G. G. Stokes to Frankland, 23 September 1891 [RFA, OU mf 01.01.0933].

26 Frankland to G. G. Stokes, 4 October 1891 [Cambridge University Library, Stokes papers, Add. MSS 7656/F374].

27 G. G. Stokes to Frankland, 7 October 1891 [RFA, OU mf 01.01.0938].

28 G. J. Burch, *Nature*, 1885, **31**, 272–5.

29 G. G. Stokes to Frankland, 13 October 1891 [RFA, OU mf 01.01.0930].

30 G. G. Stokes to Frankland, 22 November 1891 [RFA, OU mf 01.01.0931].

31 Frankland to Stokes (note 15).

32 G. G. Stokes to Frankland, 3 May 1895 [RFA, OU mf 01.07.0532].

33 Frankland to G. G. Stokes, 4 May 1895 [Cambridge University Library, Stokes papers, Add. MSS 7656/F376].

34 *Experimental Researches*, p.918.

35 At that time there was no conception of proteins as a class, still less of their structure. 'Albumen' was a term loosely applied since the early 1800s to what we should call water-soluble proteins of medium molecular weight.

36 *E.g.* L. Playfair, *On the food of man in relation to his useful work*, Royal Society of Edinburgh, 1865 (lecture delivered 3 April, 1865).

37 A. Fick and F. J. Wislicenus, *Phil. Mag.*, 1866, **31**, 485–503.

38 *E.g.*, E. Smith, *Phil. Trans.*, 1859, **149**, 681–714.

39 Another response to this work, in addition to Frankland's, was by J. B. Lawes and J. H. Gilbert who referred to earlier work of theirs in which experiments on pigs had shown elimination of nitrogen was largely independent of muscular activity or its absence: *Phil. Mag.*, 1866, **32**, 55–64.

40 *Experimental Researches*, p.938.

41 Frankland, *Phil. Mag.*, 1866, **32**, 182–99.

42 H. McLeod, *Diary*, reprinted as *Chemistry and theology in mid-Victorian London*, ed. F. A. J. L. James, Mansell, London, 1987: entry for 8 June 1866; lecture reported *Chem. News*, 1866, **14**, 126 *etc.*

43 C. Darwin to J. D. Hooker, 25 September [1866] [Cambridge University Library, Darwin Papers, DAR 115/300B].

44 J. von Liebig to F. Wöhler, 7 March 1868, in W. Lewicki (ed.), *Wöhler und Liebig, Briefe von 1829–1873*, Jürgen Cromm Verlag, Göttingen, 1982, pp.243–5.

45 MS poem in unknown hand entitled 'Muscular power; or the ascent of the Faulhorn, A Diuretic Ditty' in envelope marked in Frankland's writing 'Lord Neave's Songs' [RFA, OU mf 01.03.0640].

46 E. McCollum, *A history of nutrition*, Houghton Mifflin, Boston, 1957, p.127.

47 J. Tyndall to Frankland, 1 August 1856 [Royal Institution Archives, Tyndall papers, 9/E7–8/3; also RFA, OU mf 01.02.0305].

48 J. Tyndall to Frankland, 4 August 1856 [Royal Institution Archives, Tyndall papers, 9/E7–8/4; also RFA, OU mf 01.02.0309].

49 J. Tyndall to Frankland, 'Saturday night' [August 1856] [Royal Institution Archives, Tyndall papers, 9/E7–8/5; also RFA, OU mf 01.02.0312].

50 Tyndall , *The glaciers of the Alps* (note 2), p.30.

51 F. Cunningham, *John David Forbes, pioneer Scottish glaciologist*, Scottish Academic Press, Edinburgh, 1990.

52 See (a) J. S. Rowlinson, 'The theory of glaciers', *Notes & Rec. Roy. Soc.*, 1971, **26**, 189–204; (b) *idem*, 'Tyndall's work on glaciology and geology', in W. H. Brock, N. D. McMillan and R. C. Mollan (eds.), *John Tyndall: essays on a natural philosopher*, Royal Dublin Society, Dublin, 1981, p.113–28; (c) Cunningham (note 51), pp.255–78.

53 J. Tyndall and T. H. Huxley, *Proc. Roy. Soc.*, 1856/7, **8**, 331.

54 Frankland was a member of Council of the Royal Society from 1857 to 1859.

55 T. H. Huxley to Frankland, 24 October 1859 [Royal Institution Archives, Huxley papers, 14/D5.1].

56 T. H. Huxley to Frankland, 2 November 1859 [Royal Institution Archives, Huxley papers, 14/E1].

57 T. H. Huxley to Frankland, 'Monday morning', [Royal Institution Archives, Huxley papers, 14/E1.3].

58 Frankland to T. H. Huxley, 3 November 1859 [Royal Institution Archives, Huxley papers, 14/E2.1].

59 Rowlinson (note 52(b)), p.123.

60 Frankland, *Phil. Mag.*, 1864, **27**, 321–41.

61 Sophie Frankland to William and Margaret Helm, 1 August 1864 [PCA, OU mf 03.01.0163].

62 Edward and Sophie Frankland to William and Margaret Helm, 26 August 1864 [PCA, OU mf 03.01.0165].

63 Frankland, *Diary*, entry for 15 August 1864, *Sketches*, 2nd ed., p.351.

64 See J. D. Burchfield, *Lord Kelvin and the age of the earth*, Macmillan, London, 1975.

65 *Experimental Researches*, p.960.

66 J. D. Hooker to C. Darwin, 9 February 1864 [Cambridge University Library, Darwin Papers, DAR 101/189–92].

67 J. D. Hooker to C. Darwin, 16 February 1864 [Cambridge University Library, Darwin Papers, DAR 101/183–5].

68 C. Darwin to J. D. Hooker, 22 February [1864] [Cambridge University Library, Darwin Papers, DAR 115/221B].

69 C. Darwin to J. D. Hooker, 27 March 1864 [Cambridge University Library, Darwin Papers, DAR 115/225B].

70 See F. P. Campbell, 'The meteorological researches of Sir Edward Frankland, KCB, FRS (1825–1899)', *Weather*, 1987, **42**, 383–7.

71 Frankland, *J. Chem. Soc.*, 1860, **13**, 22–50.

72 Frankland, *J. Chem. Soc.*, 1862, **14**, 113–14.

73 W. Farr to Frankland, 3 October 1865 [RFA, OU mf 01.03.0088].

74 Maggie Frankland to family, 26 September 1873 [PCA, OU mf 03.01.0296].

75 Maggie Frankland to family, 9 November 1873 [PCA, OU mf 03.01.0407].

76 Maggie Frankland to Fred and Sophie (II) Frankland, 30 October 1873 [PCA, OU mf 03.01.0353].

77 Maggie Frankland to Edward, Fred and Sophie (II) Frankland, 30 September 1873 [PCA, OU mf 03.01.0303].

78 L. P. Casella to Frankland, 16 December 1873 [RFA, OU mf 01.03.0518].

79 Maggie Frankland to family, 21 and 26 December 1873 [PCA, OU mf 03.01.0484 and 0488].

80 *Experimental Researches*, p.961.

81 *Ibid.*, pp. 999–1032; *Proc. Roy. Soc.*, 1882, **33**, 331–40.

82 Frankland, *Proc. Roy. Soc.*, 1878, **28**, 238–41.

83 Maggie West (*née* Frankland), *Diary*, entry for 10 February 1882 [MBA].

84 Frankland, *Nature*, 1882, **26**, 380–3; *Proc. Roy. Soc.*, 1884, **10**, 17–27.

85 J. Paget to Frankland, 4 January 1883 [RFA, OU mf 01.04.1218].

86 G. A. Smyth to Frankland, 3 March 1884 [RFA, OU mf 01.04.1495].

87 Frankland to H. Sowerby Willis, 12 December 1886 [Wellcome Foundation Archives, 67597].

88 Frankland to Sophie Colenso, 26 November 1886 [PCA, 03.01.0747].

89 H. Baden-Powell to Frankland, 17 March 1887 [RFA, OU mf 01.04.0287].

90 Copy-letter, Frankland to *Morning Post*, n.d. [1887] [RFA, OU mf 01.04.0289].

91 Frankland to J. H. Nicholson, 3 October 1869 [MBA].

92 On With see H. C. King, *The history of the telescope*, Charles Griffin & Co., London, 1955, pp.271–2.

93 These are mainly in the RFA, OU mf 01.08.0302 – 0405.

94 Again there is extensive correspondence, including memoranda and letters from H. Grubb and With to Frankland from 1880 to 1882 [RFA, OU mf 01.03.0347–0370, 0402–06, 0434–0440 *etc.*].

95 Frankland (note 60).

96 T. A. Hirst, *Diary*, 31 January 1864, p.1663, reprinted in W. H. Brock and R. M. MacLeod, *Natural knowledge in social context: the journals of Thomas Archer Hirst F.R.S. (1830–1892)*, Mansell, London, 1980.

97 *Experimental Researches*, p.960.

98 Draft letter, Frankland to F. Galton, April 1874 [AFA].

99 Frankland to T. H. Huxley, 16 February 1869 [Imperial College Archives, Huxley papers, 16/251].

100 G. R. Kirchhoff, *Berlin Monatsberichte*, 1859, 662–5, 783–7.

101 W. Huggins and W. A. Miller, *Phil. Trans.*, 1864, **154**, 413–35; W. Huggins, *ibid.*, pp. 437–44.

102 *Experimental Researches*, p.909.

103 Frankland to J. Tyndall, *Phil. Mag.*, 1861, **22**, 472.

104 N. Lockyer, *Proc. Roy. Soc.*, 1868, **17**, 91.

105 Frankland to N. Lockyer, 19 September 1872 [Exeter University Library, Lockyer papers].

106 Note in Frankland's writing, with reference to *Rep. Brit. Assoc.*, 1872, p.lxxiv [RFA, OU mf 01.02.0771].

107 Frankland and N. Lockyer, *Proc. Roy. Soc.*, 1869, **17**, 288–91, 453–4; 1869, **18**, 79–80.

108 J. J. Day to H. E. Armstrong, 27 March 1869 [Imperial College Archives, Armstrong papers, C-246].

109 Maggie Frankland, *Diary*, entry for 4 April 1869 (note 83).

110 *Experimental Researches*, p.910.

CHAPTER 15

Power

1 The Chemical Society – and Banquo's ghost

The Chemical Society in the 1860s and 1870s was a lively affair with a steady if unspectacular growth. It was comparable in size with the Linnean and Royal Astronomical Societies, though about half that of the Geological Society and only one sixth that of the Zoological Society.[1] The designatory letters F.C.S. acquired a talismanic significance for some members.[2] It was a model for the handful of other national chemical societies that had sprung up in its wake. Only in the north-east of England was any serious attempt made to reproduce its virtues in a provincial context. The Newcastle Chemical Society[3] was founded in 1868, had its own journal and its own lively discussions which raised issues of national importance but did not at the same time hesitate to criticise the London chemical establishment. However it was the Chemical Society of London that provided the only national forum for chemical reporting and debate.

Frankland's connections with the Chemical Society have been frequently mentioned in the previous narrative: his election as a Fellow in 1847, his attendance at meetings, the reading of papers and the lively discussions that often followed, and the lasting friendships forged with other Fellows. We have also seen something of the efforts made to induce him to accept the Presidency, first in 1869 and then, with more success, in 1871. However it is now timely to consider a further aspect of his connection with this oldest of all national chemical societies. This concerns his use of the various positions he occupied within the Society as a power-base for achieving desirable ends. For in fact the exercise of power presents us with the most quixotic of all the traits of Frankland's complex personality. The thesis of this chapter is that, although Frankland was extremely reticent when it came to acquiring institutional power, once he got it he used it to great effect and (at times) with something akin to ruthlessness. But the very exercise of that power was rarely an end in itself, nor was it always quite what it seemed to be. There was an underlying, hidden agenda which may become clear as the argument proceeds. But this is least evident, and the issues were apparently most straightforward, in his leadership within the Chemical Society of London.

As may be seen from Table 15.1 Frankland joined the Council in 1850, while he was still at Putney, but did not continue after moving to Manchester in 1851. Once back in London, however, he had a virtually continuous spell of 15 years in one capacity or another. As Foreign Secretary he kept up a continual stream of correspondence with distinguished chemists all over the world, often writing about Foreign Membership, awards by the Society, *etc*. His Presidency proved to be relatively uneventful. It took place in the calm between two stormy periods in the Society's history, and was marked most visibly by final preparations for a move from Somerset House to Burlington House, completed in November 1873. One interesting exchange of letters occurred over a decision of the Royal Society to charge £100 for 600 copies of its *Proceedings*. Frankland wrote in some indignation, announcing that he was summoning a special meeting of Council. Clearly his chemical colleagues deemed prudence the proper course, however, and resolved to accept the offer for the current year 'with the intimation that possibly the Society will be unable to continue the arrangement for a further period'.[4] Frankland's next communication meekly acknowledges the situation, does not convey the 'intimation' decreed, and even thanks the Royal Society for its 'liberality in supplying the Proceedings on such reasonable terms'.[5]

While Frankland was President the Society gained another hundred Fellows (as compared with 60 in 1869–71 and 23 in 1867–9); numbers continued to increase at about this rate for the next couple of presidencies. In his two Presidential Addresses Frankland departed from normal practice and, on each occasion, expressed grave concern for the national state of chemistry. In the first he regretted the small number of research papers presented in

Table 15.1. *Frankland and the Chemical Society*

Year	Frankland's rôles	Presidents
1850		Phillips
	Council	
1851		Daubeny
1852		
1853		Yorke
1854		
1855		Miller
1856		
1857		Playfair
1858	Council	
1859		Brodie
1860	Vice-President	
1861	Foreign Secretary	Hofmann
1862		
1863		Williamson
1864		
1865		Miller
1866		
1867		de la Rue
1868	Vice-President	
1869		Williamson
1870		
1871	President	Frankland
1872		
1873		

the previous session, attributing lack of progress to the attitude of the universities that ignored research in the granting of degrees.[6] In the second he returned to the same theme, asserting that 'until a profound change is made in the awarding of prizes and the granting of degrees in this country, we shall look in vain for any substantial improvement in the presentation of experimental investigation'.[7] His words were portentously important.

One other event was to have prophetic significance. It occurred on 31 May 1872 when Frankland was taking the chair at a dinner given by the Society in honour of the Faraday Lecturer, the Italian chemist S. Cannizzaro. At the proper moment Frankland rose to speak:

> In the course of his speech, the President drew attention to the increasing importance of chemistry in relation to the wants of communities, and pointed out how great would be the usefulness of an Institute which would be to chemists what the (Royal) Colleges of Physicians and Surgeons were to the medical profession, the Institution of Civil Engineers to civil engineers, and the Inns of Court to the legal profession.[8]

It appears that at that stage Frankland was envisaging a new function for the Chemical Society, though in the event a new Institute of Chemistry was to spring up as an independent body. The first historian of that Institute locates its 'real origin' in that speech by Frankland in 1872. In the subsequent history of the Institute Frankland was to emerge as a leader of considerable force and authority.

Perhaps this may help to explain a most extraordinary situation revealed by the minutes of the Chemical Society Council. Although Frankland ceased to hold office as President in 1873 *he continued to attend meetings of the Council for the next eight years.* Like Banquo's ghost he was often at the feast, though there is no evidence as to the part he played. His attendance was admittedly sporadic and irregular for some of that time, but in 1876 and 1877 he was present more often than not. On 17 December 1874 he took the chair as Vice President, a position taken by most immediate Past Presidents, and for a couple of years that may have explained his presence. However it was the practice for two Vice Presidents to retire each year and be replaced; yet this never seems to have happened to Frankland.[9] No resolution creating exceptional privileges has been found and it appears that this highly irregular position was connived at by successive Councils and Presidents for some unspecified reason, possibly connected with his part in the professionalization movement. Frankland cannot have been coerced into attendance for so long and it is likely that he was reluctant to give up all reins of power when he stepped down as President and welcomed further opportunity to exert power and influence within the Society.

2 The Institute of Chemistry – and 'technical informality'

The 1870s were, as we have seen, a momentous decade for the Frankland family. They were no less significant for the practice of chemistry in Britain and for the welfare of its practitioners. For these were the years in which chemistry was to be at last recognised as a *profession*, at first chiefly by those within it but gradually by those outside. The movement's focus was to be the formation of the first organisation anywhere in the world that purported to be a professional institution for scientists: the Institute of Chemistry. Founded in 1877 its first President was Edward Frankland.

The rise of a professional consciousness in chemistry and the concomitant foundation of the Institute have been extensively studied[10] and require little elaboration here. The ambiguities residing in the word 'chemist' had long been known. An 18th century directory introduces a list of 39 such persons with the comment that 'The number of those who pretend to be Chemists almost exceeds belief; but the following are really Artists, having all regular laboratories'.[11] Nor was there any novelty in applying the term 'profession' to a chemist's calling. In 1799 Humphry Davy had claimed 'philosophy, chemistry and medicine are my profession',[12] while over 30 years earlier John Wesley had met a poor man at Marshalsea who was 'by birth a Dutchman, a Chemist by profession'.[13] In neither case did the term convey the cluster of ideas that marked the Victorian idea of a profession, and eventually found expression in the Institute of Chemistry.

At least three causes may be identified for the movement that led to its formation, and in each of them Frankland was closely involved. First, there was *a series of problems within the Chemical Society*. The black-balling tactics of the Wanklyn faction in 1867 were revived in the mid-1870s. The conspirators may have been different but the issues were essentially the same: the nature of the Fellowship. Should Fellows be created by a small band of their friends on Council, or should they be elected by a majority of members on the basis of their chemical attainments? Did the letters F.C.S. signify chemical attainment or did they not? This time one of the conspirators was the analyst C. E. Cassal, of whom it was written: 'He was a, if not the, prime mover' in battles at the Chemical Society over lax methods of admitting new members. 'The shocking attack he made on the existing system and all who upheld it did much to decide the issue'.[14] Again the popular chemical press had a field day and the Victorian habit of writing indignation letters was enjoyed to the full. Evidence for a direct connection between these activities and the foundation of the Institute will be seen later.

A second triggering mechanism was provided by *the general advance of professional consciousness* in Victorian England. The civil and mechanical engineers had enjoyed the benefits of their institutions since 1818 and 1847 respectively. More recently others had followed their example:

1863	The Institution of Gas Engineers
1868	The Institution of Chartered Surveyors
1869	The Iron and Steel Institute
1871	The Institution of Telegraph Engineers
1871	The Institution of Electrical Engineers

The Chartered Surveyors[15] may well have served as a rôle model for the chemists in some respects. All these cases from other professions cannot have passed unnoticed, for a sudden surge in the demand for chemical analysts was creating both new anxieties and new confidence among chemists. One of them wrote a long and influential article for *Chemical News* entitled 'On the necessity for organisation among chemists, for the purpose of enhancing their professional status'.[16] He was C. R. A. Wright, then lecturer in chemistry at St. Mary's Hospital, Paddington.[17] Many others joined the chorus, including N. S. Maskelyne who summed up the prevailing sentiment in the words 'chemistry is a profession as well as a science'.[18]

Of course no one knew this better than Frankland. In a MS fragment bearing the heading 'Topics for address' Frankland stressed the absence of special training for chemists before 1845 and the rise of the chemical expert, first called in for cases of poisoning, then for actions relating to pollution ('nuisance') as with the HCl emissions from alkali works. He was doubtless alluding to the effects of the two Adulteration Acts (1860 and 1872) which established the office of Public Analyst, and the two Alkali Acts (1863 and 1874) which prescribed chemical monitoring of all kinds of effluents from industrial premises. He had had much experience in his Manchester days of these matters even before legislation greatly increased the demand for such chemical services. There were also new tasks for analytical chemists in agriculture, railways and explosives industries. With much of this Frankland was very familiar. Perhaps the most illuminating example (in all senses) comes from the gas industry. In preparing its Sale of Gas Act of 1860, with tough new demands for quality control, the Government was advised, not by a chemist, but by the Astronomer Royal, George Airy. Subsequently official enquiries were directed to an engineer (R. Willis) and a physicist (C. Wheatstone). To their credit they both expressed some embarrassment at the marginalisation of chemists, Willis remarking that the subject 'belongs rather to chemists than mechanics', and Wheatstone suggesting the services of Frankland as well as the gas engineer George Lowe.[19] Frankland obliged, confirming the opinion of his friend Samuel Clegg that 'the accurate analysis of gaseous mixtures is one of the most delicate operations of modern chemistry', owing much to the work of Bunsen who had taught Frankland his methods.[20]

It is likely that Frankland had long been aware of the 'unprofessionalised' state of chemistry, if only by contrast with the status enjoyed by the older professions. From his youth he must have been deeply impressed by the profession of law represented by his natural father, medicine exemplified by Dr Christopher Johnson and pharmacy embodied in his allegedly tyrannical master Stephen Ross. By contrast the chemists were a disorganised lot and now, in the 1870s, were beginning seriously to suffer for it.

The new-style professions had four or five characteristics in common. First there was an agreed minimum standard of competence, a goal that Frankland constantly held before the eyes of his students, the readers of his books, teachers and above all Government officials. The idea of training, followed by appropriate monitoring or testing, was fundamental to Frankland's whole educational strategy. No one in England was better placed to foster and promote such ideals in a community struggling towards professional status. Secondly, a professional body could expect its members to enjoy certain levels of remuneration and a recognisable career structure. Again Frankland stands out prominently as a passionate advocate of proper financial rewards. The question of fees alone forms the theme of at least 50 of his surviving letters. And, as we have seen (p. 170) he sought to establish his personal rights in such matters within the context of general codes of practice. For a third aspect of Victorian professional institutions Frankland may seem less well prepared: the acceptance of certain obligations to society along the lines of the Hippocratic oath for medicine. Turning a blind eye to industrial pollution in his Manchester period was, even then, hardly an exercise in social responsibility. Yet his water analysis work reveals a man acutely aware of the rights of individual consumers and of his duty to protect them. His position of unofficial Government water analyst placed him in strong contrast to those who were hired by water companies and heightened his own resolve to place the public interest first. Then there was, in most professional groups at least, a strong sense of corporate identity, and this touched a nerve very deep in Frankland's personality. Never fully at ease with himself, and dogged by the stigma of his own illegitimacy, he longed for social acceptance even though he may rarely have acknowledged it. The desire to 'belong' was manifest in his membership of the Athenaeum, his eager participation in social as well as scientific functions of the Chemical Society and Royal Society, and still more in his devotion to the X-Club. And finally a professional institute could expect recognition from its peers and from the Government. Frankland's exertions to get Government and other authorities to recognise chemists long pre-dates his Presidency of the Institute of Chemistry.

If the rise of professionalism in general was one contributory cause of the foundation of the Institute of Chemistry, a third and related one might be termed *professional encroachment*. The unpleasant truth was beginning to dawn on chemists that they themselves might be the victims of other professional groups. To take but one example, the second Adulteration Act decreed that Public Analysts should be competent in the medical, chemical and microscopical examination of drugs. But, as was widely recognised, there was 'no recognised standard' for determining such competence.[21] An early assessor was not a chemist but John Simon, then medical Officer to the Local Government Board. The pages of the *Chemical News* re-echoed with indignant denunciations of attempts to usurp a chemical rôle by doctors, engineers, pharmacists or simply quacks. Again Frankland was fully aware of the dangers. The MS note referred to already contains the significant phrase 'Encroachment of the professions of Medicine & Pharmacy upon that of chemistry'. He should have known from his own experience of water analysis. By 1877 Frankland had completed a series of conflicts with the physician Henry Letheby and was about to engage battle with his successor C. M. Tidy.

Chemists in England saw themselves also as victims of a unique and peculiar type of professional encroachment, originating in the ambiguities lying within the word 'chemist'. The censuses had lumped manufacturing chemists within the 'industrial class' that included artisans of all kinds. On the other hand those people practising pharmacy, the 'chemists and druggists', were allotted to the professional class. This was bad enough, but the 1868 Pharmacy Bill granted pharmaceutical chemists a professional monopoly and restricted the term 'chemists' to those who had passed the examinations of the Pharmaceutical Society. This meant that, in law, not even the President of the Chemical Society could be described as a 'chemist'.[22] Frankland's memo contains an obscure witticism contrasting the different ways in which the word 'Chemist' is understood by Wilkie Collins and by the Board of Trade, *etc.*[23]

In response to these pressures the chemical community began to take action that led eventually to the Institute of Chemistry. From the beginning Frankland was in it up to his neck. Or, as Pilcher more elegantly wrote, 'He was intimately associated with the early work of the organisation, and may be considered . . . to have been practically the Founder'.[24]

The first steps seem to have been taken in 1875 when an Organisation Committee was set up by a small group of six activists, four of whom were included (and indicated by asterisks below). Edward Frankland was among their nominees, the complete list being:

Dugald Campbell, analytical chemist, London
Michael Carteighe*, examiner to the Pharmaceutical Society
Edward Frankland, Professor at Normal School of Science, South Kensington
W. N. Hartley*, Demonstrator at King's College, London [Hon. Secretary]
Frederick Manning*, former Assayer at the Hong Kong Mint
Theophilus Redwood, Professor at the Pharmaceutical Society
Thomas Stevenson, Lecturer at Guy's Hospital
R. V. Tuson, Professor at Royal Veterinary College
Augustus Voelcker, consultant agricultural chemist, London
J. A. Wanklyn, Public Analyst, Buckinghamshire
C. R. A. Wright*, Lecturer at St. Mary's Hospital, London.

This committee appears to have been a fairly clandestine affair and did nothing until provoked by the letter from Wright in the *Chemical News*. Its response was to issue a circular announcing a meeting chaired by F. A. Abel, then President of the Chemical Society.[25] Neither the place nor time was specified, so presumably a series of meetings was contemplated. The circulation list has survived as a hand-written document and contains 124 names, all except four being Fellows of the Chemical Society. As 60 of the recipients were, or had been, office-bearers it is not unreasonable to suppose considerable Chemical Society involvement, in which case no one on the Organisation Committee was a more likely link than the past-President, Edward Frankland.

A meeting duly took place on 27 April, all except five of the 46 present being on the original list of invitees.[26] The first motion was proposed by Odling and seconded by Frankland: 'That it is desirable that an organisation of Professional Chemists be effected'. It was carried unanimously. Other resolutions followed, none more important or far-reaching than the proposal

> that a committee be appointed for the purpose of conferring with the Council of the Chemical Society with a view to ascertaining how far that Society may be able and willing to carry out a scheme for the organisation of Professional Chemists and that this committee be requested to report to the adjourned meeting.

The composition of this new body was, to a man, identical with that of the original Organisation Committee. How far this was 'rigged' in advance it is impossible to say, but the coincidence is too striking to be overlooked. That the whole proceedings were kept conspirtorially secret is clear from the flood of letters to the *Chemical News*, only a few of which displayed hints that the writers suspected something was afoot. Then, a week later, the Chemical Society experienced a renewed outbreak of black-balling, two of the activists having been present at the Organisation meeting on 27 April. One of them (E. Neison) was soon to be involved in detailed discussions about a new organisation.[27] There was a close link between the black-balling activities and the urge for professional status. Next day the Organisation Committee foregathered.

The rendezvous was Frankland's private room at South Kensington. Unsurprisingly he was elected to the chair. The meeting proposed a scheme for establishing an 'Institute of Professional Chemists', with a Board of 21 Examiners including various officers and nominees of the Chemical Society. This was analogous to the practice of the Pharmaceutical Society of which the proposer, Carteighe, was himself an Examiner. The suggestion was then sent to the Chemical Society Council for consideration. That body met a few days later and appointed its own subcommittee to examine the Organisation Committee's proposals. There appears to be some uncertainty as to its composition. According to the minutes of the Council meeting[28] the small group was to consist of Crookes, Gladstone, Heaton, Russell, Tuson and Wright, all of whom were present. However the document produced did not bear the signatures of Heaton and Russell but was signed by the other four and also by Frankland (who had been absent at the original meeting of Council). Once again normal democratic procedures appear to have been by-passed and once again Frankland's position seems to be highly anomalous. Tuson, Wright and Frankland were, of course, members of the Organisation Committee itself so they had a curious mediating rôle.

The Report recommended creation of a special class of Fellows of the Chemical Society.[29] Meanwhile a letter to *Chemical News* from a solicitor employed by the Organisation Committee[30] advertised its intentions and triggered a further spate of correspondence. In response to the report of their subcommittee the Chemical Society Council resolved to take steps to establish an organisation of chemists, but with 'the full control of the Council over the proposed organisation'. At an adjourned meeting it agreed to recommend 'a distinct class of Fellows be created to be termed

<u>practising</u> Fellows of the Chemical Society'.[31] Frankland was present at both meetings, though, as we have seen, there appeared to be no legal reason for his being there. One can only conclude that he was determined at all costs to be a party to whatever agreement was reached.

The Council issued a printed Proposal on the following day.[32] For the next few months legal advice was taken by both sides and, holidays over, the Organisation Committee met again. It was now clear that 'insuperable difficulties' lay in the path suggested by the Council, and Frankland was given (or volunteered) the task of conveying that information to them.[33] He did so in the following letter to the President:

> I am requested by the Chemists' Organisation Committee to convey to yourself and to the Officers and Council of the Chemical Society their thanks for the cordiality with which you and they have endeavoured to institute a scheme for the organisation of Professional Chemists within that Society.
>
> The Committee regret that the obstacles in the way of carrying out, within the Chemical Society, the wishes of the promoters of the organisation scheme, appear after mature deliberation, to be so formidable as to render it undesirable to prosecute the attempt further, but they trust that any independent scheme which may be inaugurated will be carried out in friendly alliance with the Society and will receive the sympathy and support of its Fellows.[34]

The nature of these 'obstacles' or 'insuperable difficulties' has been discussed at length elsewhere.[35] There were undoubtedly formidable legal problems in changing regulations for Fellowship under the existing Charter, questions of a semi-autonomous institute within a Society whose Council demanded 'full control', and, as recent events had shown all too painfully, the Society itself was in a position of some instability. Given goodwill, determination and sound legal advice all these problems could no doubt have been solved in 1876 rather than in 1980 (when amalgamation of Institute and Society did take place). But there was also another consideration: the ambitions of Edward Frankland. Now, for the first time since 1861, he no longer held formal office within the Chemical Society. His work with the Rivers Commission was at an end. He clearly thought that his research days were almost over or else why should he publish, in 1877, his complete *Experimental Researches*? And the 1870s had brought him more than a few mid-life personal crises. It is probably not too far from the mark to suggest that the creation of

Fig. 15.1 Sir Frederick Abel, President of the Chemical Society 1875–77.
(Author.)

the Institute of Chemistry was in considerable measure a direct outcome of Frankland's need to exercise creative leadership.

So the Organisation Committee produced a Report admitting defeat in their efforts to secure action within the Chemical Society and proposing the establishment of a 'new and independent body'.[36] Notice was circulated that the adjourned meeting of the Chemical Society would take place on 4 November.[37] The meeting concluded that an Association should be formed 'independent of the Chemical Society', and nominated yet another committee, with 51 names.[38] The proposer was Frankland, the seconder Voelcker. That committee (or 28 members of it) assembled a few days later and the chairman was not Abel, President of the Chemical Society, but Frankland. Things were going his way. A new draft scheme had to be produced and the task was entrusted to a small cabal of seven men: Abel, Carteighe, Frankland, Hartley, Neison, Voelcker and Wright. On 27 February 1877 the full committee received their recommendations for 'The Institute of Professional Chemists of Great Britain and Ireland' in the form of a printed pamphlet.[39] The recommendation was accepted almost in its entirety. It was decided to propose as new Fellows the existing committee, together with 48 other names (of whom 46 were on a list prepared by Frankland).[40] To no one's surprise he was elected President of the new Institute. Vice Presidents included several of his old cronies, including Crum Brown (of notation fame), Angus Smith from his Manchester period and Robert Galloway his former fellow-apprentice in Lancaster. Every one of those elected had been named in a memorandum which Frankland had obviously prepared in advance of the meeting.[41] Other officers were also appointed.

If Frankland imagined that a new institution, unencumbered by tradition, would be easy to run he was soon disappointed. Thus it was singularly unfortunate for the infant Institute that it should have immediately fallen foul of the Pharmaceutical Society. The latter objected to the breach of its own monopolistic right to the word 'chemist', and so informed the Privy Council. Letters flew around from minor dignitaries, legal experts and the great and mighty who looked after the pharmacists' interests. It is mildly ironic that a letter of conciliation should have had to be written by that former apprentice from a chemist's shop Edward Frankland. It is even more remarkable that the confident agents of a new crusade should have had to climb down with such ignominious speed. Yet such was the force of objections by the Pharmaceutical Society that Frankland thus addressed himself to the President:

> With reference to the objections raised by the Council of your Society against the registration of this Institute, and forwarded by the Privy Council to our Committee, I beg to state that they have formed the subject of careful deliberation on our part.
> The committee, anxious to meet the views of your Council as far as

Fig. 15.2 Edward Frankland, first President of the Institute of Chemistry.
(Author.)

possible, recommends that the name of our Institute be changed to
'Institute of Chemistry' and that the clause (Section 57 Sub-section a)
empowering the Council to grant certificates to Fellows and Associates
be struck out of the articles.

Under these circumstances the Committee hopes that the Council
of your Society will, at its meeting on Wednesday next, see its way to
the withdrawal of the objections previously raised.

It is perhaps unnecessary to add that the promoters of the Institute

do not contemplate and have not contemplated any interference with
the rights and privileges of the Pharmaceutical Society.[42]

Frankland was rewarded with a gracious assurance from its
President that 'on the part of the Council of the Pharmaceutical
Society there exists every desire to facilitate the object which your
Committee have in view, so far as is consistent with the duties
entrusted to the Council'.[43] He did not add – as he might have done
– 'but don't dare to call yourselves chemists!'

Then there was the knotty problem of nominating members of
Council. What might have been expected as a signal honour was not
always received with unalloyed happiness. Thus the honorary
secretary W. N. Hartley wrote to Frankland regarding a list of
possible names. He indicated that Andrews, Debus and de la Rue
have 'objected to the whole thing'. Playfair would probably be
sympathetic but unable to attend. Of Harcourt 'I am not quite
sure'. On the other hand J. M. Thomson and Maskelyne and
Marshall Hall were thought to be favourable.[44] He was not
completely right; for example Warren de la Rue had already agreed
to serve on Council.[45]

On this matter great indignation was displayed by Williamson,
who complained that a Council had been elected at the committee
meeting *without any notice having been given of an intention to do this.*
Worse still, his name had been included and must be removed
forthwith.[46] What Frankland had emolliently called a 'technical
informality' Williamson described with good reason as a 'substantive
irregularity'.[47]

Not even membership was automatically welcome. Thus Roscoe
wrote a tart letter, regretting his election as a member without his
consent. He was willing to be elected 'if you wish to have my name
as such. But for my own part I do not care about it'.[48] A week later
he performed a neatly executed somersault, expressing 'much
pleasure in accepting the nomination'. In the interim he had met
Frankland in London, who had presumably exerted his charm and
possibly held out a hint of a promise. The hitherto disinterested
Playfair now added that he would like to become a member of
Council.[49] However there were to be some compensations for the
inevitable snubs to the new organisation. Thus Benjamin Brodie
was 'very willing to become a Fellow' – but had mislaid the
application form.[50]

At the first Annual Meeting of the Institute, on 1 February 1878,

Censors were to be elected. This simple procedure should have been uncontentious, but it was not. Frankland's old adversary, C. M. Tidy smelt a *fait accompli*, and wrote to him:

> I do feel some notice should have been given those engaged in chemical practice of taking part in the arrangements. This was not done, and feeling as I do that there has been something behindhand in the proceedings have felt compelled to withdraw. I regret to do this, but I think there has been a certain cliquism . . .[51]

He had indeed identified the great weakness of Frankland's administration. It was a feature of many power struggles in and out of the Chemical Society: a tendency to ride roughshod over proper democratic processes, to take short cuts, to form small pressure groups of activists, and to manipulate individuals behind the scenes so as to ensure success.

Once established the Institute was to discover Frankland omnipresent on Council and every Committee except that dealing with finance: Examinations, Nominations, Parliamentary, Articles and Professional Charges Committees. Having rejected numerous existing courses as qualifying for Associateship (including the Natural Science Tripos at Cambridge) the Examinations Committee requested the President to draw up a new syllabus (who better?). This was accepted a week later with only minor changes.[52] Frankland continued as President until 1880, when he was succeeded by Abel, but was Vice-President from 1880 to 1882, from 1884 to 1888 and from 1891 to 1894.

3 The Royal Society – 'Frankland won't do'

Frankland was elected FRS in 1853.[53] Pride in his membership of Britain's most distinguished scientific society was reinforced by the award of its Royal Medal in 1857[54] and by an invitation to deliver the prestigious Bakerian Lecture two years later. However the coveted Copley Medal was awarded almost at the end of his career,[55] in 1894. Its lateness was deplored by *The Times*, at that time thundering against a perceived over-emphasis on biology.[56] There was, as we shall, see rather more to it than that. Naturally enough the Medal brought warm congratulation,[57] and in his speech of acceptance Frankland paid graceful tribute to his previous colleagues Kolbe, Bunsen and Liebig, expressing gratitude for 'the highest reward which a man of science can receive'.[58]

As in the Chemical Society and the Institute of Chemistry the power-hungry Frankland found himself in numerous positions of influence within the Royal Society. To appreciate the significance of his moves one has, however, to go beyond the personal ambitions of one man and see the Royal Society in a wider context. It has indeed been customary to regard it as an organisation dedicated to the pursuit of value-free science and, at least after the reforms of 1847, pulling harmoniously together in the cause of scientific truth. Gone were the days of amateur science with all its party-spirit. The Society's members were, as a recent account describes them, 'all scientists now'.[59] On this view of 'objectivist' science and its pursuit any appearances of acrimony within the Society are seen as minor episodes, regrettable no doubt, but quite untypical. However, this emollient view is no longer sustainable in the light of new evidence. As Andrew Harrison has convincingly shown,[60] the Royal Society in the period of Frankland's connection with it (the last half of the century) was riven with strife and contention. In fact there was a major ideological divide between two classes of members, bitterly opposed to each other. They may be termed the mathematical physicists on one hand and the scientific naturalists on the other. It is perhaps too much to assert that the Society was split from top to bottom by their disputes, but that is only because neither party was completely uniform and neither side held unanimously to a specific policy. Writing of the Royal Society at this time Harrison has observed:

> When the train of objectivist assumptions is taken to its utmost and the behaviour and thought of each scientist is conceived of as an analogue of rational Nature, then the hubbub of fierce controversy and personal acrimony which marks Victorian science throughout its form and content becomes all but incomprehensible.

It was in the Royal Society that the effects of that embodiment of scientific naturalism, the X-Club[61], may be most clearly seen. All members except Spencer were Fellows and all exerted a mighty influence, or at least attempted to do so. What may be easily forgotten, however, is that their efforts were usually a continuation of what individuals had been doing long before the Club came into existence. Nowhere was this more true than in the matter of scientific refereeing.[62]

The process of refereeing, essential for all modern academic publication, is part of a system of checks and balances that goes some way to weed out careless, flawed and unworthy contributions. But it can also play a pivotal rôle in the success or failure of a young scientist's career, and may perpetuate certain institutionalised values. That it was abused in the early days of science is not open to doubt. But the idea that, following the reforms of the Royal Society in 1847, scientific objectivity alone prevailed in the refereeing process has been decisively falsified by Harrison. A classic case is that of Tyndall's objections to a paper submitted in 1855 by James Joule, an established scientific figure but blemished in Tyndall's eyes by association with William Thomson (and also with Christianity). Three years later Busk and Huxley 'consigned to the archives' a paper by the Plymouth Brethren author P. H. Gosse, and in 1862 Busk insisted on drastic modifications in a paper by the non-evolutionary naturalist J. Bowerbank, even though his fellow-referee did not agree. These questionable incidents have been fully documented and in no case depended on genuine scientific objection.[63] Not surprisingly Frankland also found himself heavily involved in the business of being an academic referee.

From his Manchester years Frankland had acted as a referee for papers submitted to the *Proceedings of the Royal Society* and *Philosophical Transactions*. He pulled no punches and was able to dispense what he saw as academic justice for a period exceeding 40 years. It was an invisible but nevertheless important way for him to exercise considerable power within the councils of the Royal Society. Frankland seems to have been innocent of the worst skulduggery, partly because he refereed chemical papers which were, in principle, less of a threat to the scientific naturalists than those in other fields. He was one of a small group of referees appointed by the Royal Society Secretary in charge of papers. Thus in 1862 he was one of only 17 persons refereeing 166 papers (out of 262 altogether). The group included the X-Club members Huxley, Tyndall and Busk. The only other chemists were Miller and Williamson. Frankland himself was concerned with nine papers that year.[64] He continued this service for many years more; amongst his few rejections was the paper by the unfortunate Phipson (p. 251).[65]

Anonymous refereeing from a distance is all very well but, though supposedly mediating Society policy, is hardly likely to influence it. More enticing prospects were at hand. As soon as he moved to London Frankland found himself a member of the Royal Society Council (from 1857 to 1859). He served again in each of the following three decades: 1865–7, 1875–7 and 1886–8. Here, on Council, Frankland began to learn something of the politics of patronage. At the level of Fellowship he was already involved in endless proposals and sponsoring. Former associates like Clowes sought Frankland's support for their candidature for Fellowship.[66] Others besought his aid for their friends, as when Darwin wrote on behalf of Meldola.[67] In fact Frankland had nominated so many chemists that he declined to be 'godfather' to more than two at a time until their candidature was successful (*i.e.* 'at the head of the list'). But he was prepared to sign lower down in the list of sponsors.[68] Medals also invited creative patronage.

In 1859 Frankland proposed Williamson for the Royal Medal[69] and was rewarded for his pains three years later. Other nominations included Guthrie (Rumford), Perkin and Gladstone (Royal), and Newlands (Davy). Interestingly it was Williamson, not Frankland, who proposed Kolbe in 1884 and Kekulé in 1885.[70] As we have already seen (p. 427) the conspiracy to deprive Forbes of a Copley Medal was engineered by those later to be conspicuous in the X-Club (Huxley and Tyndall especially) and Frankland was conveniently in a position to be their accomplice on Council. A successful campaign to secure the Copley Medal for Darwin, involving much pressure from X-Club members Hooker, Busk and Lubbock, did not coincide with Frankland's membership of Council.[71] However in 1887 Frankland delightedly informed Huxley that he had successfully persuaded the Royal Society to award its Davy Medal to J. A. R. Newlands, a non-Fellow, for his work on the periodic classification of the elements. It was, he wrote, ' a great triumph of mine over Williamson who fought against it to the bitter end'.[72] Hooker was still hard at work pulling strings in 1897, writing to Frankland (now Foreign Secretary) to solicit his support for the award of a Royal Medal to the engineer and botanist Sir Richard Strachey.[73] His ploy was successful.

Between 1870 and 1885 when the X-Club seems to have been most active, its members contributed 51 out of 315 (16%) man-years of membership, but Frankland himself accounted for only a modest 3 to that total. The Council of 1875, to which he did belong, had X-Club members in three of the five executive positions: Hooker, Spottiswoode and Huxley as President, Treasurer and Junior Secretary respectively. In Frankland's final term as Council member Huxley had given up all pretence of not being an arch-manipulator and continued to send letters to effect his own policy. In 1886 he wrote to Frankland:

> I am sorry to hear that there was a balance of opinion against the scheme for admission of colonial members to the RS. Whether the scheme proposed is the best that can be devised is an open question – but I have no doubt as to the desirableness of some such action and I think the Royal Society will look uncommonly foolish if after the appeal made to it – it is obliged to say it can do nothing.[74]

Two years later he wrote in support of a proposed federation with colonial scientific societies of which, to his dismay, his friends on the X-Club did not entirely approve:

> I am very sorry you are all against Evans' scheme. I am for it. I think it a very good proposal, & after all the talk, I do not want to see the Society look foolish by doing nothing. You are a lot of obstructive old Tories & want routing out. If I were only younger and less indisposed to any sort of exertion I would rout you out finely![75]

A number of minor committees claimed Frankland's talents from time to time. He was co-opted as a member of the Statutes Review Committee. Its Report in 1890 was noteworthy for its conservatism; it kept the *status quo* in recommending no increase in the number of Fellows and of Foreign Members, though did allow some detailed changes in procedures for reading and publishing papers. Its conclusions were largely unanimous so Frankland presumably went along with the consensus.[76] At about the same time he was on a Committee appointed to arrange for a portrait in oils of the retiring President, Sir George Stokes.[77] A further invitation came to serve on Section C of the Committee for managing the Government grant (Chemistry and metallurgy).[78]

This was all very well, but the minutiae of statutes, the giving of small grants and the business of oil paintings were hardly the stuff of scientific politics, especially if there was any feeling that time was running out. Then, out of the blue, came a letter from the Secretary, Michael Foster. It expressed the 'most earnest hope' that Frankland would accept the office of Foreign Secretary of the Royal Society without hesitating. The previous holder of that office had just been elevated to President, and there was some urgency. 'But if you do hesitate please accept it even if for one year only'.[79] The following day brought a letter from the new President, Joseph Lister, expressing similar hopes and assuring him that 'the duties are of the very lightest description'.[80] Frankland was now nearly 71, but he accepted the invitation, to the delight of many of his friends. Among the congratulatory letters was one bearing the not unwelcome news that, as Foreign Secretary, he was now an *ex officio* member of the convivial Royal Society Philosophical Club.[81]

Although the official records of the Society lend support to the view that the task was not very onerous, Frankland seems to have thrown himself with some energy into the business of communication with many foreign scientists and academies. He would write with news of various honours conferred by the Society, and received appropriate acknowledgement. Thus he heard from his old friend (and relative by marriage) Johannes Wislicenus, who wrote to thank him for Foreign Membership[82] and for the Davy Medal.[83] Letters were also received from Lenard[84] and Röntgen[85] for their Rumford Medals, from Grafsi for the Darwin Medal,[86] from Gegenbaur for the Copley Medal,[87] and from Moissan for his Davy Medal.[88] Frankland writes to the Académie des Sciences, Paris, to offer a Joule studentship of £50.[89] His correspondence network was now extended to a far wider range of scientists than before, in terms of both subject and geography. However he did not limit his 'official' letters to residents abroad. A revealing missive to the astronomer Sir William Huggins was an 'early note' telling him of his award of the Copley Medal, and indicating the vote was unanimous.[90] One wonders what divulgence of this 'inside information' had to do with the job of a Foreign Secretary.

Nor did Frankland restrain himself from socialising on the grand scale. Never a stranger to arranging dinners at the Athenaeum, on 29 April 1896 he sent out invitations to a dozen or so chosen guests for a dinner 'to meet the President of the Royal Society'.[91] On the table plan he seated himself between the President and Secretary, Lords Lister and Rayleigh. It was a characteristically splendid affair, with ten courses whose climax was probably *Filet de Boeuf Piqué à la Portugaise*. Woe betide the official whose wine was not up to standard! Frankland was already an experienced organiser of such events as Anniversary Dinners, Soirées and similar functions.[92] If all this sounds like petty posturing it must be remembered that Victorian scientists loved this sort of occasion, meritocratic heirs to the ancient customs of the aristocracy. A grand dinner worked wonders for their egos and reminded them of the new hegemony of science. But in Frankland's case there may have been more to it than mere custom.

Fig. 15.3 Sir George Stokes, President of the Royal Society 1885–90. (Author.)

Perhaps Frankland was still smarting from the award of only the most junior office in the executive. After all, his rival Williamson had been appointed to the post 22 years before. Some of his friends had expressed the opinion that he would be worthy of the highest office[93], but this was not to be. Surely his friends on the X-Club could have exerted their well-tried manipulative skills and secured his election as President? In fact the problem lay within that exclusive circle itself. First, there was disagreement over the decision by the President of the Royal Society, Stokes, to continue in office while sitting as Member of Parliament for Cambridge University. Frankland, along with Hirst and Lubbock, saw merit in the idea of 'a member for the Society' but Hooker remarked to the sympathetic Huxley that at that suggestion 'my blood ran cold and my soul sank within me'.[94] A serious fault-line was developing within that apparently monolithic structure, the X-Club. As time went on more serious divergences appeared; the scale and scope of Frankland's commercial activities became too much for even the liberal souls of some X-Club colleagues. As we shall see (Chapter 16) Frankland's money-spinning activities were a continuing cause of large embarrassment to Huxley who, at just this time, expressed himself in a dilemma over the new Presidency of the Royal Society. He wrote to Hooker:

> Who have you to suggest? The only thing I am clear about is to keep out traders on the one hand and mere noblemen on the other. . . Stokes won't do . . . as President he means stagnation or retrogression. Williamson won't do – he means crotchets and impracticability in excelsis. Frankland won't do – Biologists and Chemists and Mathematicians aside what do you think about Tyndall?[95]

'Frankland won't do'. In that pregnant but unexplained phrase his fate was sealed. Without Huxley's support he had no chance of achieving the Presidency. Neither the Copley Medal nor the position of Foreign Secretary was his until the X-Club was for all practical purposes dead. Just before both honours had been received an open letter was issued from the executive to all Fellows on the propriety of using 'FRS' in signed certificates and statements. It requested that all Fellows of the Royal Society should 'endeavour to express any statements relating to matters of trade or manufacture in such terms that no suspicion of mercenary motives or commercial partisanship can possibly attach to them'.[96]

In that note was encapsulated the Royal Society's horror of 'trade' in all its grubby manifestations. It was a shot across the bows of all engaged in commercial activities. It just might have been a veiled rebuke to the former President of both the Chemical Society and the Institute of Chemistry. Had it not been for this perception of his life-style he might conceivably have acquired a greater distinction still, Presidency of the Royal Society itself.

Notes

1 The Chemical Society was larger than is sometimes supposed. Its oft-quoted membership data as given by Leone Levi in 1868 are seriously inaccurate (*Rep. British Assoc., Trans. of Sections*, 1868, 169–72). The Society is quoted

as having (in 1867) a membership of 192 whereas official statistics indicate 499 (*The Jubilee of the Chemical Society of London. Record of Proceedings, together with an account of the history and development of the Society*, London, 1896, p.186); the published membership list in the *Journal* has 490 names for one specific point in the year.

2 The black-balling activities of the 1860s, previously noted, were connected with attempts to preserve the value of the F.C.S. ascription.

3 W. A. Campbell, 'The Newcastle Chemical Society and its illustrious child [the SCI]', *Chem. & Ind.*, 1968, 1463–6.

4 Minutes of the Chemical Society Council, 1 February 1872 [Royal Society of Chemistry Archives].

5 Frankland to Royal Society, 26 January and 12 February 1872 [Royal Society Archives, MC.9.313 and 326].

6 Frankland, Presidential Address, *ibid.*, 1872, **25**, 341–62.

7 Frankland, Presidential Address, *J. Chem. Soc.*, 1873, **26**, 772–88.

8 R. B. Pilcher, *The Institute of Chemistry of Great Britain and Ireland: History of The Institute 1877–1914*, Institute of Chemistry, London, 1914, p.24.

9 Minutes of the Chemical Society Council (note 4).

10 Pilcher (note 8); and C. A. Russell, N. G. Coley and G. K. Roberts, *Chemists by Profession: the origins and rise of The Royal Institute of Chemistry*, Open University Press/ Royal Institute of Chemistry, Milton Keynes, 1977.

11 T. Mortimer, *The Universal Director*, London, 1763, pp.18–19.

12 Humphry Davy to Grace Davy (his mother), 1799; here he was denying that he had 'turned poet' [Rolleston collection; on microfilm at the Royal Institution].

13 J. Wesley, *Journal*, entry for 7 January 1768, in *The works of the Rev. John Wesley*, London, 1809, vol. iv, p.344; the afflicted individual referred to was released from a debtors' prison through the good offices of Wesley and a gift from a friend, after which he was 'put in a way of living', presumably at chemistry.

14 Obituary of Cassal, *Analyst*, 1922, **47**, 103.

15 F. M. L. Thompson, *The Chartered Surveyors: the growth of a profession*, Routledge, London, 1968.

16 C. R. A. Wright, *Chemical News*, 1876, **33**, 27.

17 On Wright see *Proc. Roy. Soc.*, 1894, **57**, v–vii.

18 N. S. Maskelyne, *Chem. News*, 1876, **33**, 246.

19 *Papers relating to the Sale of Gas Act*, P.P. 1860, LIX, p.9.

20 S. Clegg, *A practical treatise on the manufacture and distribution of coal-gas*, London, 2nd ed., 1853, p.45.

21 A. F. Marreco, *Trans. Newcastle Chem. Soc.*, 1871–4, **2**, 220 (16 March 1874); the Newcastle Chemical Society was remarkable for its clear perception of the problem and its full debates on it.

22 Pilcher (note 8), p.42.

23 Frankland, MS 'Topics for address'; of the early laboratories he cited only the Birkbeck Laboratory, that at the Museum of Practical Geology, and the Royal College of Chemistry [RFA, OU mf 01.04.1386].

24 Pilcher (note 8), p.50.

25 Printed notice in papers relating to 'The Organisation of the Chemical Profession' [Royal Society of Chemistry Archives].

26 Minutes in 'Organisation' papers (note 25).

27 A detailed MS manifesto for a movement to create 'a new grade of Fellows of the Chemical Society' exists on notepaper headed with Neison's address (37 Fellowes Road, London NW) [RFA, OU mf 01.04.1383].

28 Minutes of Chemical Society Council, 18 May 1876 [Royal Society of Chemistry Archives].

29 MS 'Report of the subcommittee appointed to consider the suggestions of the Organisation of Chemists Committee' [RFA, OU mf 01.04.1404; and also insertion into Minutes of Chemical Society Council, 15 June 1876].

30 J. Pettengill, *Chem. News*, 1876, 33, 240–1 (9 June).

31 Minutes of Chemical Society Council, 15 and 22 June 1876 [Royal Society of Chemistry Archives].

32 Proposals of Council of Chemical Society, 23 June 1876 [RFA, OU mf 01.04.1390 and 1414 (another copy)].

33 Minutes of Organisation Committee, 13 October 1876 ('Organisation' papers (note 25)).

34 Frankland to F. A. Abel, 16 October 1876 in Minutes of Chemical Society Council [Royal Society of Chemistry Archives]; also copy-letter [RFA, OU mf 01.04.1381].

35 Russell *et al.* (note 10), pp.144–8.

36 Report of Committee [RFA, OU mf 01.04.1388].

37 Notice of meeting [RFA, OU mf 01.04.1389]

38 Report of meeting on 4 November 1876; a separate MS draft of committee nominees has only 40 names (including Frankland), so presumably the others were added at the meeting [RFA, OU mf 01.04.1392 and 1402].

39 A copy exists in the 'Organisation' papers (note 25), and another in the Frankland Archive (heavily endorsed in his handwriting) [RFA, OU mf 01.04.1416]. Slightly different versions exist in MS (annotated by Frankland) and in print [ibid., 1398 and 1412].

40 Frankland, undated MS note [RFA, OU mf 01.04.1396].

41 Frankland, undated MS note [RFA, OU mf 01.04.1420]; he includes his own name as a Vice-President.

42 Copy-letter, Frankland to President of Pharmaceutical Society, 5 June 1877 [RFA, OU mf 01.04.1394].

43 J. Williams to Frankland, 11 June 1877 [Royal Pharmaceutical Society Archives, LPC, pp.117–18].

44 W. N. Hartley to Frankland, 21 November 1876 [RFA, OU mf 01.04.1409].

45 W. de la Rue to F. A. Abel, 9 November 1877 [RFA, OU mf 01.04.1403].

46 A. W. Williamson to Frankland, 26 February 1877 [RFA, OU mf 01.04.1219].

47 A. W. Williamson to Frankland, 2 March 1877 [RFA, OU mf 01.04.1220].

48 H. E. Roscoe to Frankland, 1 January '1877' [1878 from context] [RFA, OU mf 01.04.1408].

49 H. E. Roscoe to Frankland, 8 January 1878 [RFA, OU mf 01.04.1385].

50 B. C. Brodie to Frankland, 7 November 1878 [RFA, OU mf 01.01.0562].

51 C. M. Tidy to Frankland, 1 February 1878 [RFA, OU mf 01.04.1422].

52 Institute of Chemistry Minutes for Examinations Committee 1 and 8 November 1878 [Royal Society of Chemistry Archives].

53 C. R. Weld to Frankland, 3 June 1853 [RFA, OU mf 01.01.0215].

54 He hoped that the effect of the Medal would be 'to stimulate me to further and more successful labours': Frankland to Royal Society, 9 November 1857 [Royal Society Archives, MC5.308].

55 M. Foster to Frankland, 1 November 1894 [RFA, OU mf 01.04.1581]; Frankland was nominated by W. A. Tilden, Medal Claims Book 1873–1909, p.122 [Royal Society Archives].

56 *The Times*, leading article, 1 December 1894, p.9.

57 F. R. Japp to Frankland, 4 December 1894; W. P. Wynne to Frankland, 9 November 1894; T. E. Thorpe to Frankland, 1 November 1894 [RFA, OU mf 01.04.1562, 1556 and 1547].

58 Frankland, typescript and MS notes [RFA, OU mf 01.04.0038 and 0041]; similar expressions of thanks are in his letter: Frankland to Secretary, Royal Society, 14 May 1895 [Royal Society Archives, MC16.213].

59 M. B. Hall, *All scientists now: the Royal Society in the nineteenth century*, Cambridge University Press, Cambridge, 1984.

60 A. J. Harrison, 'Scientific naturalists and the government of the Royal Society 1850–1900', Open University Ph.D. thesis, 1988.

61 For a general account see Ruth Barton, '"An influential set of chaps": the X-Club and Royal Society politics 1864–1885', *Brit. J. Hist. Sci.*, 1990, 23, 53–81.

62 For a recent examination of the refereeing process see H.–D. Daniel, *Guardians of science: fairness and reliability of peer review*, trans. W. E. Russey, VCH, Weinheim, 1993.

63 Harrison (note 60), pp. 95, 100, 98.

64 *Ibid.*, p. 87.

65 Referee's report, 10 January 1863 [Royal Society Archives, RR5/185].

66 F. Clowes to Frankland, 9 February 1896 [RFA, OU mf 01.07.0447].

67 C. Darwin to Frankland, 8 February 1882 [RFA, OU mf 01.08.0426].

68 Frankland to H. E. Armstrong, 29 December 1893 [Royal Society Archives, MM10.94].

69 Frankland to Royal Society, 20 June 1859 [Royal Society Archives, Sa. 581].

70 Medal Claims Book 1873–1909 [Royal Society Archives].

71 M. J. Bartholomew, 'The award of the Copley Medal to Charles Darwin', *Notes & Records Roy. Soc.*, 1976, 30, 209–18.

72 Frankland to Huxley, 5 November 1887 [Imperial College Archives, Huxley papers 16/272].

73 J. D. Hooker to Frankland, 19 October 1897 [RFA, OU mf 01.02.0515].

74 T. H. Huxley to Frankland, 8 December 1886 [RFA, OU mf 01.02.0481].

75 T. H. Huxley to Frankland, 3 February 1888 [RFA, OU mf 01.02.0414]; also in L. Huxley, *Life and letters of Thomas Henry Huxley*, Macmillan, London, 1908, vol. iii, p.58.

76 Draft Report of Statutes Revision Committee, October 1890 [RFA, OU mf 01.01.0978].

77 Royal Society printed Appeal [RFA, OU mf 01.01.0979].

78 M. Foster and Lord Rayleigh to Frankland, 8 February 1893 [RFA, OU mf 01.01.0808]. He concluded that, for chemistry, small grants were adequate; 29 out of 54 grants of £100 or less had proved to be satisfactory: Frankland, undated MS note [RFA, OU mf 01.01.0812].

79 M. Foster to Frankland, 8 November 1895 [RFA, OU mf 01.03.0769].

80 J. Lister to Frankland, 9 November 1895 [RFA, OU mf 01.03.0770].

81 T. E. Thorpe to Frankland, 15 November 1895 [RFA, OU mf 01.04.1726].

82 J. Wislicenus to Frankland, 18 April 1897 [RFA, OU mf 01.04.1476].

83 J. Wislicenus to Frankland, 9 November 1898 [RFA, OU mf 01.04.1280].

84 P. Lenard to Frankland, 11 November 1896 [RFA, OU mf 01.04.1686].

85 W. C. Röntgen to Frankland, 15 November 1896 [RFA, OU mf 01.04.1297].

86 B. Grafsi to Frankland, 12 November 1897 [RFA, OU mf 01.04.1511].

87 C. Gegenbaur to Frankland, 14 November and 31 December 1896 [RFA, OU mf 01.04.1508 and 01.07.0314].

88 H. Moissan to Frankland, 16 November 1896 [RFA, OU mf 01.04.1481].

89 Frankland to Académie des Sciences, 18 February 1896 [RFA, OU mf 01.07.0550].

90 W. Huggins to Frankland, 3 November 1898 [RFA, OU mf 01.03.0651].

91 Dinner invitation and table plan, 29 April and 21 May 1896 [RFA, OU mf 01.02.0854 and 0927].

92 See the voluminous correspondence with H. Rix in New Letter Books [Royal Society Archives]; his private papers contain nearly 80 pages of menus, table-plans, invitations *etc.* for the period 1894–1896 alone [RFA, OU mf 01.02.0853–0931].

93 *E.g.* H. E. Armstrong (Maggie West (*née* Frankland), *Diary* for 29 October 1901 [MBA]; Armstrong, 'First Frankland Memorial Lecture', *Chem. & Ind.*, 1934, 459–66 (464)).

94 J. D. Hooker to T. H. Huxley, 23 March 1888 [Imperial College Archives, Huxley papers, 3/316].

95 T. H. Huxley to J. D. Hooker, 30 June 1883 [Imperial College Archives, Huxley papers, 2/250].

96 M. Foster and Lord Rayleigh, open letter to Fellows of the Royal Society, 1 June 1893 [RFA, OU mf 01.01.0997].

CHAPTER 16

Retirement years

1 Reigate Hill

Like many other Victorian scientists Edward Frankland gradually felt the urge to move out of London. Partly this was on account of the growing pollution in the centre but it also reflected the extension of the commuter belt to mid-Surrey with better rail travel available. There was more than a little of status symbolism in the quest for a semi-rural retreat. In 1879 he wrote to his daughter Sophie to say that he was considering Wimbledon, Beckenham and Reigate.[1] The choice soon narrowed to Reigate, with encouragement from B. C. Brodie who already lived there (at Brockham Warren) and wrote with details of several properties.[2] By the middle of 1880 he had settled on *The Yews*, a charming house in its own grounds in a quiet road off Reigate Hill. It was duly inspected and approved by daughters Sophie[3] and Maggie.[4] By the late summer of 1880 the removal from Lancaster Gate was complete. *The Yews* was to be the domestic base for Edward Frankland's activities for the rest of his life.

Reigate Hill is a well-known feature of the southern scarp of Surrey's North Downs. At its foot lies the small town from which it is named, and a road climbs steeply up it towards Epsom Downs, six miles away. Very near the top of Reigate Hill there is a turning to the left, Beech Road. To this day it is a quietly select avenue, with a few large houses on its northern side; almost the first one encountered is *The Yews*. Over half its original land has now been sold off, but the mid-Victorian house taken by the Franklands remains immediately recognisable. Its rear garden slopes up the hill and beyond it are the remains of chalk and lime workings and, almost at the crest, the supposed track of the old Pilgrims' Way to Canterbury.

By the 1881 Census the household was well established. In addition to Edward and Ellen Frankland, Maggie (28) and Percy (22) were still formally members of the household, as were the two little girls Dorothea (4) and Helga (3). There were two housemaids, a parlourmaid, a cook and a nurse.[5] Clearly the Frankland family was not suffering from financial hardship. Moreover this was the first house that Frankland had purchased; up to that time he had always rented accommodation. In moving out of London to relatively opulent surroundings he was participating in the same trend that took Tyndall to Hindhead and Huxley to Eastbourne. It has been argued that these moves illustrate the idea that social mobility was as much a cause as a result of the pursuit of scientific naturalism by X-Club members (or at least by those members who started life at a financial disadvantage).[6] However other London academics sought similar rural retreats, as when Brodie moved to Brockham and A. W. Williamson to Hindhead. From much further north Roscoe had been lured to near Ranmore and J. W. Swan to Warlingham, while Nasmyth had gone even further afield to Penshurst in Kent. It is noteworthy that moves from London were nearly always to the south, probably because of the services provided by the South Eastern & Chatham Railway. From nearby Redhill the train journey to London Bridge took only about 45 minutes in the 1880s.

One might have expected that Frankland would have had little time for enjoying the new house, as his duties at South Kensington became increasingly onerous and his consultancy work increasingly demanding. However, with the energies of a man half his age, he quickly made a start on the process of home-improvement. First (and typically) he thought of the domestic water supply. This was obtained from a well that he sunk near the house, and softened by Clark's process, specially appropriate in chalky areas where much of the hardness is due to dissolved calcium bicarbonate. The calcium is precipitated out as the carbonate by the careful addition of calcium hydroxide solution. For this purpose tanks were placed on the hillside above the house. Then there was the question of energy requirements. Holding strongly to the view that much environmental damage was caused by the combustion of coal,

Fig. 16.1 *The Yews*, Reigate Hill: final home of the Frankland family.
(J. Bucknall Archives.)

Frankland arranged for his domestic fireplaces to burn coke or
anthracite (thus greatly reducing the bill for chimney-sweeps).
Following the example of Crookes in London and W. G. Armstrong
at *Cragside* in Northumberland he supplied the house with electric
light, making many of the fittings himself. The electricity was
generated by a steam-driven dynamo, the waste heat being
conducted to some of the living rooms and also to the extensive
hot-houses.

These hot-houses were a great pride and joy, as was his large
garden (in which he was assisted by at least one gardener).
Gardening had been a hobby since Haverstock Hill days, but now
he had more space and (in due course) time.[7] He became known in
the area for the excellence of his chrysanthemums, sometimes
winning first prize at local garden shows, in one year taking no less
than twelve prizes.[8]

Two other features of *The Yews* marked it out from its
neighbours, each reflecting the scientific passion of the owner. As
early as 1881 a dinner-guest at Reigate was H. Grubb, a supplier of

astronomical equipment from Dublin. He was to be the architect of a new observatory. [9] It was soon erected, with a magnificent $12\frac{1}{4}$ inch reflecting telescope. Some aspects of Frankland's astronomical observations have already been noted (p. 433). The structure must have been quite large for many years later Frankland narrowly escaped serious injury when falling through the trap-door. Fortunately his fall was broken by his secretary, Jane Lund, who happened to be on the stairs below, and he suffered only minor abrasions.[10]

The other structure was a fully equipped chemical laboratory which became his private research facility when he retired from South Kensington. Here his interests in electric lighting led to a series of investigations into storage batteries.[11] He also published on aspects of microbiology and water supply, particularly when water analysis was moved from Grove House to *The Yews*, at which time Frankland's assistant W. T. Burgess also transferred to Reigate.[12]

Life was not all work, however. Apart from the questions of family-rearing and social entertaining, there was also the lure of the great outdoors. Frankland acquired a tricycle. The peak year for using it was 1886. Writing to Grace, his daughter-in-law, he announced:

> I intend to try Brighton (32 miles) some day. The road from Burford Bridge to Leatherhead is exceedingly pretty. Indeed, the whole country around here is now most lovely, & the weather perfect, neither too hot (like Naples in March) nor too cold.[13]

One day he cycled 24 miles before lunch, and generally managed at least 50 miles a week on his tricycle. However he had to admit 'the shine was rather taken out of my exploits' by hearing of one 77-year old man who cycled from London to Brighton and back in one day (108 miles). On at least one occasion his tricycle was used to transport himself between Reigate and Maggie's house in central London.[14]

There were of course difficulties. Both his sons-in-law experienced broken bones through cycling accidents in that summer of 1886. By October he wrote 'the roads are now getting bad for cycling owing partly to mud & partly to new macadam'.[15] And there were social constraints:

> My machine continues to afford me much pleasure, but my excursions are much interfered with by dinner invitations, for I have only time to ride in the evening.[16]

Fig. 16.2 Edward Frankland posing for a photograph in his laboratory.
(J. Bucknall Archives.)

However it testifies to the fitness of the man that, while fellow
members of the X-Club were languishing away through illness, real
or imagined, Frankland was on his cycle bowling merrily along the
Surrey roads in his early 60s. That he was free to do so was made
possible by strange events at South Kensington that led directly to
his retirement.

Table 16.1. *Students at South Kensington Chemistry Laboratories, 1868–81*

	1868	1869	1870	1871	1872	1873	1874	1875	1876	1877	1878	1879	1880	1881
spring	45	45	41	48	49	117	98	104	101	94	103	101	79	83
summer	45	45	38	50	48	71	99	95	90	87	105	94	74	77
winter	45	48	42	50	115	92	97	88	92	101	95	90	82	85
TOTAL	*135*	*136*	*121*	*148*	*212*	*280*	*294*	*287*	*283*	*282*	*303*	*285*	*235*	*235*

2 Abnormal happenings at a Normal School

Under Frankland's leadership the successor to the Royal College of Chemistry at South Kensington experienced both prosperity and adversity. The new laboratories were far more suitable for teaching purposes than their predecessors at Oxford Street, and a steady stream of students passed through them to take influential positions in the world outside. Student numbers reached a maximum in the late 1870s (Frankland's figures[17]) (Table 16.1 opposite).

On the other hand there were real difficulties. Working men were more reluctant to travel to South Kensington than to a central location. As he had feared, demands from both education and bureaucracy saw to it that research would be extremely difficult. Moreover, student numbers were less encouraging than they should have been. It became clear that changes would be needed and that the institution's status would need to be redefined. Here were also departments of physics, mathematics and biology (but not mining science) still under the umbrella of the Royal School of Mines. It was decided that these departments, together with chemistry, should constitute a new body, to be called the Normal School of Science. The name, suggested by T. H. Huxley who became its first Dean, was intended to create the impression of that paradigm of scientific and technical education, the great École Normale of Paris. In fact the choice was singularly unfortunate, for to most Britons it conveyed anything but that idea and in due course it became the Royal College of Science. The Normal School came into existence in 1881. At this time a new Assistant Secretary at the Department of Science and Art was appointed, with responsibilities as Director for Science at the South Kensington Museum. The civil servant concerned was J. F. D. Donnelly (1834–1902), Lieut. Colonel in the Royal Engineers and a power to be reckoned with.[18] He became secretary and permanent head of the Department of Science and Art in 1884.

News of the impending changes was conveyed to Frankland from Donnelly in a letter sent to his holiday address in Norway (and presumably another copy to his home).[19] It was to prove of more than passing interest. First, Donnelly summarised the new institutional arrangements:

> The Science School at South Kensington will be organized and equipped as a Normal School primarily for the instruction of Teachers

and of students of the Industrial classes selected from the Science Schools in the Kingdom; other students will also be admitted to it while there is room for them on the payment of fees fixed at a scale sufficiently high to prevent the School competing unduly with other institutions which do not receive State aid.

Donnelly conveyed the hope that Frankland would find it possible to accept the post of Normal School Professor of Chemistry in place of his present one. The salary would have been £800 p.a. with addition of fees up to a total amount of £1200. However,

as it appears that the emoluments you have received from the offices you have held of Chemist and Lecturer on Chemistry have on the average of the three years ending 30th September, 1879, amounted to £1609 a year, your salary will be made up to that amount.

Where Donnelly obtained the titles Frankland was supposed to have held must remain a mystery. But the latter was well satisfied with his financial offer. Future duties were left gloriously vague, though courses for teachers and working men were enjoined. But there was this caveat:

Suffice it to say that it is understood that your whole time is engaged by the State in the same way as that of ordinary Civil Servants of the Crown.

Frankland declined the request to reply by telegraph, as he had no papers with him in Norway and he particularly wanted to discuss one issue with Donnelly in person. He wished to establish that 'there would be no objection to the continuance of the water-analysis laboratory on its present footing'.[20] Presumably this was settled to his satisfaction for he formally accepted the post a month later.[21] No doubt the Lords of the Committee of the Council on Education breathed a collective sigh of relief.

The question of payment for water-analyses and other services to members of the public remained to vex the School authorities. Frankland later recalled how, when the Rivers Commission was wound up, the Department of Science and Art and Local Government Board jointly agreed to close their very expensive laboratory. Future monitoring of public water supplies should be conducted under Frankland's direction at South Kensington, the Department providing the apparatus and Frankland providing chemicals and the salary of an assistant. He continued:

Soon after the re-organisation of the Science Schools I was requested by the Department to remove this work from South Kensington and to provide a private laboratory for its performance. This request I at once complied with at great cost to myself, neither asking nor receiving any compensation.[22]

That was the origin of the Grove House laboratory, in 1882. Its foundation might reasonably have been expected to calm disquiet about the conduct of the chemical laboratories at South Kensington, but sadly that was not to be the case for long.

In the summer of 1884 Huxley, in his capacity as Dean, wrote to Frankland with news of a most unwelcome development:

> I was summoned to the Privy Council office yesterday afternoon to see the Lord President and the Vice President about the affairs of our School, and more especially, about those of the Chemical Department, with regard to which, I am sorry to say, considerable dissatisfaction was expressed.
>
> Lord Carlingford and Mr. Mundella remarked upon the serious diminution of the number of students in the Chemical Laboratories. They did not think that this falling off is, to any large extent, to be accounted for by the competition of other schools, and by the comparatively high rate of our fees, to which I drew their attention. But they informed me that complaints had been made to them by influential persons that a fair share of your time and energies is not given to the work of the school and that their Lordships were of opinion that the School suffers in consequence.
>
> I was desired to make you acquainted with these views of the Lord President and the Vice President in any way I thought fit; and I have adopted the form of a private letter, as the least unpleasant method of discharging a duty, which I need not tell you I should very gladly have escaped.
>
> However, as the thing had to be done, I have thought it best to err, if at all, on the side of plain speaking.[23]

As was his habit when challenged on matters financial, Frankland gave a knee-jerk response. From hastily scribbled notes on the back and front of the envelope[24] that had contained Huxley's letter he fired off a return salvo within hours.[25] Although it would be tedious to rehearse all the conflicting arguments in this debate, some passages in Frankland's letter reveal much of interest about his values and life-style at that time. He begins on a soulful note, assuring his old friend that his letter 'gave me as much pain in the reading as I am sure it gave you in the writing'. He then went on to elaborate complaints beyond those that Huxley had conveyed, which suggests that they were common enough knowledge. They were, he said,

1. That I do not give a fair share of my time and energies to the

work of the School and that there is a serious diminution of the number of students in the Chemical Laboratories in consequence.

2. That I have a private laboratory and devote an undue amount of my time to it.
3. That I take private pupils and thus injure the School.
4. That I give evidence before Parliamentary Committees and elsewhere to such an extent as to seriously interfere with my duties at the School.
5. That I do not make original investigations.

To these charges he briefly replied:

1. The attendance book shows that I do give a fair share of my time to the School, and so far as it is in my power to do so, my best energies are devoted to the interests of the School during that time. There has been no alteration in this respect since the laboratories were removed to South Kensington, and therefore the diminution in the number of students cannot be due to the cause assigned.
2. During the whole time that my private laboratory has existed I have not spent in the aggregate one hour in it between 10 a.m. and 4 p.m.
3. I have never had a private pupil in my life.
4. Although, like almost all professors of chemistry, I occasionally give evidence before Parliamentary Committees & Royal Commissions, the occasions are very few and the time taken from my duties in this way insignificant. During the year ending this day, for instance, only 19 hours.
5. That I have an ardent desire to make original investigations is proved by the whole of my past career; but I have found, to my great disappointment and sorrow, that I cannot make them at South Kensington.

He appealed to the attendance book signed daily on entering and leaving the building, admitting, however, 'that, owing to my residence out of town, I do not usually arrive at South Kensington until $10\frac{1}{2}$ or even 11 a.m., but on the other hand I never go out for luncheon, my work is carried on continuously throughout the day and I not infrequently stay an hour or two after 4 o'clock.'

There then follow another ten pages of self-justification in the kind of detail one might expect today from an academic unlucky enough to be the subject of a university's formal complaints procedure. In Frankland's case one wonders whether the detail was intended as a smoke screen to obscure the principal grounds for complaint or whether it simply reflected his own inability to see the wood for the trees. On balance the latter seems more likely. Of course he made some telling points: the constant interruptions from 'British & Foreign Professors & Teachers', to be expected by a scientist of his age, experience and reputation; the responsibility for his subject in general; the routine chores of giving interviews, writing testimonials and advising students; the large number of successful pupils; the sheer difficulty of conducting sustained research given demands of bureaucracy and public enquiry. All these points could have been made equally well in 1996. The curious thing is that Frankland did not seem to think it worth stressing the benefit brought to the College (and indeed the country) by his externally-directed activities. Perhaps that kind of perception had to wait until the twentieth century. The most remarkable feature of all in the long and rambling letter is the reference to his water analysis laboratory at Grove House, in existence since 1882. He repeats his assertion: 'I have not in the aggregate spent a single hour in it between 10 a.m. and 4 p.m. during the Sessions of the School'. This may well have been true, but he failed to mention any quick visits to see his assistants, all holiday periods, work done in the evenings before dinner in town, and above all the labours of assistants performed on his behalf.

There is little doubt that something of the extent of Frankland's 'extra-mural' consultancy activities was widely known in London and beyond. While it may have been technically the case that most of them did not impinge on his ten-to-four day at South Kensington, the fact remains that his reputation as 'an accomplished chemical pluralist' had stuck, and with good reason. His surviving corre-spondence discloses a mass of consultant work in water analysis and many other areas. It was natural to assume that it was undertaken at the expense of his duties at the Normal School of Science, and in one sense it was. No man, especially at his age, has infinite energies and those expended on profitable external activities inevitably diminish those available for paid employment. He ended his long letter on an appropriately apocalyptic note:

> I feel strongly that the bulk of my daily duties at South Kensington is
> not of a kind that ought to occupy so much of my time. It is most
> uncongenial to me; but, as I have already explained, I feel unable to
> shake myself from it. On this account I am very decidedly of opinion
> that a younger professor who would be free from the disturbing
> influences of connection with thousands of past students, and of
> conferences with foreign colleagues, would be of far more value to the
> School than I can now possibly be. And, entertaining this opinion, it is
> my intention to take an early opportunity of retiring from my
> appointment, & devoting the remainder of my life to original work in
> my own private laboratory.

Perhaps he realised that his 12-page defence was inadequate,
composed as it was in the white heat of indignation. Within a week
he was writing again to Huxley. This time he sounded a different
note, deflecting attention away from all questions of loyalty and
financial probity. There was, he implied, a hidden agenda for his
opponents that had nothing to do with money. 'The only valid
charge that can be brought against me is that I do not teach
elementary chemistry in the laboratory', and this is 'the real origin
of the complaints'. However he added with unusual truculence that
'all the Lords in creation could not induce me to do it now'.[26] To
these letters Huxley replied that 'the completeness of your defence
is the best consolation I can find for being the means of troubling
you', and promised to pass on the information to the authorities.[27]

At this point another figure entered the fray. Four days later
Huxley wrote to Frankland,[28] enclosing a letter from J. F. D.
Donnelly. This powerful official had pointed out the declining
number of students and raised the question of technical analyses
being performed in the South Kensington laboratories. Frankland
replied to Huxley that most students had finished practical work by
February and therefore visitors (as MPs) would discover 'a very
melancholy appearance' to the laboratories that gave an unfair
impression of overall numbers. As for commercial analyses, these
had been banished to his private laboratory in 1882.[29]

Frankland had entirely missed the point, by intention or
accident. So Donnelly wrote to him again, explaining that the
concerns related to analysis *for payment*, not to technical analysis as
such.[30] Right on cue, Frankland replied within 24 hours. He
stressed the desirability of technical students having experience of
technical laboratory work, much as medical students need to deal
with patients. Were these analyses to be done without fee they

would be flooded with work to the detriment of practising chemists in the Society of Public Analysts or the Institute of Chemistry.[31] To this Donnelly observed that patients do not as a rule pay to be treated by medical students, asking 'why cannot you purchase the objects to be analysed?'[32] Frankland promptly retorted that 'Patients at Hospitals are, as a rule, unable to pay' (unlike rich industrialists, by implication), and that, although artificial mixtures could of course be prepared, 'what the students want is the real thing'.[33] However he would try this plan if Donnelly recommended it (*i.e.* gave him instructions). News of these exchanges was duly forwarded to Huxley[34] who lay low until the storm was past.

By the next year Frankland had reached 60, a time of life which Huxley had often declared was that at which men of science should be strangled lest age should harden them against the reception of new truths, and make them clogs upon progress.[35] He had become tired of teaching, was still fully involved in water analysis and needed time for other activities. Above all even he must have grown weary of the constant wrangling over fees. He cannot have forgotten his veiled threat to Huxley. Accordingly he sought advice from Donnelly, whose reply cannot be accused of excessive length:

> Dear Frankland,
> Here is the form for suicide.
> Yours truly,
> J. F. D. Donnelly[36]

That day, 16 April 1885, Frankland sent the form to Huxley, formally tendering his resignation. His action received a rather longer letter from Donnelly, expressing the 'high appreciation' of the Lords of the Committee of the Council on Education for his services to science.[37] Perhaps scenting some problems ahead over the size of his pension, Frankland wrote to his old mentor Lyon Playfair, now Liberal MP for Leeds (South) and until recently Deputy Speaker of the House of Commons. Playfair advised him that the Treasury could add some years' increments 'for distinguished service', though such decisions were 'not always grounded on an intelligent basis',[38] to which Frankland replied: 'I was in no doubt as to My Lords saying I had been a good boy . . . but I suppose everybody fears the Treasury'.[39]

His forebodings were fully justified. A few days later he received a note from Donnelly,[40] enclosing a copy-letter from the Treasury to the Vice President of the Committee of the Council on Education.[41] It informed the Council that Frankland would be granted an annual pension of £360, with effect from 1 October 1885.

Never a man to let the grass grow under his feet where money was concerned Frankland sent an instant – and lengthy – response to Donnelly.[42] He conveyed the impression of a person in deep shock. The news had taken him 'entirely by surprise' because he had expected a pension of not £360 but £724. Pensions were then calculated on the basis of years of service (n), being $n/60$ of final annual salary. Since Frankland had been formally appointed in 1868, and since he received a salary of £1609, he had not unreasonably expected a figure of £$(27/60 \times 1609)$, which is £724. At least that is how he represented his case.

Incandescent with fury, Frankland had immediately stormed down to the Treasury, only to be informed by their solicitor that the calculation was on the basis of a salary of £800, the remaining £809 being students' fees and therefore not pensionable. Now, to Donnelly, he played his trump card: when his salary was reassessed in 1881 *both elements had been consolidated into a single whole*, so that even if there were no student fees he would still receive the same amount, £1609.

Donnelly brought this information to the attention of the Department of Science and Art, obtained their support and wrote on their behalf to the Treasury, urging reconsideration of Frankland's case.[43] It took the Treasury bureaucrats over two months to respond. Eventually a reply was despatched to the Vice President of the Committee of the Council on Education,[44] a copy of which was forwarded to Frankland.[45] His appeal was rejected.

Always reluctant to give up without a struggle, Frankland made one more effort. He decided to write to the recently ennobled Lord Iddesleigh. Formerly Sir Stafford Henry Northcote, he had been Leader of the Opposition in the House of Commons and was now Chancellor of the Exchequer. Such a man would surely see the justice in Frankland's claim. He had only three weeks previously offered Huxley a Civil List pension of £300, and possibly Frankland knew of this; Donnelly certainly did.[46] Again he approached Lyon Playfair, this time with a request that he would intervene with Iddesleigh in person. A note from the MP[47] conveyed an invitation to see him at home any morning between 10 and 11, but a further letter was less encouraging: 'your letter to

Lord Iddesleigh is too much of an administrative question for me to forward'.[48] Undeterred, Frankland wrote to Iddesleigh directly about his pension, though was permitted to enclose a letter of introduction from Playfair. He rehearsed the old arguments, particularly stressing that he had never heard until recently the basis on which his pension was to be calculated. With breathtaking disregard for both his present financial prospects and his recent research record he wrote that, unless his Lordship could reverse the decision, 'I shall now have to seek remunerative employment instead of devoting the rest of my life to original research as intended'.[49] In due course a reply arrived from Iddesleigh (addressed from 10 Downing Street) in which the noble lord declined to intervene, satisfied that Frankland's misapprehension was not due to 'any advice which was given to you officially'.[50] Technically this was correct, though one may be permitted to query the ethics of a situation in which an employee could be given so little information on such an important topic. The correspondence – and the controversy – concluded with a polite acknowledgement by Frankland,[51] a surviving envelope bearing the date 10 December 1885 and the words 'Last letter from Treasury', and a letter from Frankland to Donnelly thanking the Lords of the Committee of the Council on Education for their kind but unavailing efforts.[52]

A remarkable feature of these controversies is the rôle played by Huxley. An old friend and ally, a fellow-member of the X-Club, Huxley was caught between personal loyalties to Frankland and institutional loyalties to the School. Moreover, the constant repetition of the charges against Frankland was beginning to take its toll and exert an influence on him that he regretted.[53] Nor was Huxley the only unwilling player. Donnelly shared many of his misgivings, for in terms of general outlook it is quite correct to label Donnelly a 'Huxleyite'. Together they sought damage limitation and made every effort to avoid unpleasantness. Worst still, this was the very issue on which the X-Club was beginning to break apart, and which the Royal Society so lamented: the pursuit of science for monetary gain. The incident of Frankland's pension was to exert an enduring and adverse effect on the relationship between himself and Huxley.

3 'Remunerative employment'

'Remunerative employment' had formed the substance of Frankland's plaintive prediction if he were to receive a 'reduced' pension. Whether he had ever intended to reduce his profitable activities may be open to question. The fact was that after 1885 he continued much as before. It is now quite certain that a serious financial ambition underlay much of Frankland's consultancy practice. He refers to money far too often in his private papers for this not to have been the case.

Frankland's first attempts to diversify his commercial interests relate to a discovery he had made 30 years previously . Early in 1885, before his actual retirement, he entered into discussions with the Belgravia Dairy Company of South Kensington concerning the marketing of the 'artificial human milk' that he had invented in Manchester. For some years such a product had been sold by a rival firm, the Aylesbury Dairy Company, with leaflets acknowledging its inventor. Frankland had suggested to his neighbours in South Kensington that they could undercut their rivals and also have a special advantage in advertising. They could announce that:

Artificial Human Milk
that has been prepared under the
special supervision
of
Professor E. Frankland
the
Inventor
can only be obtained from
The Belgravia Dairy Company, Ltd.

At 6d. a pint it could be sent fresh to customers with each delivery of milk, or bought over the counter at one of the Company's depôts. Regrettably, within a year of its launch the Company Secretary informed Frankland with considerable understatement 'I am afraid it is not a very heavy success', the royalties up to that time amounting to the princely sum of 5/6d. Not surprisingly Frankland pulled out of the enterprise with considerable speed.[54] He did not, however, abandon attempts to get royalties from some of his other inventions. Having effected some improvements in the design of electrical accumulators he suggested to a north London electrical works that he would allow them to manufacture under his patent for a royalty of 20% of the net selling price, with £100 to be paid in advance[55]. They declined, on the basis that the up front payment was unacceptable.[56] Some satisfactory arrangement was later made with the Jarrow Chemical Company.[57] Meanwhile he had acted as a witness in litigation over carbon electrodes between Edison & Swan and Woodhouse & Rawson, earning himself £100 from the latter firm.[58] His friend J. W. Swan was on the other side, but Frankland was used to putting his personal feelings aside on such occasions. He unsuccessfully tried to recruit Crookes to the case, but the latter preferred a holiday in Ireland[59] to the week demanded in London courts.[60]

When litigation or speculative ventures in manufacturing failed there was always the surer ground of chemical analysis. In the ten years up to his retirement on 1 October 1885 his water analysis laboratory had conducted 1066 analyses. In the following decade the number was 3135, virtually a *threefold* increase in turnover.[61] Of course assistants and maintenance had to be paid for, and these expenses formed a prominent part of the rhetoric used to justify his receipts. Thus by 1893 the normal cost of a single analysis was five guineas,[62] and he complained that the £350 received annually from the Local Government Board was 'totally inadequate' in order for him to pay all his laboratory expenses.[63] This might have been true if all his work had been done in their service, but in fact it was not. There were three main sources of analytical business that came his way in the years after 1885.

First were the long-standing analyses conducted for the Local Government Board. The arrangements dated back to the days of the Rivers Commission and were conveniently free of formal documentation. It seems that, in conformity with the Act of 1866, he was required to report on the quality of Metropolitan waters, for which he received a salary of £150, and also of any others the Board wished him to undertake. The latter responsibility earned him a further £200 (thus accounting for the £350 that figures in his complaint).[64] Yet in 1893 Frankland filed an additional claim of £86.2.0 for 41 samples, so inadvertently precipitating an official examination[65] into the basis of his fees. It appears that nothing had been put in writing since 1875, after which the Grove House laboratory had been set up, and Frankland was quick to stress the loss-making potential of such a venture. When the Local Government Board queried his claim[66] he wrote a characteristically long and detailed reply, mentioning that the 41 samples had been provided by the Water Examiner, not by an LGB Inspector as heretofore. This, he said, 'is an entirely different matter', necessitating 'arduous work' often conducted on Sundays.[67] And he was charging only half the usual fee, hardly covering the cost of analysis. Perhaps the magnanimity of this gesture melted the hearts of the Treasury Lords for this time his pleas fell on receptive ears and the extra fee was paid.[68]

A second official opportunity came with service rendered to the Royal Commission on Metropolitan Water Supply, from 1892. This Commission showed what Hamlin has called 'the transformation of Edward Frankland',[69] for his bacteriological researches had convinced him of the effectiveness of filters and thus of the safety of the resultant water. He thus seemed untypically on the side of the water companies, who indeed had paid him to undertake microbiological analysis (yet another source of income). Though officially representing the London County Council he was now able to exert a moderating view and commend certain aspects of the water companies' work. It gave rise to considerable acrimony, Sir Alexander Binnie alleging that when Frankland gave evidence 'the payment influenced his report'. The fact that Binnie was forced to a hasty withdrawal of that charge by letter[70] and by means of printed apology (p. 396)[71] cannot obscure the general feeling of disenchantment with 'expert' witnesses who receive substantial remuneration for their pains. In this case Frankland[72] said he received £192.3.0, though official records[73] stated a figure of £801.10.0.

In the third place, and most importantly, Frankland undertook all manner of analyses for individuals and companies. Taking the year 1887 as a random example we find a total of 383 analyses recorded. If all experiments related to public water supplies, official civic bodies *etc.* are eliminated there remain 199 (52%) of all analyses in the private sector, ranging from schools and private dwellings to a large number for the London and South Western Railway. The fees in 1870 for examining the waters from Shanghai[74] were £75.12.0, while those in 1896 for Eastbourne[75] were £257.5.0, just two out of hundreds of analyses conducted for non-Government agencies. Nor did it follow that earning capacity was limited to the analyses recorded in the laboratory books. Writing to Armstrong concerning the latter's investigation at Knaresborough sewage works, Frankland advised that for three days' work one should expect 100gn., in addition to analysis fees, travel costs and hotel expenses.[76] It is impossible to conceive that Frankland did not deliberately spend much of his retirement in distinctly lucrative employment.

It would, however, be distinctly unfair to imply that all Frankland's analytical work came through diligent marketing of his services. Much of it was on the basis of personal recommendation and unsolicited requests for help, whether from individuals or from Government departments. A glimpse of correspondence behind the scenes that led to his involvement is afforded by a letter to the judge Sir Arthur Kekewitch from H. H. Fowler, President of the Local Government Board. It concerned a dispute about water supplies at Walton-on-the-Hill and West Derby. He had consulted Major Tulloch, Chief Engineering Inspector of the LGB, who 'suggests, as the only man whom he personally knows to be capable and competent for the work, Dr. Edward Frankland'. Fowler concludes 'Major Tulloch cannot of his personal acquaintance suggest another name and I therefore refrain from recommending anyone besides Dr. Frankland'.[77] There must have been many other cases like that.

Some impression of Frankland's activity level may be judged from these remarks to his friend Consul Andersen of Christiania, explaining why he could not join him on holiday in 1893:

> I must definitely abandon all hope of fishing at Hegre next year. There are several Bills in Parliament imperatively claiming my attention. Also a most important lawsuit about smokeless powder, involving millions of money, in which I have promised to help the Government. And, lastly, there is cholera, which, even now, is giving me an intolerable amount of work & which will, no doubt, be rampant in Europe & perhaps in some parts of England next summer.[78]

4 A house divided

The 1880s were a decade in which the Frankland family underwent profound change. It was the end of an era when Frankland's mother and step-father were taken ill and eventually died. For so many years they had exerted a strongly supportive rôle for the Frankland

Fig. 16.3 William Helm in later life. (J. Bucknall Archives.)

household without, it seems, getting in their way. For people of
their age and time they were surprisingly mobile. Since leaving
Lancaster in the late 1840s they had lived (often for brief periods) at
Manchester, Disley, the Isle of Wight, Watford and Walton-le-Dale.
By 1863 they had settled at Leyland, near Preston, and this was
their home for most of the next 20 years, William Helm being
churchwarden from 1864 to 1870. The Frankland children often
spent summer weeks at Leyland, and grandparents were frequent
visitors when needed in London. They celebrated their golden
wedding in 1880. Gradually their health deteriorated, Margaret
being afflicted with rheumatism and William with epilepsy. In
September 1883, two days after receiving a telegram that Ellen's
mother had died, Frankland had another from William Helm,
summoning him to Leyland where his own mother was unconscious.[79]
Though leaving immediately he arrived too late but remained for
her funeral.[80] Two months later another telegram brought the news

Fig. 16.4 Margaret helm in later life. (J. Bucknall Archives)

that William Helm had died, and once again Frankland left hurriedly for Leyland.[81] Again the funeral was reported and tributes paid.[82] The Helms were buried in Garstang, from which they both originally came.[83]

Despite this triple bereavement in 1883 the house in Reigate was not devoid of family celebration. In June 1882 there was much jubilation over the wedding of Grace Toynbee to Percy Frankland, the last of the surviving children of Edward and Sophie. From now on *The Yews* was home only to the family of Edward and Ellen. For his two oldest daughters the unhappiness associated with their father's second marriage seemed now to be largely behind them. Both were now married women and, with their husbands, would spend a few weekends at Reigate each year. On 19 November 1881 there had been rejoicing at the arrival of Esmond Colenso, Sophie's first child and Edward Frankland's first grandchild.[84] The idyllic picture suggested by these happenings might have been even more intensified by the birth, just over a year later, of Maggie's first child, Margery.[85] The coming of grandchildren gave him a new and intense delight, and years later he was declared 'the best grandfather living'.[86] Yet in fact disaster was just round the corner.

In July 1882 Sophie and Frank Colenso brought their seven month old son Esmond to *The Yews*. What then occurred must remain rather problematic, but the child was taken ill on the Sunday with a rash and at the end of the following day had died in his mother's arms, apparently from very severe diarrhoea. Three days later he was buried in Reigate and the distraught parents returned to Norwich.[87] At first sight this was simply one of countless such episodes experienced by Victorian families, but subsequent events revealed deeper under-currents.

It was in the following spring that Maggie took her own baby on a visit to *The Yews*. The welcome was truly regal, and she permitted herself the curious reflection that had half the attention been bestowed on Sophie and her child she would gladly have dispensed with her share. In the next few days she made several visits to Esmond's grave, and on one of them was accompanied by her step-mother. It was, she said, an ordeal because she considered Ellen 'in some measure responsible' for the baby's death. In particular she blamed her for 'such heartless cruelty' to her sister while the child was dying.[88] A few days later they all returned to London, and after a lecture by Siemens at the Royal Institution, Maggie was cross-questioned by her father about 'Sophie's treatment of Ellen, in as it were ignoring her existence'.[89]

Whatever be the truth in the allegations flying about, it is quite clear that family divisions had not been permanently healed, and that Ellen was the object of deep animosity from her step-daughters (who, it will be remembered, were only a few years younger than she). For the next few years there seem to have been genuine attempts to maintain some kind of relationship, whether for reasons of public appearance or of family loyalty we cannot say. But one feature of Frankland's life is deeply significant. Maggie's *Diary* had often recorded her father's frequent absence from home, and now that she was married she notes instead his many visits to the West household without Ellen. For many years he used Wimpole Street as a convenient *pied à terre*, with a room permanently set aside for him, and the visits became so frequent as to suggest an escape mechanism from life at Reigate. This is borne out by his growing habit of taking vacations alone and of Ellen's departure with the children for their own holidays by the sea.

Domestic friction (from whatever cause) cannot be held responsible for further episodes in which Frankland showed signs of alienation from both his sons. In 1887 Fred returned from New Zealand for the first time in 13 years, bringing his wife and child (Freddie) with him, and entertaining the prospect of work in England. Sadly, his father's welcome was less than ecstatic. While Maggie's diaries must always be read with circumspection, this most loyal of Frankland's children is not likely to write ill of her father. Yet she records a disappointing evening with Fred present at a family meal, adding 'it did not appear to me that Papa was really wishful that Fred shd stay in England. Fred felt this also very keenly'.[90] However she was reassured after Fred had again departed by a letter from him quoting Frankland himself, showing 'that after all Papa has a great affection for Fred & that it is only that he does not always show it in the way we want.'[91]

This illustrates the difficulty Frankland experienced over deep personal relationships. His tendency over a problem of this kind was at all costs to avoid it. The threat posed by Fred to his own professional reputation (as he saw it) was only to be removed by Fred's emigration. Now, over a dozen years later, it was still the same. At what must have been an emotional last evening with the whole family together before his son left England for a second time the only person absent was Edward Frankland; he went to the X–Club instead.[92]

Fred, however, was soon to be out of sight if not out of mind. The same was not true of Percy and just about this time he was beginning to pose a threat to his father of a quite different kind. Of his scientific competence there could be no doubt. He was his father's assistant at the Grove House laboratory, and that is where the trouble began. In 1888 Percy left to become Professor of chemistry at Dundee and had borrowed his father's lecture notes and his copy of *A treatise on chemistry* by Roscoe and Schorlemmer. Shortly afterwards there developed between father and son a row of such volcanic intensity that it is hard to believe what one reads in their correspondence. The dispute was not, as has sometimes been suggested,[93] primarily to do with chemical *versus* microbiological methods of analysis, but about money.

It seems to have started with a request from Percy to settle moneys due to him from various affairs of Grove House. After receiving several requests from his son for outstanding payments, Frankland launched at him one of those lengthy essays in self-justification for which he was becoming rather well-known.[94] He also sent a telegram demanding the immediate return of his lecture notes and book.[95] In the letter he demanded an answer to his own questions, concluding ominously:

> But if you fail to do so within a fortnight, or persist in your policy of insolent silence, our relations of father & son must cease. I shall also at the end of that time give a copy of this letter to Sam and to Fred.
> E. Frankland

However great the provocation it is hard to see this as anything but an irrational response to a trifling disagreement over less than £200. Taking a leaf from his father's book, Percy stepped up the pace of hostilities with a letter of immense length, enumerating in minute detail the injustices to which he considered he had been subject.[96] He ended ominously:

> After your most unwarrantable and libellous letter I absolutely refuse to have any further correspondence about the matter until my claim has been paid.
> Percy F. Frankland

For those with a taste for such controversy the correspondence[97] may be readily followed, though neither party emerges with much credit. The disagreement was formally about who received which fees, whose laboratory and assistant it was, whether all payments had been declared, and so on. It may also have been about who really ran the business and about the exact meaning of Frankland's 'retirement'. Once again it is hard to resist the impression that Frankland, for good reason or ill, considered himself under threat from one of his sons.

Percy moved to Birmingham in 1894 but the simmering resentment remained, each man too proud to apologise to the other. Matters were not mended by Frankland's next extraordinary move. He withdrew his nomination of Percy to the Athenaeum[98], an act of apparent spite that added fuel to the fire. His letter was acknowledged by the club's secretary, H. R. Tedder, who insisted that his son must be informed.[99]

Efforts were made to effect a reconciliation by Frankland's brother-in-law Adolph Fick, to whom he wrote with an explanation of the rumour that 'Percy went up to his father' at a Royal Society reception. Indeed Maggie wrote in her *Diary* that at last the ice was broken.[100] She was wrong. Anxious to dispel any notion that he had spurned an advance by his son Frankland offered the explanation that 'we accidentally met at the entrance of the Hotel & I had shaken hands with him before I realised who he was'. He then adds: 'I ought to have hesitated to take the hand which had deprived me of my property & skill', but obviously thought better of it for that sentence is crossed out in the draft letter.[101] In a separate account of his troubles written to Sophie he adds that 'Percy seized my hand & said I will at all events congratulate you on the receipt of the Copley medal. The whole thing was over before I realised who it was who had taken my hand'.[102] One wonders whether he seriously thought that anyone would believe him.

Never a woman to mince her words (and in the eyes of some of the family a contributing factor to the continuing feud) Grace Frankland had written to Ellen complaining that not only had Frankland prevented his son's election to the Athenaeum, he had also deprived him of the coveted Fellowship of the Royal Society. Ellen replied that, on the first count, Frankland considered Percy 'would probably share his feeling as to the extreme unpleasantness and undesirability of their belonging to one and the same [club]',[103] and that, on the second, despite rumours from some 'members of Council', Frankland had in fact done nothing at all.[104]

There seems little doubt that the two women were trying – however ineffectually – to bring about a reconciliation between their husbands. By a sad irony their goal was to be achieved in a wholly unexpected way. In January 1899 Ellen Frankland died after a short illness. Her death triggered an impulse in Percy to write once more to his father.

In a letter that went far beyond conventional expressions of condolence Percy displays a degree of sympathetic understanding for his father that few could have expected. He had always a greater respect for Ellen[105] than his sisters seemed to have shown and now he writes 'I do not think that the relationship of stepmother to stepson could well have been more free from what tradition supposes to be all but inevitable', denying that he had ever received a harsh or unkind word from her. Indicating that Grace and he would like to be at the funeral he offered to stay away from the house if his father so wished. He concluded:

As I have not been able to resist the impulse of writing, I should like to take this opportunity of saying how deeply I regret the long estrangement which has existed between us, & the share which I have had in causing it, & would express the hope that you may be prepared to forget and forgive what is past.

In any case I should like you to know that unfortunate circumstances which have divided us during the past ten years cannot prevent my sharing even though at a distance your present sorrow.

Your very affectionate son.

Percy[106]

In a postscript he promises to return the missing lecture notes tomorrow.

After Ellen's funeral Frankland wrote to Percy in a conciliatory tone.[107] He said 'My dear wife always regretted the estrangement & only a few days before her death expressed the hope that some means could be found of putting an end to it'. He hoped that the misunderstandings in Grace's letters to Ellen would be withdrawn, and, for the first time in years, signed himself

I am, dear Percy,
    Your affectionate
        Father

It is gratifying to report that the last six months of Frankland's life were unclouded by the estrangement from his younger son.

5 Recognition at last

To a modern generation that may not have even heard his name one of the most curious features of Edward Frankland's life was the way in which contemporary institutions fell over themselves in attempts to honour him. As retirement loomed nearer he could look back to his Oxford Honorary DCL in 1870 and a range of minor honours. In the next few years that number would be augmented considerably. Table 16.2 lists his honorary memberships or fellowships for which documentation still survives.

He was particularly proud of the last distinction, since the Académie had only eight places for Associées Étrangères, the Britons hitherto being Kelvin and Lister.[121] In 1884 Frankland received the honorary LLD from Edinburgh. On 11 July 1887 he took the oath as a newly appointed JP for Reigate. This was largely an honorific appointment, though he did attend a number of Quarter Sessions and adjournments for the next two years (after

Table 16.2. *Frankland's Honorary Memberships*

1865: Honorary Member, Societas Naturae Scrutatorum Helvetorum, Geneva[108]
1866: Corresponding Member, Académie des Sciences, Institut de France[109]
1869: Honorary Member, Manchester Literary and Philosophical Society[110]
1873: Honorary Member, Deutsche chemische Gesellschaft zu Berlin[111]
1876: Corresponding Member, Scientific Academy of St. Petersburg[112]
1877: Honorary Member, American Chemical Society[113]
1879: Honorary Member, New York Academy of Sciences[114]
1887: Honorary Member, Trinity Historical Society, Dallas[115]
1893: Honorary Member, Association Internationale pour le Progrès de l'Hygiene, Brussels[116]
1893: Honorary Fellow, British Institute of Public Health[117]
1893: Honorary Member, Institution of Civil Engineers[118]
1894: Honorary Member, Asiatic Society of Bengal[119]
1895: Foreign Member, Académie des Sciences, Institut de France[120]

which Surrey County Council took over those responsibilities of the Reigate borough).[122] His greatest honour was yet to come.

On 17 June 1897 he received the following letter from the Prime Minister, Lord Salisbury:[123]

> My dear Sir
> It affords me great satisfaction to be authorized to inform you that the Queen has been pleased to direct that you should be appointed Knight Commander of the Order of the Bath, on the occasion of the approaching jubilee, in recognition of the services you have rendered to the community and of the eminent position you occupy in the scientific world. I am extremely glad to be the instrument of acquainting you with Her Majesty's gracious intentions.
> Believe me
>   Yours very faithfully,
>   SALISBURY

Frankland's prompt reply was signed 'Yours very gratefully', thanked his Lordship for his kind words and reminded him that he had been the bearer of good news on a previous occasion, – when informing Frankland of his Oxford DCL.[124]

The family were of course delighted, though the information was extracted from him only 'after great pressing'.[125] In general the scientific community were pleased that one of their number had received such a signal honour. Many contemporaries would have echoed the opinion of *Nature* that the award was for 'water analysis',[126] a misconception that has been repeated down to the present time.[127] Yet colleagues, family and historians would have been extremely surprised to learn of the real reasons for the award.

The question was first raised by Oliver Lodge, then Professor of Physics and Mathematics at Liverpool. Late in 1896 he wrote to A. J. Balfour, First Lord of the Treasury the following letter:[128]

> If not too unpardonably interfering with what does not concern me I want to make a suggestion based on the following facts.
>
> Crookes and Frankland have both recently declined the Presidency of the British Association on the ground of age & health. Both were really entitled to this recognition long ago by their scientific discoveries.
>
> Frankland was kept out of it, so gossip says, by the jealousy of Williamson – for many years an honorary official of the B.A. Crookes has no doubt suffered severely in reputation through dabbling with new & unorthodox forces & through the consequent attacks of Carpenter.
>
> Both are to a certain extent men of the world, & of coin.
>
> May it not be possible to think of a simple knighthood for, one or both of them, in connexion with 60-year honours?
>
> Possibly the recent water controversy may make the case of Frankland undesirable & personally I do not feel his case so strong as that of Crookes.
>
> Trouble not to reply, but take it as an anonymous suggestion among (doubtless) others.
>
> Faithfully yours,
>
> Oliver Lodge

A copy of this letter was sent from Balfour's office to Salisbury's private secretary, marked:

> Rt. Hon. A. Balfour
> Private Secretary
> _____
> Encl. Mr. Oliver Lodge
> Recomds.
> Professor Crookes or
> Professor Frankland for
> Knighthood, both having
> declined Presidency of British Assocn.[129]

The final decision was that both men should be honoured. In Frankland's case it was not to be 'an ordinary knighthood' but a KCB.

From this correspondence it thus emerges that:

1. Frankland was considered less worthy than Crookes;
2. He was not knighted for water analysis, his activities in that area actually being a disadvantage;
3. He was honoured for having declined the Presidency of the British Association.

It is hard to find any other interpretation of the correspondence than that his KCB was, in effect, some kind of consolation prize. The importance of the BAAS connection cannot be stressed too strongly. This was the forum which Lodge had made specially his own, and one by which Lord Salisbury had been particularly influenced; indeed he was himself its President in 1894. It is ironic that an organisation with which Frankland had so little patience in his early life should have unwittingly lent him support towards its end. But for whatever reasons the award was given it was a fitting climax to a life's work of distinguished research in so many aspects of chemistry. The investiture at Windsor Castle was on Thursday, 9 December 1897 in 'Levée dress'.[130] Socially, Frankland had arrived.

Notes

1 Frankland to Sophie (II) Frankland, 7 September 1879 [PCA, OU mf 03.01.0579].
2 B. C. Brodie to Frankland, 29 August, 5 and 25 September 1879, 27 June 1880 [RFA 01.04.1469, 01.02.0654, 01.04.1466 and 1614]. Brodie died shortly afterwards.
3 Frankland to Sophie (II) Frankland, 30 July 1880 [PCA, OU mf 03.01.0587].
4 Maggie Frankland, *Diary*, entry for 20 July 1880 [MBA].
5 1881 Census [PRO, RG11/797].
6 A. J. Harrison, 'Scientific naturalists and the government of the Royal Society 1850–1900', Open University Ph.D. thesis, 1988, p.429.
7 *Sketches*, 2nd ed., pp.445–8.
8 Frankland to I. C. Andersen, 29 November 1892 [JBA, OU mf 02.01.0472].
9 Maggie Frankland, *Diary*, entry for 2 April 1881 (note 4).
10 Maggie West (*née* Frankland), *Diary*, entry for 27 March 1897 (note 4).
11 Frankland, *Proc. Roy. Soc.*, 1893, **35**, 67–70, 1890, **46**, 304–8; *Electrician*, 1894, **13**, 471–3.
12 Frankland to Percy Frankland, 18 December 1889 [RFA, OU mf 01.04.0214].
13 Frankland to Grace Frankland, 29 August 1886 [RFA, OU mf 01.01.1129].
14 Maggie West (*née* Frankland), *Diary*, entry for 4 June 1887 (note 4).
15 Frankland to Sophie Colenso, 27 October 1886 [PCA, OU mf 03.01.0741].
16 Frankland to Grace and Percy Frankland, 7 September 1886 [RFA, OU mf 01.02.0747].
17 Frankland, MS table 'Attendances in Lab.' [RFA, OU mf 01.03.0488].
18 On Donnelly see *DNB*, M. Reeks, *History of the Royal School of Mines*, Royal School of Mines, London, 1920, pp.126–30, and W. H. G. Armytage, 'J. F. D. Donnelly: pioneer in vocational education', *Vocational Aspect*, 1950, **2**, 6–21.
19 J. F. D. Donnelly to Frankland, 19 July 1881 [RFA, OU mf 01.03.0556; copy-letter, 01.04.0284].
20 Copy-letter, Frankland to J. F. D. Donnelly, 25 July 1881 [RFA, OU mf 01.03.0560].

21 Copy-letter, Frankland to J. F. D. Donnelly, 25 August 1881 [RFA, OU mf 01.03.0562].

22 Draft letter, Frankland to T. H. Huxley, 21 June 1884 [RFA, OU mf 01.03.0471].

23 T. H. Huxley to Frankland, 19 June 1884 [RFA, OU mf 01.03.0510]; copy-letter at Imperial College Archives, Huxley papers, 16/266.

24 Envelope [RFA, OU mf 01.03.0508–9]

25 Frankland to Huxley (note 22).

26 Frankland to T. H. Huxley, 27 June 1884 [Imperial College Archives, Huxley papers, 16/261].

27 T. H. Huxley to Frankland, 29 June 1884 [RFA, OU mf 01.03.0513].

28 T. H. Huxley to Frankland, 3 July 1884 [RFA, OU mf 01.03.0504].

29 Frankland to T. H. Huxley, 4 July 1884 [RFA, OU mf 01.03.0506].

30 J. F. D. Donnelly to Frankland, 8 July 1884 [RFA, OU mf 01.03.0459].

31 Draft letter, Frankland to J. F. D. Donnelly, 9 July 1884 [RFA, OU mf 01.03.0462].

32 J. F. D. Donnelly to Frankland, 12 July 1884 [RFA, OU mf 01.03.0466].

33 Draft letter, Frankland to J. F. D. Donnelly, 14 July 1884 [RFA, OU mf 01.03.0468].

34 Frankland to T. H. Huxley, 20 July 1884 [Imperial College Archives, Huxley papers, 16/263].

35 L. Huxley, *Life and letters of Thomas Henry Huxley*, Macmillan, London, 2nd ed., 1908, vol ii, p.418.

36 J. F. D. Donnelly to Frankland, 16 April 1885 [RFA, OU mf 01.03.0341].

37 J. F. D. Donnelly to Frankland, 12 May 1885 [RFA, OU mf 01.04.1086].

38 L. Playfair to Frankland, 12 June 1885 [RFA, OU mf 01.03.0337].

39 Frankland to L. Playfair, 14 June 1885, [RFA, OU mf 01.03.0339].

40 J. F. D. Donnelly to Frankland, 30 July 1885 [RFA, OU mf 01.04.1090].

41 C. G. Barrington to Vice President of Committee of Council on Education, 27 July 1885 [RFA, OU mf 01.04.1088].

42 Frankland to J. F. D. Donnelly, 1 August 1885 [RFA, OU mf 01.04.1091].

43 J. F. D. Donnelly to the Secretary, H M Treasury, 15 August 1885 [RFA, OU mf 01.04.1095].

44 Copy-letter, R. E. Welby to Vice President of Committee of Council on Education, 27 November 1885 [RFA, OU mf 01.04.1097].

45 G. F. Duncombe to Frankland, 8 December 1885 [RFA, OU mf 01.04.1101].

46 L. Huxley (note 35), pp.416–7.

47 L. Playfair to Frankland, 17 December 1885 [RFA, OU mf 01.04.1733].

48 L. Playfair to Frankland, 19 December 1885 [RFA, OU mf 01.03.0345].

49 Copy-letter, Frankland to Lord Iddesleigh, n.d. (actually 22 December 1885) [RFA, OU mf 01.04.1104].

50 Lord Iddesleigh to Frankland, 6 January 1886 [RFA, OU mf 01.04.1286].

51 Copy-letter, Frankland to Lord Iddesleigh, 7 January 1886 [RFA, OU mf 01.04.1289].

52 Copy-letter, Frankland to J. F. D. Donnelly, 14 December 1885 [RFA, OU mf 01.03.0344].

53 T. H. Huxley to J. D. Hooker, 3 June 1884 [Imperial College Archives, Huxley papers, 14/46].

54 A. Robinson to Frankland, 20 and 25 February 1885, 8 and 11 January 1886 [RFA, OU mf 01.03.0825, 0827, 0835, 0829]; R. M. Soutter to Frankland, n.d., 11 January 1886 and 11 March 1887 [*ibid.*, 01.03.0833, 0830 and 0818]; Frankland to A. Robinson, 12 January 1886 [*ibid.*, 01.03.0838]; Frankland to R. M. Soutter, 12 March 1887 [*ibid.*, 01.03.0820].

55 Frankland to Eidsforth & Mudford, 19 November 1885 [RFA, OU mf 01.04.1177].

56 Eidsforth & Mudford to Frankland, 24 November 1885 [RFA, OU mf 01.04.1179].

57 Frankland to F. L. Rawson, 6 June 1887 [RFA, OU mf 01.01.0030].

58 *Ibid.*

59 W. Crookes to Frankland, 11 May 1886 [RFA, OU mf 01.04.1443].

60 Frankland spent almost all that week in London, sleeping in Maggie's home (Maggie West (*née* Frankland), *Diary,* entry for 11 May 1886 (note 4)).

61 Water analyses notebooks [RFA, OU mf 01.09].

62 Frankland to J. Bellamy, 9 April 1893 [RFA, OU mf 01.07.0083].

63 Frankland to Local Government Board, 18 December 1893 [RFA, OU mf 01.07.0880].

64 R. E. Welby (Treasury) to Local Government Board, 1 September 1893 [RFA, OU mf 01.07.0866].

65 F. Mowatt (Treasury) to Local Government Board, 28 November 1893 [RFA, OU mf 01.07.0878].

66 Sir W. E. Knollys (LGB) to Frankland, 12 December 1893 [RFA, OU mf 01.07.0877].

67 Frankland to Local Government Board, 18 October 1893 [RFA, OU mf 01.07.0872].

68 F. Mowatt (Treasury) to Local Government Board, 12 February 1894; S. A. Provis (LGB) to Frankland, 19 February 1894 [RFA, OU mf 01.07.0868 and 0869].

69 C. Hamlin, *A science of impurity: water analysis in the nineteenth century,* Hilger, Bristol, 1990, p.281.

70 Sir A. R. Binnie to Frankland, 20 May, 1898 [RFA, OU mf 01.03.0859].

71 Printed sheet and typed copy [RFA, OU mf 01.04.0474 and 01.03.0856].

72 Frankland to Sir A. R. Binnie, 13 May 1898 [RFA, OU mf 01.03.0868].

73 Binnie (note 70).

74 J. Pook to Frankland, 17 October 1870 [RFA, OU mf 01.07.0728].

75 Frankland, MS note, 31 December 1896 [RFA, OU mf 01.04.0475].

76 Frankland to H. E. Armstrong, 26 July 1893 [Royal Society Archives, MM.10.98].

77 H. H. Fowler to Sir A. Kekewich, 27 April 1893 [Birmingham University Library, L.Add 2786].

78 Frankland to Andersen (note 8).

79 Maggie West (*née* Frankland), *Diary,* entry for 15–17 September 1883 (note 4)).

80 *Preston Guardian,* 27 September 1883, p.3.

81 Maggie West (*née* Frankland), *Diary*, entry for 26 November 1883 (note 4).
82 *Preston Guardian*, 28 November 1883, p.5.
83 Garstang parochial records [Lancashire County Record Office, DRB 2/88].
84 Maggie Frankland, *Diary*, entry for 19 November 1881 (note 4).
85 *Ibid.*, 25 November 1882.
86 *Ibid.*, 31 October 1898.
87 *Ibid.*, 2, 3 and 6 July 1882.
88 *Ibid.*, 20 April 1883.
89 *Ibid.*, 27 April 1883.
90 *Ibid.*, 21 October 1887.
91 *Ibid.*, 30 January 1888.
92 *Ibid.*, 5 January 1888.
93 By the present author among others; *e.g.* C. A. Russell, 'Percy Frankland: the iron gate of examination', *Chem. Brit.*, 1977, 13, 425–7.
94 Frankland to Percy Frankland, 29 November 1889 [RFA, OU mf 01.04.0265].
95 Copy of telegram, Frankland to Percy Frankland, n.d. [RFA, OU mf 01.04.0155].
96 Percy Frankland to Frankland, 12 December 1889 [RFA, OU mf 01.04.0219].
97 A long series of letters between the two men covering 1889 and early 1890: Frankland to Percy Frankland [RFA, OU mf 01.04.0146, 0152, 0155, 0161, 0163, 0165, 0214, 0265, 1172]; Percy Frankland to Frankland [*ibid.*, 01.04.0219, 0251, 0253, 1169].
98 Frankland to Tedder, 10 May 1893 [RFA, OU mf 01.04.0239]; he 'has no seconder'.
99 H. R. Tedder to Frankland, 11 and 30 May 1893, 19 and 28 January 1894 [RFA, OU mf 01.04.0240, 0241, 0243 and 0244].
100 Maggie West (née Frankland), *Diary*, entry for 30 November 1894 (note 4).
101 Draft letter, Frankland to A. Fick, 7 October 1896 [RFA, OU mf 01.04.0249].
102 Copy-letter, Frankland to Sophie Colenso, 15 November 1896 [RFA, OU mf 01.04.0209].
103 Copy-letter, Ellen Frankland to Grace Frankland, 19 June 1895 [RFA, OU mf 01.04.0199].
104 Copy-letter, Ellen Frankland to Grace Frankland, '12th Nov.' [probably 1895] [RFA, OU mf 01.04.0202].
105 Grace Toynbee had of course been a neighbour of the Grensides at Wimbledon.
106 Percy Frankland to Frankland, 22 January 1899 [RFA, OU mf 01.04.0185].
107 Frankland to Percy Frankland, 6 February 1899 [RFA, OU mf 01.04.0181].
108 Certificate of membership, 23 August 1865 [RFA, OU mf 01.03.0526].
109 Certificate of membership, 2 July 1866 [RFA, OU mf 01.03.0554].
110 J. Baxendale to Frankland, 7 April 1869 [RFA, OU mf 01.03.0522].
111 A. W. Hofmann to Frankland, 17 December 1873 [RFA, OU mf 01.01.0542].
112 Certificate of membership, 29 December 1876 [RFA, OU mf 01.03.0549].
113 G. F. Barker to Frankland, 24 May 1877 [RFA, OU mf 01.04.0299].
114 Certificate of membership, 2 June 1879 [RFA, OU mf 01.03.0539].
115 D. W. Austin to Frankland, 29 July 1887 [RFA, OU mf 01.03.0552].

116 Certificate of membership, 9 May 1893 [RFA, OU mf 01.03.0542].

117 W. R. Smith to Frankland, 25 October 1893 [RFA, OU mf 01.07.0556].

118 J. Forrest to Frankland, 8 December 1893 [RFA, OU mf 01.01.0778].

119 C. R. Wilson to Frankland, 15 March 1894 [RFA, OU mf 01.03.0523].

120 M. Berthelot to Frankland, 4 June 1895 [RFA, OU mf 01.03.0547].

121 Frankland to I. C. Andersen, 30 June 1895 [JBA, OU mf 02.01.0486].

122 Quarter Sessions Order Book, Michaelmas Session 1886 [Surrey County Record Office, QS2/1/98 and 99].

123 Lord Salisbury to Frankland, 16 June 1897, in *Sketches*, 2nd ed., p.450.

124 Frankland to Lord Salisbury, 17 June 1897 [Hatfield House, MSS 3M/Y6].

125 Maggie West (*née* Frankland), *Diary*, entry for 21 June 1897 (note 4).

126 Obituary notice of Frankland, *Nature*, 1899, 60, 372.

127 *E.g.* the present author's *Lancastrian Chemist*, p.4; and C. Hamlin, 'Edward Frankland's early career as London's official water analyst, 1865–1876: the context of previous sewage contamination', *Bull. Hist. Medicine*, 1982, 56, 56–76 (76).

128 Transcript of letter, Oliver Lodge to A. J. Balfour, 21 November 1896 [Hatfield House, MSS 3M/Class E].

129 Transcript of letter, J. S. Sandars (Private Secretary to A. J. Balfour) to Hon. S. McDonnell (Private Secretary to Lord Salisbury), 25 November 1896 [Hatfield House, MSS 3M/Class E].

130 Sir A. W. Woods to Frankland, 30 November 1897 [JBA, OU mf 02.02.1753].

CHAPTER 17

The last journey

On Friday, 14 July 1899, Edward Frankland left London for yet another holiday in Norway. This time he was weary from overwork and still recovering from the shock of Ellen's death six months before. According to his daughters he seemed to have aged many years in that brief period.[1] His destination was the health resort of Golaa in Gudbrandsdalen, high in the mountains and with spectacular views. He travelled alone to Christiania, arriving in Norway on the Sunday evening. Here he met Jane Lund, his secretary, who had gone ahead a fortnight earlier.

Jane Lund was a young lady who played an important part in Frankland's final years. He had known her for some time, as secretary of his friend I. C. Andersen, Consul in Christiania, and companion on fishing expeditions in Norway since the 1870s.[2] Writing to the Consul in 1892 regarding a possible visit to Hegre, Frankland remarked: 'I suppose you will again take Miss Lund with you? She is such an excellent housekeeper & makes the time pass so pleasantly'.[3] That she was a personable young woman is suggested by a comment from Andersen about a possible suitor now being engaged to someone else: 'You need not now be any more concerned for the safety of Miss Lund'.[4]

For the last four years Frankland had gone to Norway with at least one new objective. He had for some time determined to write some kind of autobiographical sketch for the entertainment of his family, and a Norwegian holiday would give the leisure and freedom from distraction that could not be found at home. However there was a serious difficulty. Dogged by failing eyesight (for which cataract was pronounced responsible[5]), he recognised

Fig. 17.1 Frankland in 1894. (*Sketches.*)

that the only way was to employ an amanuensis. By a happy chance his friend was prepared to lend him Jane Lund for one month in 1896:

> Her duties are to write for me, six hours a day, all my letters and reports and their copies, and spend the rest of the time at the autobiography. Nothing could exceed the neatness and accuracy of the

> fair copy which she brings out . . . The morning is generally devoted to
> dictation and the afternoon to copying . . . Miss Lund takes much
> interest in the work and this makes it quite a pleasure to me. In fact it is
> great fun to go back to one's childhood in this way; but whether it will
> be any pleasure or fun to anyone else to read it is quite another matter.[6]

In fact he greatly under-estimated his own literary skills, for, despite inevitable inaccuracies, the *Sketches* are a model of elegant writing, and have a charm and grace that are all their own.

This turned out to be so satisfactory an arrangement that Jane became Edward Frankland's secretary, replacing Blanche Gough who had probably produced the typed letters emanating from *The Yews* in the late 1890s, but had died in September 1896.[7] Miss Lund duly came over to Reigate with Frankland in early 1897, and thereafter became almost one of the family, being present at family celebrations, accompanying them on walks, to the theatre, the opera and even to church.[8] This could well have been an expression of Frankland generosity to a visitor from overseas. However it has been suggested that Jane Lund became Frankland's mistress,[9] so it is necessary to consider how far such an allegation may be sustained.

Our most intimate glimpses of the two are afforded by Maggie's *Diary*. Through 1897 and 1898 she does record a number of incidents that could bear the most innocent interpretation or could be taken as guarded expressions of disapproval. She records the incident, already mentioned, of Jane happening to be on the observatory stairs as her employer fell through the trap-door and landed on her;[10] on another occasion 'we found Papa and Miss Lund looking for sunspots in the observatory' (even though the sun was then totally obscured by clouds!).[11] Several times it is recorded that Edward and Jane left Maggie's house alone; they were both found together on a visit to the Royal Mint[12] and together they took Maggie to the theatre[13]. Jane herself seems to have been friendly with both Maggie and Sophie and not infrequently stayed with them. On Frankland's 74th birthday, when Ellen was desperately and terminally ill, Maggie discovered her father and Jane writing in the study.[14] After Ellen's death Frankland and his secretary were more often together for longer periods. An entry in the visitors' book at the Freshwater Bay Hotel in the Isle of Wight confirmed that they had booked in there for a brief holiday in February.[15] In May they came to stay with Maggie for five days,[16] and later accompanied Fred and his wife to the laying of the foundation stone of the Victoria & Albert Museum. Edward Frankland was in Court dress, and he and Jane stayed the night with Maggie.[17] The following day the three of them decamped to Southport 'where Papa is taking me for a week's holiday'.[18]

None of these events could possibly be taken as unequivocal evidence that Jane was Edward's mistress. However, such references have to be read with caution. Maggie, of all the children, was most intensely loyal to her father and it is unlikely that she would have recorded acts of unequivocal adultery, though her strong religious belief would have left her in no doubt as to the propriety or otherwise of such a situation. Yet all that she did record may suggest to certain people something more than good companionship. Above all there is the undeniable fact that for several years Jane and Frankland spent a month or two together in Norway, and after Ellen's death they must have often been alone in Reigate. But *on the basis of the data so far available* no firm conclusions may be drawn and it is most unwise to assume that Jane was anything other than a loyal and valued friend.

Once in Norway, and having met up with Jane, Frankland wrote cheerfully to his assistant William Burgess reporting a pleasant time, though winter clothing had to be donned. Burgess was acknowledging that letter by 31 July.[19] A week earlier Frankland had despatched a letter concerning a water supply near Nuneaton. Though written from Norway, it was addressed from his water analysis laboratory, and related to a sample collected after he had left Britain.[20] Presumably it used data supplied by Burgess. The intention was to continue work on the autobiography, and a start was made on the chapter on 'Travelling', though in the event it reached only to 1855.[21] One morning about a fortnight after their arrival at Golaa Frankland was taken seriously ill. The only account we have of the next few days is from his daughter Maggie who would have received it directly from Jane Lund:

> He was sitting down to the table to dictate to Jane when he said he did not feel up to it that day & moving his head from side to side felt a sudden sharp pain in his neck & he seemed fainting so Jane got him to lie down on the sofa & his face became gray & perspiration was all over his head. Jane wanted to run for the doctor but he begged her not to leave him so all she c^d do was to give him some smelling salts & then when he seemed better she ran to the larger house & begged for a hand bell & that the doctor sh^d come.

Fig. 17.2 Golaa, Norway: the cottage where Frankland died (a sketch by his daughter Maggie). (*Sketches.*)

The treatment applied was mustard poultice to the neck, but though it eased the pain Frankland remained feverish and delirious through the night. For the next ten days his condition did not improve, he slept much, ate little and (as is so often the case in such circumstances) reminisced a great deal about his childhood and his mother. On 5 August his condition deteriorated further. Aware of his oldest daughter's responsibilities to her own children he refused at first to allow Jane to summon them, though he relented after several days. On 6 August Maggie received letters from Miss Lund telling of his illness, and two days later a telegram 'Come immediately'. Unfortunately the journey was so long that Sam and Maggie West arrived just too late. Two days before they could reach Golaa Edward Frankland died, during the evening of Wednesday, 9 August. During the morning he had been enquiring when Maggie and Sam should arrive, and following their journey in his mind. Then from 10 a.m. the pains in his back reached such a level that sedation was indicated and at 2.30 p.m. the doctor administered opium. 'He fell into a sleep from which he never awoke and breathed his last at about 6.30 in the evening'.[22]

The symptoms of his illness are so vaguely described that the exact cause of death must remain in doubt. Maggie tells us that during this time Jane nursed him devotedly and scarcely left his side, day or night. She prayed the Lord's Prayer in Norwegian with him each evening. His daughter recalled a remark he had made before his illness:

> I am not frightened to die, only frightened of suffering pain. I have tried to lead a good life. My God may not be the same as you think yours. My God I imagine much more powerful and great.[23]

In that imagination – or faith – he died.

When Maggie and Sam arrived there were the predictably sad farewells, Maggie placing in her father's black Norwegian coffin flowers from his children and grandchildren. After they had left the room 'dear Jane' was the very last to leave.[24]

The funeral took place at Reigate on 22 August. The cortège was met by the Rev. Professor Thomas Bonney who also preached the sermon. Bonney was then Professor of Geology at University College London, an authority on Alpine geology and a preacher at St. Peter's, Vere Street, whom Frankland had been known to appreciate. His address has not been reported, though the music was: Chopin's *Preludes* in E♭ and D♭, Mendelssohn's *Funeral*

March, and (at the end) Spohr's 'Blessed are the departed' and Beethoven's *Funeral March*.[25] All the adult members of the family were present, except Fred (now in New York) and Sophie (apparently in Cassel). After Jane Lund and William Burgess the mourners were headed by Lord Lister, President of the Royal Society. The Chemical Society and the Institute of Chemistry were represented by their secretaries (Dunstan and Pilcher respectively), while closer friends included Roscoe, McLeod and Armstrong.[26]

Edward Frankland was buried with his second wife, Ellen, in St Mary's Churchyard, Reigate. A six-foot high Celtic cross in brown granite was subsequently erected bearing the words:

IN LOVING MEMORY OF

EDWARD FRANKLAND

K.C.B., D.C.L., Ph.D., LL.D.

FOREIGN SECRETARY OF THE ROYAL SOCIETY

BORN AT CHURCHTOWN, LANCASHIRE

18 JAN: 1825

DIED AT GOLAA, GUDBRANDSDALEN, NORWAY

9 AUG: 1899

AND WAS INTERRED HERE IN REIGATE

ALSO OF

ELLEN FRANCES

SECOND WIFE OF THE ABOVE

BORN IN LONDON 25 JAN: 1848

DIED AT REIGATE 20 JAN: 1899

The funeral over, close friends and family returned to *The Yews* for the reading of the will. The first reaction was of blank astonishment. Unknown to any of them, Frankland had left a very considerable sum, approaching £140 000. In today's terms he would have been a millionaire several times over. The will, drawn up in the previous May, was a complex document whose provisions were likely to cause much controversy. They did. To Percy was bequeathed *The Yews*, its land and outbuildings, furniture, his scientific apparatus and library, clothes, consumables *etc.* To Dorothy and Helga, children of his second marriage, he left £5000 each. The other three children of his first marriage (Fred, Maggie and Sophie) were to receive equal shares of the residue; that meant between £30 000 and £40 000 each. The huge disparity in treatment of his six children inevitably gave rise to bitter acrimony, even on the day of the funeral.[27] It is necessary only to say that Percy, spurred on by Grace, took the side of Dorothy and Helga against the other three, and that for many years to come the family was once again deeply divided against itself. Minor bequests included £500 to 'my faithful assistant William Thomas Burgess' (if still in his employment) and 'to my Secretary Jane Lund who has been of great service to me in my professional and other work the sum of two thousand pounds free of duty whether she is or is not in my employment at my death'.[28]

Far more enduring was Edward Frankland's legacy to chemistry, though, for certain strange reasons, it has been less than generously remembered until recent years. To him are due important elements of the theory of valency, and even the familiar term 'bond'. He is the founder of that burgeoning branch of chemical science known as organometallic (and again we owe him the word). It is not too far from the truth to hail him also as one of the founders of synthetic organic chemistry in general. From these researches emerged dozens of new compounds, from ethane to heterocyclic bases. He was co-discoverer of helium and pioneer in stellar spectroscopy. He was the first to make thorough analyses of gases from different kinds of coal, and the first to measure the calorific value of foodstuffs.

Much of his consultancy work was of great value to his contemporaries though, being less systematic, has not left indelible marks on subsequent science. His miscellaneous work includes various anti-pollution measures, development of milk substitutes for infants, studies of illuminating power and combustion rates, and much work on gas analysis in general. Most important of all was the immense body of analyses and research on public water supplies. Some of his work, notably on glaciers and astronomy, was falsified in his own lifetime. No one can be right all the time.

Probably as enduring as the research elements in his chemistry were what might be called the 'social dimensions' of the subject. For advances in chemical communication posterity is more indebted to Frankland than to any other English chemist of his time. It is an unconscious tribute from posterity that 'Frankland's notation' is of such universal use today that no one bothers to call it by that name. He is one of the pioneers to whom we also owe many elements of modern pedagogic practice, not least the provision of teaching laboratories, the proper construction of demonstration experiments and the systematic training of chemistry teachers. Nor were his bequests only in the area of communication. Frankland, more than any other person, was responsible for the foundation of the Institute of Chemistry, and thus the first professional institution for scientists anywhere in the world.

Setting aside all the obituary hype and wildly exaggerated rhetoric indulged in by a few admirers, the fact remains that Frankland was a scientist of great stature, the most prominent chemist in Victorian Britain. Lyon Playfair put it well in a private letter to him: 'I consider that you are the leading chemical investigator of the country'.[29]

This immensely impressive legacy of scientific achievement inevitably raises fundamental questions about the man behind it. To enquire into the well-springs of his conduct, to ask 'what made him tick?', or to try to identify a specific world-view that explains everything, is an exercise in frustration. It probably is for all human beings, who are far more complex creatures than most historians would like to imagine. For one as reticent as Frankland the task may seem well-nigh hopeless. The data are too well-concealed. Fortunately, however, he gave one (and only one) written exposition of his own self-assessment, and that was in a letter to the statistician Francis Galton, who was anxious to compile an analysis of those factors, social and genetic, that favoured the development of scientific genius. It was eventually published as *English men of science*,[30] and two drafts of Frankland's response to his questionnaire have survived.[31] It is a singularly honest document, some of his assessments making much sense of his life-style over many years:

> Considerable energy of body & power of enduring fatigue, but desire for rest & capacity for enjoying almost perfect idleness for weeks. Early rising often tried but not successfully. Usually rise at $7\frac{1}{4}$ & go to bed for past two years at 11, formerly at 12 or 1 o'clock. Fond of travelling & mountaineering. Cannot walk more than 30 miles per day on level road, but in 1859 ascended Mont Blanc & remained 22 hours on summit with very little fatigue. Sleep required up to 1870 – 8 hours. Afterwards 7 hours.
>
> Considerable energy of mind, but more fixity of purpose. Having undertaken to do a thing, the feeling that it must be done & if possible within the time specified overwhelms everything else. Retentiveness of memory for things seen & phenomena observed very great, but for

> book knowledge very small (comparatively). . . . I cannot say what
> peculiarities have contributed to my professorial success. I seem to
> have drifted into it rather than to have exerted any effort. I have held 9
> appointments, but only applied for one of them.

Thus, on his own admission at least, he was not guided through life by any grand strategic plan, and tended to take things as they came. While such simple-minded pragmatism may well apply to choice of jobs, especially in an age when scientific career patterns scarcely existed, it overlooks the most obvious motif of all (possibly just because it is *so* obvious). Frankland's whole life-style, from his early schooldays to the day of his death, was dominated by science in general and by chemistry in particular. It was not the only driving-force, particularly in his later years, but it was the dominating one and, moreover, the ostensible means for achieving other desirable ends.

A single-minded devotion to science (or anything else) is bound to generate all manner of social tensions. Such was the case with Frankland. It is instructive to glance briefly at three areas of potential conflict. First, there was the fact that pursuit of science in the formative years of his career was frankly incompatible with notions of social progress, and especially so in England. Frankland was not in the least concerned with notions of class structure in a Marxian sense, but his illegitimacy and the known social status of his half-brother may have impelled him to a drive towards social advancement. Yet the more he pursued chemistry the more it became evident that it would lead neither to wealth nor to social kudos. Then came the moment of revelation in Marburg: 'the German division of Society into classes, which is here effected according to <u>mind</u> and not <u>money</u>' (p. 72). This German idea of a meritocracy of the intellect was to be with him all his life. Though doubtless pilloried on occasion in his own country as academic snobbery, it gave him a self-image that made the pursuit of science a worthy thing in itself, at least as meritorious as the Gorsts' engagement with law and diplomacy. So deeply were German ideals ingrained in his mind that he confessed to Galton 'German is now almost like my native tongue & I mostly think in the language'.[32]

Other potential areas of tension with science could occur with family or with religion. In the latter case there appears to be some inconsistency, in that his reactions to the science/religion issue in theory seem very different from his reactions in practice. Like other members of the X-Club he appears to have been committed to a 'warfare model' for the interaction between science and Christianity, eulogises Darwin over against his clerical detractors and decries all attempts by the church to influence the conduct and conclusions of science. Yet he writes 'the [orthodox] religious creed taught me in my youth has had no deterrent effect whatever on the freedom of my researches'.[33] If, in practice, Frankland's science caused few problems of a religious kind, the same cannot be said of his family relations. Time after time the family holidays were interrupted by his own excursions with an allegedly scientific objective, or by visits to clients or to scientific meetings. Even his honeymoon was combined with work for a coal-owner near Tenby.[34] As time went by, and as the urge for power, fame and coin increased, the family for whose very welfare he was working became in practice less important in his life, and his frequent absences associated with the work of the Rivers Commission inevitably took their toll of relationships. By the 1870s this was beginning to show, and the stability of his first family was shattered by the death of Sophie and his subsequent remarriage. Again he had the chance to start a new family but there is little evidence to suggest an involvement comparable to that in the Manchester days. For one thing he was older, he was far more busy, there were even more demands from the world of science that had to be satisfied. It can be argued that the 'deportation' of Fred, and the long-standing breach with Percy, were cases where scientific values triumphed over those of an average Victorian family. The first son was a scientific embarrassment because his chemical skills were defective, while the younger son was a rival, not merely on matters financial, but also in respect of the arguments over the relevance of bacteriology to water analysis.

Given that scientific values were an all-pervasive, dominating influence we may next enquire why this should have been so. At least five reasons suggest themselves, each reflecting a different view of science itself.

First, there was the view of *science as a means of describing nature*. Whether Frankland's late reflections on his early life are always accurate representations of what happened in the 1830s and 1840s there can be no doubt that certain attitudes, when spoken of approvingly, represent values that he still held at the end of his life. In his *Sketches* he speaks repeatedly about his early observations of nature, and to Galton he wrote 'From the age of 3 had always a great

love of the new and the marvellous & much curiosity about facts'.[35]
He confesses a debt to his early reading of children's books like
Sandford and Merton and still more to a schoolmaster, James
Willasey, for helping him to develop his observational skills. This
typically Victorian – almost Gradgrindian – devotion to *facts*,
though specially marked in the early years, remained with him
always. It may be remarked that this is often seen as a characteristic
of early organic chemistry, a sphere that Frankland made specially
his own. If ever there was a time when Frankland pursued science
in this way it must have been in those idyllic years with Duppa in
the laboratories at the Royal Institution.

Then, secondly, science for Frankland meant *a means of
understanding nature*. His youth was marked by puzzlement as well
as curiosity: *why* did things happen as they did? This of course
involved scientific theorising, and it must be said that Frankland
was not at his best when elaborating hypotheses. That is one reason
for his slow enunciation of the theory of valency, and for his lack of
recognition in that area. He he was far too much an empiricist to
promulgate theories before he felt the pressure of observation was
too strong to be resisted. Ironically, where he did theorise publicly
without adequate experimental data, he invariably burnt his
fingers, as in his theory of glaciers for example.

Frankland's third conception of science sounds almost Marxist,
as a means of controlling nature. However it owes nothing to Marx
but much to the utilitarianism of his early environment. This was
the philosophy commonly espoused by the Mechanics' Institutes,
implicit in the applications of chemistry to such diverse occupations
as calico-printing and textile bleaching, and explicit in its application
by the Johnsons of Lancaster to sanitation, medicine and agriculture.
Frankland remarks 'at the age of 6, having read something about
the restoration of drowned persons to life, I had the most
unwavering conviction, that if I could get at the putrid carcase of a
dog floating in a neighbouring pond, I could restore it to life.'[36] This
was to become the dominant use of science in Manchester days,
partly from necessity but much more from the convictions so
confidently expressed in his Inaugural Address. Here, natural
theology is not used as a religious apologetic, nor even to justify the
social *status quo*, but rather to enhance the image of chemistry as a
means of controlling nature. The belief that science can genuinely
be used to benefit mankind, though sneered at by those whc find

'social control' a more comfortable theory, was rarely more demonstrable than in the work of Edward Frankland in purifying and monitoring domestic water supplies.

Fourthly, science can be pursued as a *method to accomplish certain social ends*. In some cases it *may* be used to diminish social unrest, to maximise profits, to achieve specific political goals, and so on. The question is often controversial and now rather *passé*. In the case of Edward Frankland it is doubtful if science was ever employed in this quasi-political manner, with the important exception of his adherence to the doctrines of the X-Club. Here science, or what passed for it, was relentlessly used in the promotion of scientific naturalism and the denigration of organised religion. Of that there can be no doubt. But it was not so much the practice of science as the rhetoric of its philosophy that was intended to fulfil these purposes. The chemistry of Edward Frankland was quite irrelevant.

Finally, science can be seen as *a means of personal advancement*. In Frankland's case this became increasingly important with the passage of time. It involved the acquisition of money and the pursuit of power. Each of these is worth some extended comment.

In pursuit of the first objective, financial achievement, Frankland was outstandingly successful (though it was much too vulgar a topic to be discussed in the columns of scientific journals). Frankland had many advantages in this area, starting from the annuity awarded by his natural father to his mother. With the aid of this she was at least able to send him to 'expensive' schools, but probably was also able to launch him into a career with slightly more ease than some of his contemporaries in Lancaster. But we should not exaggerate. Frankland was keen to stress that

> my pecuniary position has been achieved entirely by my own exertions. My late dearly beloved wife brought me £100, but beyond this every shilling I possess has been earned by my own work.[37]

So he would work like a beaver, from the hard years in Manchester to his final weeks in Norway. As a former colleague remarked, when turning aside from research 'at the urgent request of clients it was only a case of *reculer pour mieux sauter*. We must live, he used to say, with something between a smile and a sigh.'[38] Yet from all the examples that we have examined it is perhaps hard to recognise the strength of purpose that led to such financial success. The pluralism that marked his career goes some way to accounting for a more than adequate income, but there were also fees for lectures,

Table 17.1. *Some Scientific Estates*

Name	Death date	Re-sworn estate (£)
Lord Avebury	1913	315,579
William Spottiswoode	1883	187,078
Edward Frankland	**1899**	**136,472**
George Busk	1886	46,933
Joseph D. Hooker	1911	36,861
John Tyndall	1893	22,122
Herbert Spencer	1903	18,635
Thomas Archer Hirst	1892	14,640
Thomas Henry Huxley	1895	9,290
Alexander W. Williamson	1904	33,511
William Crookes	1919	30,057
Benjamin C. Brodie	1880	30,000
Lyon Playfair	1898	10,538
Robert Galloway	1893	3,541

patents, and above all analyses. His books brought in substantial royalties and, in later years, he invested in stocks and shares. Frankland's stockbroker was Col Inglis, next door neighbour at Reigate and tenant of Rougham Hall, Suffolk.[39]

He would surely have agreed with Dr Johnson in supposing that no man is more innocently employed than when he is making money. In that case his criticism of Francis Bacon is all the more astonishing, not merely for its questionable historical accuracy but even more for illustrating the universal human capacity for seeing in others that which others perceive so clearly in ourselves:

> He appears, however, to have been of a mercenary disposition, and this degraded his scientific aspirations, so that he cared little for knowledge for its own sake, but only for the solid pudding that could be got out of it.[40]

It is interesting to compare the legacies of Frankland with other members of the X-Club and with a few other chemists[41] (Table 17.1):

Apart from the member of the landed aristocracy (Lubbock) and the Queen's printer (Spottiswoode) no fellow-member of the X-Club, or any contemporary academic chemist, remotely approached the financial success of Edward Frankland.

Then, secondly, it is worth noting briefly something of Frankland's apparent pursuit of power. This can easily be overstated. His reluctance to accept the Presidency of the Chemical Society is much more consistent with his own pragmatic self-portrait than with that of a power-hungry chemist. However by the mid-1870s his rôle in the formation of the Institute of Chemistry suggests a new sense of adventure and an actual enjoyment of the fruits of high office.

Yet even now it is doubtful if it is power as such that he sought. It is more likely to have been power (or office) as a token of social acceptance. This, more than any other trait, is increasingly apparent during the later years of his life and may have coincided with the rise to fame (and real power) of his natural half-brother. All through his life Frankland knew full well, though rarely mentioned it, that his half-brother had so many advantages denied to himself. Sir John Gorst (as he became) was a successful lawyer and politician: Conservative MP for Cambridge, Solicitor General in 1885 and Under Secretary of State for India in 1886. He was knighted 12 years before Frankland's KCB and his own son was well on the way to a scintillating career in the diplomatic corps (and another KCB). By a strange coincidence the Surrey newspaper that carried an account of Frankland's funeral reported in the following week a meeting at Levens Hall, Westmorland, of the Primrose League, at which the speaker was Sir John Gorst, and his subject 'Education'. Though the politics were far from his own Frankland would have been as familiar with the location as the subject.[42] The exploits of his half-brother were paraded across the country, even reaching an obscure provincial newspaper 300 miles away. It would have been most unusual if Frankland had not secretly nourished hopes of some kind of social acceptance.

This helps to explain the constant name-dropping in his early letters and his *Diary*. One can imagine the exhilaration and pleasure with which he communicated the news to his parents of his introduction to scientific circles in Berlin:

> Rose introduced me to a great number of professors and other eminent men, whose acquaintance will be of great service to me; the wine circulated freely, and laughter, jokes, and scientific conversation filled up the evening. We had a select party at our end of the table, consisting of Prof. H. Rose, Prof. G. Rose, Prof. Poggendorf, Prof. Dove, Dr. Rammelsberg, and Dr. Weber; these names will no doubt be strange to you, but they are some of the most renowned in the scientific world; I was introduced to all of them.[43]

Then there were the innumerable table-setting plans for dinners he threw in later life, where the good and the great were invited and carefully seated in relation to the host. Such mundane drafts speak volumes about the social values of their originator. His account of the BAAS meeting at Norwich (p. 236) tells much the same story. But it was in the X–Club that he found a route to the heart of the scientific establishment (or a vociferous faction of it), just as John Gorst had penetrated the heart of the political and diplomatic establishments.

Thus science was pursued for a mixture of reasons whose relative weights varied with time. None however became more important than in the matter of his own reputation. Why else should he publish, at great personal expense, a highly indigestible volume of his own *Experimental Researches?* That question of his fame and reputation constitutes perhaps the supreme enigma of a life that in more than one respect could claim to be enigmatic: *why, despite all his massive achievements and the publicity that went with them, is he so little known today, even among chemists?* There are, of course, some trivial and fairly straightforward reasons. Within his own subject he had the misfortune of not having a famous reaction or theory or piece of apparatus named after himself. That kind of thing keeps a man's memory alive long after his death, as witness the Daltonian atomic theory, the Perkin reaction or the Davy lamp. But there is an irony here in that at least two phenomena *were* given his name in the nineteenth century. One was the water analysis apparatus which in fact became superseded. The other was the 'Frankland notation' which, conversely, became so successful that it was universally employed and no one needed to bother with a name to describe it. However the need for further explanation is indicated by a curious manifestation immediately after his death.

Within a few days the tributes started to flow in obituary notices, short and long. Yet there were two astonishing omissions. The Chemical Society, to which so much of Frankland's life had been given, never published a Memorial Lecture for its distinguished former President. The story has been told elsewhere,[44] and in the end the *Journal of the Chemical Society* had to be content with an ordinary obituary notice (by Herbert McLeod), belated though very able;[45] a so-called 'Frankland Memorial Oration' was delivered by H. E. Armstrong in Lancaster no less than 35 years after Frankland's death.[46] Even more remarkable is the fact that the Royal Society, though promising a notice of its late Foreign Secretary, in fact never produced one.[47] One can only conclude, either that the task was too difficult or that Frankland had enemies who were anxious to deny him even posthumous honour. While both may have been true in a measure, the former explanation seems the more obvious.. The whole cloak of secrecy with which Frankland had at first to envelop his private life because of his illegitimacy was hard to cast off in later days, and he certainly had no wish for people to pry into his financial activities. As one would-be chronicler ruefully observed with masterly understatement, 'the doctor has no particular desire to receive biographical attentions'.[48] The fact that he was an intensely private man must have had much to do with his later obscurity.

Yet, as we have seen, in his own lifetime powerful forces were at work to limit his influence. These were clearly evident in the Royal Society, less obviously so in the community of scientific naturalists whose approval he so desired. They were, in short, a deep antipathy to making money out of science and a determined marginalisation of those who did. Possibly feelings of jealousy and envy were also aroused. In the world of commerce there were those whose aspirations were frustrated by what they saw as Frankland's excessive demands. In promoting science education he and his X-Club friends were taking on the whole establishment of public schools, clerically dominated universities and catholic religion. It is here, and only here, that Frankland's life may be seen to display a truly political purpose. He was always ready to use politicians and the political machine for his own personal or scientific ends. Yet the astonishing fact remains that in the vast collection of his surviving papers, of politics in its narrow and conventional sense there is hardly a trace.

His books, though popular with students, constituted a threat to authors of more traditional works, both by undermining traditions of science learnt from books and by undercutting their own sales. So Frankland had plenty of enemies, though for all kinds of reasons. It is the contention of this account that the most persistent foe of all lay within himself, a striving for acceptance and above all for financial reward. In both of these aspects of his personality he was the victim of his own circumstances, particularly his illegitimacy.

Frankland illustrates well the old Spanish proverb: 'Take what you want, said God, take it and pay for it.' To achieve success in matters financial he had to pay a formidable price, ranging from exculpation at the X-Club and late recognition by the Royal Society to (at very least) frayed family relationships and exhaustion of mind and body. His acceptance by the scientific establishment, which he sought through the X-Club, demanded not merely renunciation of theological orthodoxy but also alignment with those who, in the fullness of time, would oppose Government interference while at the same time demand Government money. In a way his was the misfortune of being the only chemist in the group. The centre of gravity of chemical research moved away to the Continent, and specially to Germany, and English chemists became a small minority in the hall of fame.

Yet by a strange turn of events the chemical community is focusing with increasing diligence on that branch of the science of which Frankland was the founder, organometallic chemistry. And, in England at least, it is now becoming perceptibly more aware of what it has missed in the last half century by virtually ignoring its own history. So, in 1982, the Royal Society of Chemistry followed an example set 80 years before by Percy Frankland in creating a Frankland prize.[49] In 1902 it was given to the Storey Institute, Lancaster, but its successor was established for distinguished work anywhere in the world in organometallic chemistry.[50] In the following year a Sir Edward Frankland Fellowship for research in the same field or in the co-ordination chemistry of transition metals was also announced.[51] Perhaps the time has come for a recognition of the towering achievements in chemistry of this self-made man from Lancaster.

Notes

1 *Sketches*, 2nd ed., p.448.
2 Andersen wrote to Frankland with a list of Norwegian girls' Christian names (I. C. Andersen to Frankland, 11 April 1878 [RFA, OU mf 01.08.0643]); presumably he selected the name 'Helga' for his fourth daughter from this list.
3 Frankland to I. C. Andersen, 15 April 1892 [JBA, OU mf 02.01.0470].
4 Frankland to I. C. Andersen, 29 November 1892 [JBA, OU mf 02.01.0472].
5 By Frankland's oculist nephew Adolph Fick: Maggie West (*née* Frankland), *Diary*, entry for 7 July 1897 [MBA].
6 Frankland to Ellen Frankland, 9 August 1896, in *Sketches*, 2nd ed., p.367.
7 Maggie West (*née* Frankland), MS autobiographical note, n.d. [JBA, OU mf 02.01.0505].
8 Maggie West (*née* Frankland), *Diary*, entries for 16 and 25 January, 7 April, 23 and 24 October, 26 November 1897 (note 5).
9 W. H. Brock, *Fontana History of Chemistry*, HarperCollins, London, 1992, p.402.
10 Maggie West (*née* Frankland), *Diary*, entry for 27 March 1897 (note 5).
11 *Ibid.*, 25 October 1897, 19 January 1898.
12 *Ibid.*, 26 January 1898.
13 *Ibid.*, 17 May 1898.
14 *Ibid.*, 18 January 1899.

15 *Ibid.*, 25 May 1903.
16 *Ibid.*, 1 May 1899.
17 *Ibid.*, 17 May 1899.
18 *Ibid.*, 18 May 1899.
19 W. T. Burgess to Frankland, 31 July 1899 [RFA, OU mf 01.04.0405].
20 Frankland to W. M. Sykes, 24 July 1899 [JBA, OU mf 02.01.0497]; it is the last surviving letter from him, and is in his own hand.
21 The rest of the chapter (and the book) was completed by the editors' insertion of many of his letters.
22 Maggie West (*née* Frankland), *Diary*, entries for 31 July – 12 August 1899 (note 5).
23 *Ibid.*, 9 August 1899.
24 *Ibid.*, 11 August 1899.
25 *Surrey Advertiser*, 28 August 1899.
26 *Surrey Leader*, 25 August 1899.
27 Maggie West (*née* Frankland), *Diary*, entry for 22 August 1899 (note 5).
28 Edward Frankland, Will (Probate 27 September 1899) [Somerset House; and JBA, OU mf 02.02.1854].
29 L. Playfair to Frankland, 19 December 1885 [RFA, OU mf 01.03.0345].
30 F. Galton, *English men of science*, London, 1874.
31 Draft letter, Frankland to F. Galton (a) 12 April 1874 [RFA, OU mf 01.01.0111]; (b) 'April' 1874 [AFA].
32 *Ibid.*, (b).
33 *Ibid.*
34 *Sketches*, 2nd ed., p.277.
35 Frankland to Galton (note 31b).
36 *Ibid.*
37 Copy-letter, Frankland to F. Galton, April 1874 [AFA].
38 J. Attfield, *Chemist & Druggist*, 1899, 321.
39 Frankland to I. C. Andersen, 7 October 1893 [JBA, OU mf 02.01.0480].
40 Frankland to Ellen Frankland, 28 June 1888, in *Sketches*, 2nd ed., p.365.
41 Wills in Somerset House. The most successful entrepreneurs in the chemical *industry*, such as C. Allhusen and F. C. Hills, did become millionaires.
42 *Surrey Leader*, 1 September 1899.
43 Frankland to William and Margaret Helm, 8 January 1850 (reproduced in *Sketches*, 2nd ed., pp.267–73 (271)).
44 *Lancastrian Chemist*, pp.6–12.
45 H. McLeod, *J. Chem. Soc.*, 1905, **47**, 574–90.
46 H. E. Armstrong, First Frankland Memorial Oration of the Lancastrian Frankland Society, *Chem. & Ind.*, 1934, **53**, 459–66.
47 *Year Book of the Royal Society of London*, 1900, p.143; the Presidential Address contained a brief two-page summary of his achievements (pp.146–7).
48 'Old Briar', *Cross Fleury's Journal*, 1898 (March), no. 19, pp.3–4.
49 Percy Frankland to W. French (Principal, Storey Institute), 20 December 1902 [Lancaster Public Library, MS 7676].
50 *Chem. Brit.*, 1982, **18**, 737.
51 *Ibid.*, 1983, **19**, 847.

Index of persons

Subject index